EUROPE IN THE NINETEENTH CENTURY

EUROPE IN THE NINETEENTH CENTURY

By

D. G. O. AYERST, M.A.

FORMERLY HEADMASTER OF KING EDWARD VII SCHOOL, LYTHAM;
FORMERLY SENIOR HISTORY MASTER AT
BLUNDELL'S SCHOOL

CAMBRIDGE
AT THE UNIVERSITY PRESS
1952

PUBLISHED BY
THE SYNDICS OF THE CAMBRIDGE UNIVERSITY PRESS

London Office: Bentley House, N.W.1
American Branch: New York

Agents for Canada, India, and Pakistan: Macmillan

First Edition 1940
Reprinted 1947
,, 1949
Second Edition 1952

First printed in Great Britain at The University Press, Cambridge
Reprinted by Spottiswoode, Ballantyne & Co., Ltd., Colchester

PREFACE

The object of this book is to make clear the course and nature of the most significant movements in European history between the outbreak of the French Revolution and that of the European war of 1914. Significance is, of course, a matter of point of view. I have not endeavoured to conceal my own opinions; but I have tried to explain rather than to judge, and to explain those things in particular which are now uppermost in all our minds. Wherever possible events are described through the eyes of those who caused them; and I have tried to make the makers of the nineteenth century live again in a boy's imagination.

The book is designed in the first place for school certificate candidates; but I have sometimes—as in the section called 'Characteristics'—put in what I think a reasonably intelligent boy ought to know, without considering whether examiners will try to elicit this information from him. Throughout the book I have taken the line that one reason well understood is worth a dozen 'causes' or 'consequences' learnt parrot-wise and forgotten boy-wise as soon as the examination room is closed. In the text, therefore, I have tried hard to keep a sense of historical proportion, preferring to write much about nationalism and little about the Convention of Olmütz.

Tabulated information there is, however, in abundance in the appendices, which are on an unusually large scale for a book of this kind. A biographical dictionary, for instance, gives supplementary information about every character whose career is not fully dealt with in the text. A dictionary of political terms explains such words as 'referendum' or 'sovereign', while another alphabetical index gives the causes, clauses and duration of the principal European treaties.

Special care has been given to the maps and diagrams. They have been drastically simplified to drive a particular lesson home. Opposite each map is a historical summary and explanation which should be of considerable use in revision. An historical atlas should be unnecessary, but good physical maps are naturally indispensable. For the careful draughtsmanship of the maps I am indebted to Mr J. M. Kupfer who has translated my rough sketches into what I hope may prove memorable shapes.

I am particularly grateful to Mr E. L. Woodward who read the first three quarters of the book in manuscript. He has saved me from many errors, but he is of course in no way responsible for those that remain. I owe him, however, much more than a limited accuracy: whatever understanding I may have of the nineteenth century is derived largely from his. I should also thank Mr E. C. Fry for much valuable work on a preliminary draft of the second half of the book.

LYTHAM

December, 1939

FOR THE SECOND EDITION

The biographical dictionary has been brought up to date and the epilogue has been extended by five pages for this edition. The opportunity has been taken to correct a few small errors.

NORTHWOOD, MIDDLESEX

February, 1952

CONTENTS

PART II

METTERNICH'S EUROPE

1815 TO 1852

PART IV

THE CONSOLIDATION OF NATION-STATES

EPILOGUE

The end of the story

FAMILY TREES

MAPS

FIGURES

DIAGRAM

PART ONE

REVOLUTIONARY EUROPE

1789 TO 1815

1. THE SUNSET KINGS

France, the model for Europe. In the eighteenth century all continental Europe copied France as closely as it could. Louis XIV, the great king who at the end of the previous century had made France by far the most powerful country in the world, had built himself a magnificent palace at Versailles. In the eighteenth century every king and prince from Naples to Potsdam built himself a miniature Versailles. Louis XIV had taken his nobles from their country life and local interests to live a butterfly life at his court. He was 'le roi soleil', the sun king, and their only business was like planets to revolve around him. They gained pleasure but they lost power, which was what the king wanted. The work of local government was taken from them and given to Intendants and other officers, paid servants of the king who had to do as they were told. Henceforward Louis could quarrel with any noble, no matter how distinguished, without fear of a breakdown of local government or of a rebellion led by the courtier in disgrace. A non-resident landowner, known to his tenants only as a grasping rent collector, has no following. The other kings of Europe copied as far as they could this method by which Louis XIV had made himself the only political power in France.

In art and manners, as well as in government, France set the fashion for Europe. In St Petersburg and Berlin society talked French more than Russian or German; and French replaced Latin as the language of diplomacy, which it long remained. Frederick the Great of Prussia devoted almost as much energy to writing French verse, which he did badly, as to governing his country, which he did well. Only music escaped the French domination. The language of opera was Italian and the best composers were German. The music of the 'Frenchified' courts of Germany ('courts' for Germany was not one state but many) was the work of Mozart or Haydn.

Concentration of power. The concentration of political power in the hands of the king was used on the whole for the good of the people, but for four reasons it failed to satisfy their desires. First, nothing could be done without the consent of the central government and the government was far too busy to attend to a quarter of the things which needed doing. Secondly, the concentration of power was not complete. Although nobody could over-rule the king if he was determined to have his way, the thirteen 'parlements' of France could and did delay reforms almost indefinitely.

The 'parlements'. These 'parlements' were the supreme law courts, filled with hereditary judges, of the various districts of France. They were far more than ordinary law courts. They could make trade regulations of their own and no royal order was binding until it had been registered by the 'parlements', which could place many difficulties in the way of registration. They were the only real check on the king's absolute power and, as such, they were sometimes popular; but they were a check which protected abuses and hindered reforms. The royal government in the years just before the revolution tried at intervals to remove some of the worst abuses. It was opposed by the 'parlements', which thus must share some of the responsibility for the revolution. The third reason why men disliked the king's absolute power was that the wishes of the governed were not consulted; and, lastly, although the nobles' power had been taken away, their privileges had not.

In politics all except the king were virtually powerless. The richest noble had no more hope than the poorest peasant of successful opposition to the king's will. But in privileges there was no equality. It was this inequality which made the sun king's system unstable and soon led France from the glories of Louis XIV to the stormy sunset of Louis XVI's reign. The inequality of privileges both weakened the government and enraged the people.

Privileges. The taxes, for instance, were paid by those least able to bear them, for in France the nobles and the clergy were largely exempt from direct taxation. The peasant still had to pay tithes to a church, whose bishops had as likely as not ceased to believe in Christianity, and feudal dues to the 'seigneur', who had long ceased to do any of the local government work for which originally they were a payment. Then the best careers were reserved for the best blood and in most regiments only nobles could hope to be officers.

In the towns trade was still largely controlled by the guilds, which in the middle ages had enforced decent standards of workmanship and of living conditions, but which were now mainly concerned with preventing competition and making it difficult for any but the fortunate few to earn a living. It was not only the guilds which hampered traders. Each province had its own customs barriers. A consignment of wine from Provence to Paris had to pay duty forty times on its journey. In addition to the national frontiers of France there were 3600 miles of these provincial frontiers each with its own customs houses. Trade between the provinces was hindered also by the fact that there was not one law for all France. The provinces had their own local customs, so that what was legal in one was illegal in another.

In spite of these handicaps the middle class was growing richer and the peasants were buying their farms to which they were passionately devoted. 'The peasant sows his heart with his grain', as a French historian put it. Unlike the English gentry the French nobles were not to any great extent themselves farmers. They merely received money or produce in the form of feudal dues and rent from the real landed class, the peasants. The revolution when it came, therefore, did not so much redistribute the land as free it from the various payments which the farmers had to make to the lord.

Both the peasants and the middle class were conscious that the state could not get on without them; both claimed that they were, therefore, entitled to equality of treatment with the nobles and the clergy. Many of these two privileged classes agreed with them in theory, more particularly since men like Lafayette had come back from the American War of Independence full of enthusiasm for the simple equality of the young republic. They had ceased to believe that their privileges were justified, for they had read and enjoyed the books in which Voltaire had exposed the pretence on which society rested and Rousseau had expounded the virtues of a free and equal society which he imagined to exist among primitive peoples. They continued, however, to enjoy their admittedly unjust privileges just because they were too enjoyable to surrender.

The bankrupt government. The king allowed the privileged classes to retain their immunities because socially the king and his courtiers were one world. He shared their social outlook, their opinions and prejudices. The king's inclinations made him uphold his courtiers, his interest was to sacrifice them. Their privileges were fatal to the monarchy because they made it impossible for the government to avoid bankruptcy. The fact that the nobles and clergy could not be taxed meant that the king could never get sufficient revenue to pay for the government, even though in pre-revolutionary France it was not uncommon for a peasant to pay rather more than half his income away in taxes, besides being liable for the 'corvée' or forced labour at road-making. The king was therefore always in need of ready money and it had become the practice to anticipate the taxes. Instead of collecting them himself the king 'farmed' many taxes out to tax-gatherers who paid the king a lump sum and collected that and as much more as they could from the taxpayers. After the 'taille', a sort of income tax, the most oppressive tax was the 'gabelle'. Salt was a government monopoly and every family had to buy seven pounds of salt a year at exorbitant prices whether it wanted to or not. Each year some four thousand people were sentenced to prison, flogging, the galleys or exile for breaking the 'gabelle' laws.

But in spite of this severity, and although the taxation was more than taxpayers could bear, the government was unable to pay its way because there were too few taxpayers and too many expenses. The magnificent palace at Versailles cost Louis XIV £30,000,000 to build, while Louis XV spent a tenth of that sum on one mistress alone. More expensive still were the great wars which France fought against England during the eighteenth century. The government had to live by borrowing. At last the money which the government received in taxes was insufficient even to pay the interest on the money which it had already borrowed. The government was bankrupt and the revolution had come. The government was bankrupt not because France was too poor to pay its way but because those who could have afforded to pay went untaxed. It was not poverty but privilege which caused the revolution.

France's relative prosperity. Certainly it was not poverty which caused the revolution. The French middle class was more prosperous and more numerous in 1789 than in 1689; the peasants were less downtrodden than they had been. They were gradually becoming peasant proprietors instead of feudal serfs. In fact personal serfdom had almost died out in France—in this respect France was one of the more fortunate European countries—but the peasants still had to pay many feudal dues which were a relic of the old system.

The revolution came first in France and not in Austria or Prussia or Russia not because social conditions were worse but because they were slightly better. The people of central and eastern Europe were too downtrodden to dream of revolting; they were so oppressed that they could hardly even dream of a better future. The peasants of France were improving their condition; they had gained sufficient to know that there was more to be gained. They were angry because their efforts to gain more were frustrated by the government and the social system. The French middle class was energetic, frugal and ambitious. They saw almost unlimited possibilities open to them if the barriers erected by the government and the social system could be removed. The French revolution was not so much the product of despair as of a belief that with a little energy and common sense things could easily be made a great deal better.

2. THE REVOLUTION IN FRANCE

May to October 1789

The States-General. On a sunny day in May in the year 1789 Louis XVI left his palace to open the first meeting of a States-General to be held in France since 1614. The French revolution had begun. The States-General was an old-fashioned parliament divided into three houses in which sat the representatives of the clergy, the nobles and the rest—the Third Estate. The men of the eighteenth century no more expected to see a revival of the States-General than of the rule of the Pharaohs; to them it was just another of those barbarous devices of the middle ages which an enlightened and efficient monarchy had swept away. But now in the last years of the century the 'efficient' monarchy had become so overwhelmed by its financial inefficiency that it had had to resurrect the old discarded assembly. The court's views on the subject were clear. The States-General's business was to invent some way of balancing the budget—probably by a reform of taxation—and then to go away. The States-General thought differently. All through the previous winter there had been an excited election campaign: lists of grievances, 'cahiers' they were called, had been drawn up by the electors at the government's request and given to the newly elected deputies. These were their instructions, and the representatives of the Third Estate were determined not to go home until they had got their way.

The National Assembly. The first month was spent by the king, the nobles and the clergy, and the middle-class representatives in watching one another carefully. Then followed, from June 17 to June 27, ten days that shook the world. Not even a street riot was needed to destroy the proudest despotism in Europe. Three resolutions were sufficient. The first was passed on June 17. For over a month the deputies had been discussing how the States-General should vote. Was each estate to count as a separate house of parliament so that the nobles and clergy could always overrule the Third Estate by two to one? Or were all the deputies to sit and vote together so that the Third Estate could count on a majority with the help of sympathisers from among the nobles and clergy? The resolution of June 17 settled the question by adopting the second method. The Third Estate, the representatives of the great unprivileged majority of Frenchmen, declared themselves to be a National Assembly and invited the nobles and clergy to join them. The resolution went on to say that no taxes which were not freely granted by this new National Assembly would be legal. The old-fashioned States-General had disappeared; a modern all-powerful parliament had been born.

Three days later the Third Estate found that the king had locked up their usual meeting hall; they adjourned to the neighbouring tennis court (a building like a racquets court) and there they took an oath never to separate until they had given France a constitution. Another three days passed and the king hit back. He called a special meeting of all three estates, announced that he would not agree to the resolutions of June 17, and ordered the meeting to break up. The Third Estate took no notice, stayed behind and passed another resolution declaring that the deputies could not be arrested by the king. Four days later the king gave way and accepted all that the Third Estate had done. The revolution had won the first round.

Revolutionary principles. In principle the French revolution was over. The people were no longer to be responsible to an absolute king who could do what he liked with their lives and property. The king instead was to be responsible to the people represented in a parliament. Thus the victory for what is called 'responsible government' had been won. Further, the opinions of nobles and clergy were no longer to count for more than the opinions of ordinary men and women. The Third Estate had become the National Assembly. Thus the victory for what is called 'equality before the law' had been won. It remained to work out the implication of these two victories and to defend them. They had been won by moral courage, they had to be defended by physical courage. They had been achieved without the firing of a single gun or the breaking of a single head. They had to be maintained by the pitiless use of the guillotine and on a hundred hard-fought battle-fields. The old order fought back.

The first attack came very soon. Even in this crisis of government and society Louis was more interested in his hunting than in politics; but his wife, Marie Antoinette, an Austrian, had a clearer though fatally limited view of the situation. She was determined to regain by force what had just been lost by indecision—and she very nearly succeeded. The rank and file of the pre-revolutionary army, a paid professional force belonging to the king, was to a considerable extent German-speaking. At this time about a sixth of the army were foreigners. Not all the remainder spoke French for German has always been the language of Alsace. To use these foreign troops against Frenchmen seemed to most people an unforgivable, anti-patriotic action. Yet this is just what Marie Antoinette proposed to do. Only success could have excused her action and she did not succeed.

Fall of the Bastille. An army was collected at Versailles and the defeat of the revolution seemed certain. At this moment the mob of Paris intervened for the first time. A young journalist told the excited crowd that the king's German troops were coming to punish Paris. The crowd quickly helped themselves to arms from the gunsmiths' shops and hurried off to attack the Bastille, Paris's 'Tower of London'. At first they had little success although there were only a handful of troops inside. The next day they were reinforced by deserters from the French-speaking troops and with their help they soon forced the Bastille to surrender. Its brave commander, who had protested against the capitulation and tried in vain to blow up the fortress, was murdered. That evening Paris went mad with joy and celebrated its victory by murdering also the head of the old royal government of the city. The first warning of the Terror had been given.

The Bastille was only the symbol of tyranny; its capture did nothing to change the material balance of power, but it was sufficient to embolden frightened revolutionaries and to scare the hesitating king away from the use of force back to his earlier decision to recognise the revolution. It is for this reason that the day of its capture, the fourteenth of July, is still the French national holiday. The revolution had won the second round.

The nobles' attitude. Two consequences followed. The more conservative nobles, led by the Count of Artois, the king's brother, at once went into a disgusted exile and became traitors in the eyes of most Frenchmen. Three weeks later the more liberal-minded nobles themselves proposed the abolition of all their remaining privileges. Far into the night of August 4 the National Assembly was in session. It listened to reports from the provinces of the burning by impatient and infuriated peasants of the country houses in which the hated title deeds to the seigneur's feudal rights were stored. Instead of asking for police action to restore order the nobles and clergy suggested that the abuses which were the cause of the riots should be removed. A bishop proposed the repeal of the game laws; a noble moved that government employment should no longer be bought and sold; the courts of the lords of the manor were abolished and personal servitude, the condition of a serf, disappeared in a night. Before the session was over the nobles of France in a mood of incomparable generosity had voluntarily removed all those centuries' old grievances which had made a revolution inevitable. A few years before their gesture might have been in time to save France from the Reign of Terror and Europe from a generation of bitter war. As it was the revolution had already started on its violent course. They were only in time to give freely what was already in any case being taken forcibly. They had made a magnificent gesture; they might have carried out a practical reform.

Louis was incapable of keeping for long to any decision. Torn between the commands of his wife and the determination of the revolutionaries, he sympathised with both and obeyed each in turn. Before long the court at Versailles was again planning the overthrow of the revolution by force. After a dinner at the palace army officers replaced the new tricolour cockades of the revolution by the white of the Bourbons. Agitators in Paris used this insult to the national cause to inflame still further the crowds who were suffering from the high price of food that year.

March of the women of Paris. Early in October accordingly a strange procession of angry, hungry women set out for Versailles. There they demonstrated at the palace, convinced that the king and queen could give them bread and determined to make them. They were followed by the National Guard and its commander Lafayette, whose gallantry in the American revolution had made him a national hero and concealed his political incapacity. This National Guard was the middle class in arms to defend its property and its newly won power. The army was under the orders of the court, the National Guard was under the orders of the town councils. At this moment it alone stood between the king and queen and the people of Paris who were overrunning their palace. Lafayette could do nothing except prevent an attack on their persons and next day he persuaded the court to give way to the mob.

The king and queen were carried back to Paris in triumph. The crowds were still loyal and even enthusiastic in a coarse, familiar way. They brought with them, they said, 'le boulanger, la boulangère et le petit mitron'—the baker, the baker's wife and the little cook boy. Their king, so they thought, had been removed from the influence of corrupt and selfish nobles to the influence of his very loyal subjects. What had really happened was that the king had been taken away from Versailles, where he was his own master, to Paris where the revolution was master. He was no longer at the head of his troops but in the hands of the mob. The revolution had won the third round.

The workers of Paris. The people who had brought the king back to Paris were the working men and women of the capital. In 1789 they were in a mood of violent discontent. There had been bad harvests and a shortage of food was unavoidable. The shortage, however, was worse than it need have been, partly because of the extremely stupid economic organisation of France and partly because speculators, including many of the court circle, had been engaged in cornering the market. In addition there was much unemployment and the people of Paris and the other large towns were near starvation.

It was Paris which counted for most in the history of the revolution. In 1789 it already had a population of 750,000. It was in a position to dictate to the government. Before long its working class was well organised by astute politicians and could quickly be mobilised to demonstrate for or against any particular man or measure. The rest of France might disagree with the policy of Paris, but it could not make its influence felt in the same way. The provinces could only send resolutions to the Assembly, Paris could send a howling mob.

The town working class was in some ways the most powerful of the revolutionary forces, but in the long run it was the least successful. It wanted economic security, but it had not the ghost of an idea how to get it. It was not a working class disciplined by factory conditions, but a miscellaneous collection of unemployed servants, general labourers and sweated workers either in small workshops or in their own homes, where they made up goods for their employers. Such men could only denounce the rich, they could not suggest any scheme of organisation which would improve their lot. Trade unionism and socialism are the result of factory organisation, but as yet there were no large factories in France. Both the middle class and the peasants knew what they wanted and took it. The town working class only knew that they were miserable.

The peasants and the middle class. The peasants' demands were simple. They wanted, and had got, the ownership of the land they tilled, free from any obligation to pay tithes to the church or feudal dues to the seigneur. From that time they played a passive part in the history of the revolution. Intensely conservative by nature, they hated the extreme politics of the later revolutionary leaders; but their support of the revolution was guaranteed by their fear that a restoration would mean the loss of what they had won. Only in a few exceptional districts such as La Vendée, where the nobles were still squires in the English sense, did the peasants take up arms against the Terrorists.

The middle class also wanted to abolish privilege in order to gain the unrestricted right to enter any trade or profession and, once there, to make as much money as possible without interference. They wanted political power as well but, as we shall see, they lacked the practical ability to draw up a constitution which would satisfy them and also work. In the end they were glad to welcome the dictatorship of Napoleon which at least secured order.

3. THE THREE MISTAKES
October 1789 to April 1792

Mirabeau. From the outbreak of the revolution until his death in April 1791 the figure of Mirabeau dwarfed all others. Carlyle has painted his portrait for all time: the man

> through whose shaggy beetle-brows, and rough-hewn, seared, carbuncled face, there look natural ugliness, small-pox, incontinence, bankruptcy, —and burning fire of genius.... All manner of men he has gained; for at bottom it is a social, loving heart, that wild unconquerable one:—and more especially all manner of women.... In War, again, he has helped to conquer Corsica; fought duels, irregular brawls; horsewhipped calumnious barons...in the whole National Deputies, in the whole Nation, there is none like or second to thee.

Much of Mirabeau's youth had been spent in prison, partly because of his bad character and partly because his father, a man of equally poor morals, preferred to see him there. By getting a 'lettre de cachet' from the king he had been able to keep his son in prison without a trial. Thus Mirabeau had strong personal reasons for hating the old order. In 1789, although a noble, he had been elected a deputy for the Third Estate. Almost at once he had become their leader—in the affair of the tennis court, for instance—but from the time the court was moved to Paris he turned more and more to the side of the king because he was alarmed by the growing violence of the revolution. Eventually he was given a secret pension by the king, who also paid his considerable debts, in return for excellent advice which was rarely taken.

A weak constitution. The National Assembly moved to Paris with the king and soon settled down to its interrupted task of giving France a constitution. There were interminable debates as to the proper relation between the government and parliament under the new constitutional monarchy. Most of the revolutionary leaders started with four assumptions of which the first was very doubtful and the last two definitely wrong. They believed that men were naturally good and that if they were left almost entirely alone by governments they would become actually good. They believed that France had suffered from having had too strong a government and no parliament at all, and that therefore it was necessary not to give the new government too much power. They believed that the only way of preventing tyranny (the government having too much power) was to keep government and parliament absolutely separate. This is called the 'separation of powers'. Finally, they believed that in England this separation of powers existed.

Mirabeau, on the other hand, believed that especially in a revolutionary period, when big changes are being made, it is necessary to have a strong government if anarchy is to be avoided. He had lived in England and knew that her success was not due to the separation of powers, which has never existed here, but to the concentration of all power inside parliament of which the government is a part—all cabinet ministers are also members of parliament. Mirabeau proposed that this should be the rule in France. The other leaders objected. They felt on the one hand that Mirabeau was ambitious and wanted to be both minister and deputy, and on the other hand that the king was seeking a backstairs way to control the Assembly. They assumed that the king would appoint the ministers who would be, so to speak, his outposts in parliament. Mirabeau's proposal was rejected, and in November 1789 a decree was passed forbidding ministers to be also deputies. France failed to get a strong government and this failure led straight to the Terror. This was the first fatal mistake made by the revolutionary leaders.

The revolution and the church. The second mistake was the treatment of the church and here Mirabeau was as much to blame as anyone. The fantastically rich higher clergy—bishops and abbots—were almost to a man nobles by birth and counter-revolutionary by interest. Few people objected when the state took away their wealth and paid instead a moderate salary to all the clergy. Most educated Frenchmen at this time had lost any real belief in the truth or at least in the importance of Christianity. It did not occur to the revolutionary leaders that men would die for their faith as well as for their country. They were, therefore, surprised when the next move against the church lost them the support of the ill-paid parish clergy, who at first had been in sympathy with the revolution.

It was decided in July 1790 by the Civil Constitution of the Clergy that bishops and parish priests should no longer be appointed by their superiors but that in future they should be elected by the people whether they were Catholics or not. The clergy were to be the servants of the people, not the servants of God: so it seemed to many devout men. The Pope refused to have anything to do with the new scheme. The truly religious priests supported him almost to a man and became enemies of the revolution, while those who were priests mainly in order to earn a living accepted the new French revolutionary church. The revolutionary leaders learned that there were far more simple, devout Catholics than they had ever dreamed. The king, who was a sincerely religious man, never forgave the Assembly for what it had done. From this time Louis never hesitated; he became the open enemy of the men of 1789.

The flight to Varennes. The revolution had by now lost much of its popularity outside Paris. The reforms which ninety-nine Frenchmen out of a hundred wanted had been carried out in 1789; the later work of the National Assembly appealed only to one section and not the whole nation. Many prudent men also were frightened by the growing power of the mob. If the king had been prepared honestly to accept the earlier work of the revolution, he might now have placed himself at the head of a large moderate country party against the extremists of Paris. This Mirabeau urged him to do, advising him to leave Paris where he was virtually a prisoner and go to Normandy where he had many loyal supporters. The queen also wanted him to leave Paris; but she wanted him to go eastwards towards Marshal Bouillé's royalist army, towards the troops which her brother Leopold of Austria was collecting beyond the frontier, and towards the exiled nobles on the Rhine whose one thought was to secure foreign help with which to destroy all that had been done since May 1789. Mirabeau's advice was that the king should appeal to loyal Frenchmen against Paris; the queen's advice was that he should appeal to foreigners against France.

Mirabeau's advice was the best that could be given, Marie Antoinette's the worst. Mirabeau died and the queen had her way. In June 1791 the king and queen left Paris secretly to join the army on the eastern frontier. At first all went well, but at the little village of Ste Menehould the postmaster recognised the party and rode on ahead to Varennes to call out the revolutionary National Guard. There they were stopped and brought back prisoners to Paris, having failed by only a few miles to make good their escape. Everyone now knew that the king was a traitor. He was allowed for a little longer to call himself king—the French were very slow to accept the idea of a republic—but the fate of the monarchy was sealed.

The Legislative Assembly. That autumn the National Assembly finished its work and was replaced by the Legislative Assembly, the first and, as it turned out, the only parliament of the constitutional monarchy. Distrusting—and with some reason—the political competence of the working class the National Assembly had limited the right to vote to solid middle-class people. In spite of this a large number of men who were beginning to call themselves republicans were elected, partly because loyal Catholics who would not accept the new revolutionary church were not allowed to vote. The members of the National Assembly had made the serious though generous error of retiring from politics by declaring themselves ineligible for the new assembly.

The army and the war. The inexperienced new parliament hastened to make the third fatal mistake of the revolution. They went to war in the spring of 1792. The refugee nobles and Marie Antoinette had long been urging other European countries to intervene in France. The central European powers, however, had been far too interested in quarrelling over Poland to undertake a crusade to help the Bourbons. If the leaders of the revolution had been content to live at peace with them there would have been no war, at any rate for some time. In the interval France could have reorganised her army. As it was she cheerfully went to war with her army in a state of complete disintegration.

The officers of the old French army were nobles and so, although many of the men sympathised with the revolution and welcomed the weakening of discipline, the army as a whole was a counter-revolutionary force. But just as there were nobles who supported the revolution so there were revolutionary parts of the army. The artillery and the engineers were not exclusively or mainly officered by nobles. To provide sufficiently competent officers for these specialist corps it had been necessary to throw this career open to talent, and middle-class artillery or engineer officers, such as Napoleon, frequently went over to the revolution. The revolution could, and did, improvise infantry regiments. It could not have improvised artillery and engineers. Nothing however can excuse the revolutionary leaders for going to war with less than half an army to support them.

The extremists oppose war. The middle-class republican leaders had probably taken too seriously the threat of an unprovoked foreign invasion. They were certainly mistaken in thinking that by the winter of 1791–2 they were strong enough to threaten the old order across the frontier with a missionary war. They were wrong again in their belief that a declaration of war by France would be followed at once by revolutions throughout Europe. The new extremist leaders with whom we shall be concerned in the next two chapters saw more clearly the possibilities. They opposed a war because they thought that it must either restore the Bourbons to their old power or else eventually lead, as it did, to a military dictatorship. If the moderate republicans could have shared their insight and foreseen the Reign of Terror and the coming of Bonaparte they would have kept the peace. But in one thing at least Marie Antoinette and the moderate republicans were in agreement. Neither thought that the old order and the new could exist peacefully side by side in Europe. And so in April 1792 Louis XVI at the command of his ministers declared war on his relations. France was at war, and her king was on the side of her enemies.

4. TOWARDS TERROR

April 1792 to March 1793

Danton. Out of the confusion which followed the king's flight and lasted until the outbreak of the war one great figure emerged: the genial, ferocious, dominating figure of Danton. Like many other revolutionary leaders Danton was a lawyer, but there was nothing of the lawyer's fastidiousness or precision about him. He was essentially an unbuttoned man: to that, to his eloquence, and to his courage he owed his power over the people, first of the Cordeliers district—'the hotbed of sedition'—then of all Paris, and, for a few months, of all France.

Danton's first important intervention in politics had not been a success. In the month of July 1791 he had organised a demonstration on the Champ de Mars, the Hyde Park of Paris, to demand that a referendum should be held to decide what should be done with the king, who was in disgrace because of his attempted flight to Varennes. The last thing that the middle-class leaders wanted was to be dictated to by the ignorant and violent mob of Paris. Lafayette had used the National Guard to break up the meeting, some people were killed and Danton had had to take refuge in England. This 'massacre of the Champ de Mars' was the last victory of the middle class before the Terror.

Before long Danton was back in Paris organising the slum dwellers. Next time he did not fail. It was the war which gave him his opportunity. For three months after the declaration of war little happened. Then the Duke of Brunswick, the commander-in-chief of the Austrian and Prussian armies, moved slowly on Paris, issuing a manifesto in which he threatened that unless the old order was restored at once Paris would be given up to an 'exemplary and never to be forgotten vengeance'.

August 1792 and its consequences. This was Danton's cue. At one o'clock in the morning of August 10, 1792, a revolutionary committee with Danton at its head met at the Town Hall. Soon all the bells of Paris were jangling angrily as the 'tocsin' rang out to call the crowds together for the attack on the Tuileries, the palace where the king was living. Its defenders had made elaborate preparations and not only the palace itself but the bridges across the river as well were held by the king's Swiss Guards and loyal detachments of the National Guard from the rich west end of Paris. The rising might have failed if the royalist general had not been lured to report at the Town Hall, where he was murdered on the steps as he was going in. At this the defence lost heart; the troops were withdrawn from the bridges to the palace itself. In the evening the Tuileries was stormed by ferocious crowds who revenged themselves for their casualties by massacring and mutilating the heroic Swiss Guards. The king, however, had escaped before the attack took place, fleeing in fear of his life to the Assembly, which in fear of the mob promptly 'suspended' him. The experiment of constitutional monarchy was over.

For six months Danton was the government. Under his influence the local government of Paris was now reorganised and taken out of the hands of the respectable middle class. In each district there was a ward or 'section' organisation, and at the centre the Commune or town council of Paris. All these bodies now passed completely into the hands of the extreme revolutionaries. The middle class had moved the king from Versailles to Paris in 1789 in order to be his master; Danton's revolution in August 1792 made the working class of Paris the master of the middle class.

The September massacres. The new revolution, however, did nothing to stop the Duke of Brunswick's slow but apparently sure advance. By September he was outside Verdun, and when Verdun fell the way to Paris would be open. The position seemed hopeless; only Danton refused to despair. To save Paris, he told the frightened deputies, there was need of 'daring, daring and still more daring'. He alone had it. Only terror could now prevent a royalist rebellion in the capital. The answer to the fall of Verdun was the September massacres, which Danton defended if he did not plan. Again the 'tocsin' rang; but this time the people of Paris hurried not to the palace but to the prisons. Crowds lined the streets outside, while inside lynch-law committees—at one prison they were butchers by trade—dealt rapidly with all the prisoners, releasing many but killing more. Anyone who could possibly have helped a royalist rising and many who could not possibly have done so were brutally murdered. They were killed not because they were guilty but lest they should have the opportunity to become guilty. The slaughter lasted four days and when it was all over between 2000 and 3000 royalist prisoners had been killed.

Valmy. From Danton's point of view the massacre had this advantage. It committed the moderates beyond hope of pardon. Now they could never convince the allies of their respectability. By failing to prevent the massacres they had lost the power to 'sabotage' the revolution by making a frightened peace. At the price of much innocent blood Danton had prevented treason on the home front. He had preserved republican unity; but he had not defeated the allies. In fact they were not defeated that year. Verdun duly fell, but at Valmy a few days later the Duke of Brunswick received a slight check. He could have pushed forward but he preferred to leave the punishment of rebel France to next year. Next year it was too late, for by then the republic was armed. Valmy was not a victory but it was an inspiration. A French force of half-trained recruits had escaped defeat in an engagement with the best professional troops of Europe. France had been saved when salvation seemed impossible. She gained not only time, but courage. Seldom has a drawn battle been more decisive.

The Convention. During that anxious September of 1792 France had been electing its third parliament in four years. The Legislative Assembly now gave way to the Convention, which had to find a new constitution to replace the constitutional monarchy which had disappeared on August 10. The right to vote was no longer confined, as the respectable reformers of 1789 had provided, to the safe, propertied middle class. This time everyone was allowed to vote, but less than a tenth bothered to do so. Everywhere there were sweeping republican successes.

Girondins and Jacobins. The day after the battle of Valmy the Convention met in Paris. Its members were bitterly divided among themselves into Girondins and Jacobins. The former had been in power during most of the time between the death of Mirabeau and the August revolution, but now they began to lose power to the more extreme Jacobins. The Girondins took their name from the fact that most of their leaders came from the Gironde in the south-west of France. Essentially they were the provincial party, while the Jacobins were the party of Paris. The Girondins were people of the highest character and the finest personal, but the poorest political, courage. They agreed with the thinkers of 1789 that men were good and governments bad, despite all the evidence that was piling up to the contrary. They were full of schemes to save the gentle, liberal republic in which they believed; but they lacked the ruthlessness to carry out their plans at the cost of bloodshed. Themselves men of the highest dignity, they allowed the determined, brutal politicians of the extreme Left to hustle them in the most undignified way.

The Jacobins took their name from a disused Paris church which was their meeting-place. The Girondins were just a group of members of parliament; the Jacobins were a highly organised political party with an intensive life of their own outside and independent of the Convention. They were primarily the party of Paris but there were also Jacobin clubs throughout France federated to the parent club in Paris. Here at open debates the policy of the party in the Convention was decided. As the Jacobins grew more powerful, important decisions were made at the discussions in their club instead of in the Convention. The club members came to be all of one way of thinking, and victory went not to the moderate who could win the support of waverers from the other side, but to the extremist who could whip up the enthusiasm of his fellow Jacobins.

Robespierre. Much the most important of the Jacobin leaders was Maximilien Robespierre, a lawyer like Danton; but, unlike Danton, lawyer-like—cold, clear, precise, even tedious. The two men, so different in temper, were born to be enemies, but now for a time they were allies. Robespierre's was a highly abstract mind, which cared far less for men than for principles. Terror, which to Danton was a weapon to be used ruthlessly but only in moments of great danger, was to Robespierre an article of faith. Traitors must be punished; therefore all traitors must be punished. If there were a hundred traitors, then a hundred must die; if ten thousand, then ten thousand. And everyone was a traitor who did not live up to Robespierre's high standard of republican virtue. The term 'traitor' included profiteers as well as royalists. If France was sickened by the extent and duration of the Terror, this was due not to Robespierre's cruelty but to his inhuman consistency.

The struggle between the Girondins and the Jacobins was fought in the Convention and in the streets of Paris while the conflict between the revolution and its enemies was being decided on the frontiers of France. The two struggles were closely linked and each affected the development of the other. At first all went well with the war. Dumouriez, the 'hero' of Valmy, led a French army to an easy conquest or liberation of Belgium, which was then a discontented Austrian province. Other armies carried the French colours without serious opposition to the Rhine and through Savoy. France had gained in a few months those 'natural frontiers' which she had desired for many centuries.

Execution of the king. Meanwhile, when the Convention met in Paris the Girondins bitterly attacked those who had been responsible for the September massacres. The Jacobins, to distract attention, counter-attacked with a demand for the king's execution. The Girondins, moderate and humane, refused to believe that the execution of the king was a necessary or desirable consequence of setting up a republic. The Jacobins held that to pardon the king was to condemn the republic—for if the king was not guilty, so Robespierre argued, the republic was not justified. The debates were long and even, but at last on January 15, 1793, Louis XVI was condemned to death.

A week later he was taken from the Temple, the squalid prison where the royal family was now kept, and driven to the Place de la Concorde. There the guillotine, which was to be so overworked for the next year and a half, was given its first notable victim. Louis stood up to address the people, but the drums beat to drown his voice and the executioners forced him down. As the blade fell it is said that his confessor exclaimed 'Son of St Louis, ascend to heaven'. There was another point of view. Ten days later Danton told the Convention: 'You have declared war on kings, you have thrown down your glove before them and this glove is the head of a tyrant.'

Why England went to war. The challenge was taken up. In 1792 William Pitt, the British prime minister, had said that he looked forward to half a generation of unbroken peace; but the execution of Louis XVI was the beginning for England of a generation of nearly unbroken war. The earlier stages of the French revolution had seemed to Englishmen almost wholly admirable. For a hundred years the French and British ideas of government had been violently in conflict; at last the French seemed to have been converted to the British system. Obviously they were a sensible people. But in November 1792 the Convention offered the help of France to all nations struggling to be free. This was a challenge to all international order, an offer to intervene in favour of any rebellion anywhere. It seemed to threaten an endless repetition of incidents like the invasion of Belgium and the opening of the Scheldt which followed it. Since 1713 the Scheldt, Antwerp's river, had virtually been closed by treaty to navigation. Antwerp was by her position well suited to be a danger to London in trade and to England in war. Consequently the French disregard of the treaty was heartily disliked by the British government. Then in January 1793 came the execution of the king. It was this which horrified England, led Pitt to expel the French ambassador and finally brought war.

The anger of the English, though sincere, was unreflecting, and unexpected by the French. After all, the September massacres had been far more horrible but had not led to war while, besides, England herself had executed a king less than one hundred and fifty years before. To the French the revolutionary wars were a crusade against tyrannical governments in which they could count on the sympathy of oppressed peoples for, as Rousseau had taught them, 'man is born free, but *everywhere* he is in chains'. They were largely right as far as their European enemies were concerned, but seriously wrong in the case of Britain. Englishmen not only sang but believed that they were not and never, never would be slaves. It was not only the British government but the British nation which fought against France.

French defeats. The spring of 1793 was disastrous for the French. Their armies were turned out of Belgium, their commander, Dumouriez, deserted to the enemy after losing the battle of Neerwinden and western France rose in rebellion. There the Catholic peasants of La Vendée, the remote, marshy district at the mouth of the Loire, started their long struggle against the wicked unbelievers in Paris who were driving away their priests and conscripting their sons to fight for a cause they loathed.

MAP 1. THE ENCIRCLEMENT OF FRANCE, 1793

1. This map shows the dangers which the Jacobin republic in Paris had to meet in 1793. The foreign countries at war with France are shaded—Belgium at that time belonged to Austria.

2. The seven arrows show the principal military attacks which the Republic had to repel. The broken arrow (1) shows the Austrian and Prussian invasion of 1792. Arrows 2, 3 and 4 show the invasions of 1793. Arrow 4 is the English army under the Duke of York.

3. The Republic had to fight French royalists as well as foreign enemies. Arrows 5, 6 and 7, linked by the Fleur-de-lys, show the royalist armies of Normandy, Brittany and La Vendée.

4. There was also trouble from the moderate republicans (Girondins), who were in rebellion in Bordeaux, Lyons, Marseilles and Toulon.

5. Notice the part played by the British navy. It not only blockaded the two naval ports of Brest and Toulon, but was able at different times to bring help to both groups of French rebels. It occupied Toulon in alliance with the Girondins in 1793, and two years later it sent a force to Quiberon Bay to help the royalists.

Map 1. THE ENCIRCLEMENT OF FRANCE, 1793

5. TERROR

March 1793 to July 1794

The revolutionary dictatorship. Danton was in Belgium in the February of 1793, but he hurried back to Paris when the first bad news came through. The Girondins had declared a war which their incompetence was losing. France looked to Danton to save her now as he had saved her in the previous September. He wasted no time. Undisciplined terror by the mob of Paris, and by the worst elements of that mob, appeared to have saved France in the autumn of 1792. Terror, organised by the best elements of the revolution, must save France in 1793. A revolutionary dictatorship was established.

The day after his return from Belgium, March 9, 1793, Danton persuaded the Convention to set up a Revolutionary Tribunal to try political prisoners speedily and without appeal. The anarchy of the September massacres was not to be repeated, but their paralysing effect on possible traitors was to be continued. 'Let us be terrible,' Danton told the Convention, 'that the people may not be.' A month later, on April 6, came the organisation of a Committee of Public Safety with Danton at its head, the first strong government France had enjoyed since 1789. This committee was elected by the Convention to supervise the government departments and make them more efficient. It soon became the master both of the government and of the Convention. Provincial France, which since 1789 had been governed almost entirely by locally elected officials and hardly at all by Paris, was brought under close control again by 'representatives on mission', Jacobin members of the Convention who were sent down with full powers to control doubtful towns and wavering sections of the army. These three—the Tribunal, the Committee and the representatives on mission—were the organs of the revolutionary dictatorship.

Fall of the Girondins. The next step was to deal with the Girondins. Once more a revolutionary organisation was set up by the Commune with Robespierre's approval. On the last day of May all was ready. The Paris National Guard was called out. The

Convention was not attacked, but it was besieged. Outside there were always crowds and often troops; at times as many as 30,000 men crowded the streets in the neighbourhood. Inside resolution after resolution was proposed. At last a compromise was found; the Girondin deputies agreed to resign while the revolutionary committee was to be dissolved. The committee and the crowd, however, refused to accept the solution, and when the deputies left the hall they found every street blocked. Frightened and ashamed they slunk back and at eleven o'clock that night of June 2, 1793, they did what they were told—dismissed the Girondin deputies and ordered the arrest of twenty-nine of them.

Spread of civil war. Ultimately the fall of the Girondins may have strengthened the republic, but its first effects were bad. There were now two civil wars to fight instead of one. Soon there were Girondin armies fighting for a moderate, decentralised republic as well as royalist armies in La Vendée fighting for Louis XVII (the Dauphin). Lyons, Bordeaux and Marseilles, the three great towns of the south and all centres of Girondin influence, declared war against the government. The Normans raised an army to march on Paris, the army of La Vendée grew, and grew victorious.

Success had deserted Danton, and so when a new Committee of Public Safety was elected in July his name was left out. He retired more and more from politics and eventually left Paris for his country home. Robespierre took Danton's place as a leader, while at his desk in the War Office Carnot, 'the organiser of victory', continued to slave day and night to build up the armies which in a few years were to carry the tricolour in triumph to half the capitals of Europe.

The tide turns. It was not until the autumn, however, that the tide turned. Then the French armies won two victories on the Belgian frontier at Hondescoet and Wattignies, where Carnot himself took part and led a charge. In October Lyons fell after a long siege. December was a month of victories. Toulon, the home of the Mediterranean fleet, was re-taken from a British expeditionary force which had held it since August; while in the west the army of La Vendée was shattered at Le Mans. The critical year had ended in triumph. The republic was never again in mortal danger from enemy armies. The revolutionary dictatorship established by Danton in the spring had saved France. It had been severe, but at first it had not been bloodthirsty. It had not delighted in terror for its own sake. It did so now when the case for persecution was weakening, when France was nearly saved. The Reign of Terror, which may be dated from a decree of September 1793 proclaiming terror 'the order of the day', makes 1794 one of the black years of history. It cannot be excused; it can be explained.

Famine and 'assignats'. The dangers through which France had passed in 1793 strained men's nerves almost beyond bearing. It was natural that the people of Paris and their leaders should see spies everywhere and imagine that every breakdown was an act of sabotage. Traitors there certainly were, but not half as many as most people thought. Perhaps profiteers were more common than traitors; certainly poverty and hunger were greater dangers to most Parisians than counter-revolution and espionage. It was impossible to see that Paris was well and cheaply fed; it was easy to distract attention by pretending that the shortage was the work of traitors. Possibly the politicians believed that it was. Actually the economic crisis was the result of two things: the general upset caused by the invasion and the civil wars, and the inflation which the government had carried out.

The States-General had been called because the royal government was bankrupt. The revolution had not solved the financial problem. It too was bankrupt. Since the government could not collect enough money by taxation to pay its bills, it had to print paper money for the purpose. At the end of 1789 the government had taken over all church property and in return it had undertaken to pay the priests. It had gained some exceedingly valuable property, but no money. This it got by borrowing on the strength of the church lands. The borrowing took the form of 'assignats', paper notes issued by the government which gave their holder a claim on the new national property. If 'assignats' had only been issued to the value of the church lands all might have been well, but gradually the government took to printing 'assignats' whenever it wanted money. Prices rose because the amount of money had been increased without any increase of the things which money buys. By the period we have now reached, 'assignats' were worth barely a third of their nominal value. Before long they were altogether worthless.

In Paris prices rose although the Commune tried hard to control them. It bought corn, for example, and sold to the bakers below cost price. All the same there were long queues outside the shops and prices went on rising. There was much discontent. One day a deputation of washerwomen went to the Convention to complain of the price of soap and to ask for the death of traitors; the next day grocers' shops were raided. At last 9000 or 10,000 men marched from the working-class district of St Antoine to demand that the Convention should fix a maximum price for bread. They got their way, but the economic crisis continued.

The Reign of Terror. The Reign of Terror can only be understood against this background of political suspicion and economic distress. The women who daily sat knitting in the Place de la Révolution watching heads fall into the basket below the guillotine were bitter, hating and hateful because they were scared and starving. Brutal punishments in public were nothing new, though they had never been so plentiful. The 'ancien régime' had whetted people's appetite for brutality; the reign of terror satisfied it. As time went on the executions grew more plentiful. In six weeks in the summer of 1794, 1300 people were guillotined in Paris: more than in all the previous fifteen months of the Revolutionary Tribunal's existence. Paris perhaps had taken its cue from the provinces, where with less excuse far worse atrocities had been committed by 'representatives on mission'. At Lyons Fouché, a runaway priest who was to become Napoleon's chief of police, punished the city for its rebellion by mass executions by artillery. At Nantes, the capital of La Vendée, Carrier invented the republican sacrifice called 'noyades'. His victims were placed in a boat with a hole in its bottom and left to drown.

Idealists and racketeers. The Reign of Terror was not necessary to save the revolution from disaster, for it had hardly started before the foreign invasion had been checked and the civil wars won. It was necessary if the relatively small group of Jacobin enthusiasts were to make a new France in their own image. They changed the names of the month, abolished the week, replacing it by a decimal calendar of ten day units and did many other strange things which were the outward signs of the unwanted new world they were intent on making. If these enthusiasts needed the Terror, so also did the other group of extremists, the men of the Commune, led by Hébert, who wanted to extort as much money as they could in safety.

At first Robespierre, the terrorist on principle, and Hébert, the terrorist for profit, worked together. In October 1793 they secured the execution of the Girondins who had been in prison since June. Madame Roland, whose house was their favourite meeting-place, noticed a statue of liberty on her way to the guillotine. 'Ah, Liberty,' she cried, 'what crimes are committed in thy name.' But it was the weakness of her friends which had made those crimes possible. They had hated, but they had not stopped the September massacres. A fortnight before Madame Roland's death Marie Antoinette had been executed. A France which had no room for Madame Roland could have no place for her.

Fall of Hébert and Danton. In the new year Robespierre decided that he could no longer endure Hébert. The man who said 'the richest of Frenchmen should not have more than a £120 a year' could hardly tolerate for long the perfect racketeer. To get rid of Hébert, Robespierre secured the help of Danton, who had come back from the country feeling that the need for terror was passing. Robespierre attacked Hébert because he was unworthy to be a terrorist; Danton attacked him because he was a terrorist. As soon as Hébert had been executed, therefore, Robespierre turned on Danton. Hébert fell because he disgraced the Terror, Danton because he disapproved of it. Within fourteen days of Hébert's death Danton was being tried by the Revolutionary Tribunal which he himself had set up.

Of all the leaders who went to the guillotine Danton alone retained his popularity with the crowd after his fall. It was said that during his trial his great voice could be heard on the far side of the Seine; and, certainly, the uneasy crowd outside seemed likely to turn into deeds the biting words of his defence. A popular rising might save Danton, kill Robespierre. A decree was therefore hastily rushed through the Convention taking away Danton's right of defence. 'I see now', he shouted, 'that in times of revolution power falls ultimately to the greatest scoundrels.'

For nearly four months Robespierre was solitary and supreme. How far he himself was personally responsible for the fiercer terror of these months is an open question. Some people think he was largely a figure-head, but at any rate his contemporaries thought him responsible and he certainly regarded himself as all powerful. In June a new law, following the precedent in Danton's case, took away the right of political prisoners to defend themselves. Prisoners could now be tried in batches, instead of individually, and the guillotine be fed more frequently. It was the height of the Terror, Robespierre's final triumph. While his victims suffered and his colleagues wondered when their turn would come, Robespierre busied himself with his new cult of the Supreme Being, a pale compromise between Christianity and irreligion which seemed to most Frenchmen either blasphemous or ridiculous. Meanwhile his enemies were frightened but determined. The survivors of the Girondins, of the Hébertists and of the Dantonists drew together, united only by their hatred of Robespierre and their fear for their lives.

Thermidor, 1794. A great victory at Fleurus opened the way for Robespierre's fall by removing the last danger of foreign invasion and the last excuse for the Terror. Carnot took the first step. He ordered the artillery of the National Guard, who were friendly to Robespierre, to go to the front. Robespierre knew that he was in danger, but he could not believe that the Convention, which he had so long dominated, would dare to overthrow him. He went to their hall and told them that the republic had new and unsuspected enemies, that terror was now more than ever necessary. The deputies, however, were more frightened for themselves than for the revolution or the country. After long hesitation they timidly summoned up courage enough to order Robespierre's arrest; his brother bravely stood by him and was arrested too. The Commune rescued them, however, and the Convention itself was now in danger. They entrusted their defence to a clever, roguish deputy called Barras, a noble by birth, who collected troops and marched on the Town Hall.

The hour demanded action. Robespierre faltered, Barras struck. He had no difficulty in entering the Town Hall and arresting Robespierre and his party. And so on July 28,1794—the tenth of Thermidor in the new revolutionary calendar—Robespierre was executed. He lay at the foot of the scaffold waiting his turn, his head bandaged, for he was wounded. Whether he had tried to kill himself or whether one of Barras's men had shot him has never been decided.

Lifted aloft, (Robespierre's) eyes again opened; caught the bloody axe. Samson (the executioner) wrenched the coat off him; wrenched the dirty linen from his jaw: the jaw fell powerless, there fell from him a cry; —hideous to hear and see. Samson, thou cans't not be too quick. Samson's work done, there burst forth shout on shout of applause. Shout, which prolongs itself not only over Paris, but over France, but over Europe, and down to this generation. Deservedly and also undeservedly....Stricter man...lived not in that age. A man fitted, in some luckier, settled age, to have become one of those incorruptible barren Pattern-Figures, and have had marble-tablets and funeral-sermons. His poor landlord, the Cabinet-maker in the Rue S. Honoré, loved him; his brother died for him. May God be merciful to him, and to us! (Carlyle.)

6. BONAPARTE, SERVANT OF THE REPUBLIC

July 1794 to November 1799

The Dauphin's death. In June 1795 a boy lay dying in prison. His father and his mother had both been executed and his death seemed only a matter of time: a race between the guillotine and his gaoler's bullying. Recently he had been better treated, but he was very ill. The boy was the Dauphin, to royalists Louis XVII. The leaders of the Convention hated him because after Robespierre's fall there had been a great revival of royalism in France and the young boy's life was a menace to their lives. Any day a revolution might depose them and enthrone him as a constitutional king. At the same time they dare not execute him, for that would be to provoke the rising they feared. All they could do was to hope his illness would be fatal.

At last he relieved their minds by dying: royalism virtually died with him. The only sort of monarchy which France would accept was a constitutional monarchy, and there was no longer a possible constitutional king. The new claimant to the throne was the boy's uncle, one of the first and most bitter of the émigrés.

The Directory. The death of Louis XVII decided that France was to remain a republic. But what sort of republic was it to be? Would it have the moderate, respectable government France wanted or would there be room for Barras and the ex-terrorists? The Convention was drawing up a new constitution under which there were to be two houses of parliament and a government of five Directors elected by the parliament. Barras and his friends were afraid that they would not be elected to the new assembly because of their growing unpopularity, particularly with middle-class people. They protected themselves by a law providing that two-thirds of the members of the new parliament must be chosen from the members of the Convention.

'A whiff of grape-shot.' France was furious. What, after all, was the good of elections unless one could turn out people one disliked? Once again there was a rising in Paris; but this time it came from the west end, not the east. The revolutionaries counted on the National Guard, which after Thermidor had again become a safe middle-class body. Barras and the Convention relied on fifteen hundred hastily armed recruits from the slums and on the army. The odds seemed heavy against the government, but fortunately Barras remembered young General Bonaparte and Bonaparte remembered the guns of the National Guard. Captain Murat galloped off at midnight to secure them. It was five in the morning of October 4, 1795 (Vendémiaire in the revolutionary calendar), before the anxious watchers in the Tuileries heard the clatter of his return. Bonaparte's memory had turned the tide. When the revolutionaries advanced up the Rue St Honoré they were scattered, in Carlyle's phrase, by 'a whiff of grape-shot'.

Napoleon Bonaparte. The Directory, as the new government was called, soon found employment for Bonaparte. In March Carnot appointed him to the army of Italy, his first independent command. Bonaparte's earliest enthusiasm had been the liberation of his native island, Corsica. Corsican politics in the eighteenth century turned on the rivalry of France and England in the Mediterranean. One party looked to England for help, the other to France. Bonaparte belonged to the French faction and it was quite natural that he should become a French artillery officer. In 1793, when the English were holding Toulon, Bonaparte had attracted some attention by his handling of the siege artillery. Latterly, however, he had fallen into disgrace through his friendship with Robespierre's brother—after Thermidor anyone in any way connected with Robespierre was suspect. He was now again a rising man, important enough to marry Josephine de Beauharnais, a Creole beauty, a widow well known in the promoted underworld which during the Directory was the fashionable society of Paris.

Josephine's friendship with Barras might be useful to a rising soldier of twenty-six, but it was not her influence but her beauty which had captivated Bonaparte. He fell passionately in love; she, too, was fascinated by his burning energy, ambition and confidence. His appearance was memorable, but not prepossessing. If you could catch sight of him—for he was short and in 1796 very thin—you could pick him out of any crowd by his long, untidy, chestnut hair; his hollow cheeks; his sunken neck and his quick, jerky movements which were yet something of a slouch. But what men and women remembered most were his deep blue eyes, sometimes smiling, sometimes terrifying, always alive. 'He talked', writes a French historian, 'as though he were making an attack, looking straight at his interlocutor and never allowing him to get on one side of him.'

French victories. But at least a dozen generals were better known than Bonaparte in 1796. There was, for instance, General Jourdan, who had commanded the French at Fleurus in 1794. True, the victory was really Carnot's, who corrected his mistakes and led the decisive charge in civilian clothes, but the credit was Jourdan's. After Fleurus his army had been irresistible, chasing the allies through the Low Countries and capturing Mainz in 1795, just before Bonaparte got his chance in Vendémiaire. Everywhere the French arms had been victorious. A royalist expedition sent from England to revive the La Vendée revolt had been overwhelmed at Quiberon Bay; the passes between France and Italy had been captured; the Rhine had become the French frontier and Holland had been subdued. In 1795, by the Treaty of Basel, Prussia had retired from the war. Spain, too, left the anti-French coalition. Only England, Austria and Piedmont remained.

Revolutionary armies. Civilian Paris had wearily exchanged the cruelty of the Convention for the corruption of the Directory, but Jacobin enthusiasm still stirred the armies. Under Bonaparte they were soon to gain such glory that these earlier successes are sometimes forgotten. How were they won? Largely by intoxicating enthusiasm and infectious courage, but also by three practices which distinguished the French armies from others: the attack in column, living on the country and the nation-in-arms.

The attack in column was a matter of necessity. The half-trained, semi-amateur French troops had neither the steadiness nor the training for the attack in line, with its greater firing possibilities, used by the professional armies of the eighteenth century. The enthusiasm, the weight and the dash of the French columns, however, overwhelmed the staid Austrian armies. The columns were protected by a thick screen of skirmishers, who by their individual tactics—suitable to the French genius—worried the enemy formation and drew its sting before the columns came into action. But the attack in column was always a second best and failed before the English armies of the Peninsular War.

The professional armies, and particularly the Austrian army, were extremely slow. When they needed provisions from the country they paid for what they took, but they carried all they could with them. Their enormous baggage trains delayed them; but in war speed is necessary. If one army can outmarch another of equal size it will probably beat it—its extra speed is the equivalent of extra men. The French armies travelled light, living on the country, raising food supplies by requisition. They were therefore very fast. This system worked well in the fertile and, on the whole, the friendly countryside of the Low Countries, Germany and Italy. It worked badly in the hostile and barren territory of Russia and Spain.

The original French revolutionary army was one of volunteers strengthened by ex-regulars, but soon conscription was introduced. The first stage was the 'levée-en-masse', the simultaneous conscription of all men between eighteen and twenty-five. This failed because the conscripts could not be trained. France needed soldiers, not men. Later the army authorities fixed the number of men required for each year, who were then requisitioned. Conscription meant that virtually the whole nation was trained to arms; since rich and poor fought side by side it meant also that the nation was trained to equality. France's enemies, on the other hand, had small, long service professional armies, highly trained and drilled but difficult to replace. Their training and tactics were not adapted, as were the French, to the quick assimilation of large numbers of raw recruits.

The Italian campaign. The army of Italy was the worst equipped and the least successful of the French armies. The men were in rags, their boots worn out, their food scanty, their pay forgotten. General Bonaparte encouraged them: 'Soldiers,' he said, 'you are naked and ill-fed; I will lead you into the most fruitful plains in the world. Rich provinces, great cities will be in your power. There you will find honour, fame and wealth.' Thus Bonaparte appealed to the greed not the self-sacrifice of his troops. His men were to be as enthusiastic as the earlier revolutionary troops, but they were to be inspired by professional spirit not by crusading zeal.

They needed their courage, for the plan of campaign for 1796 went wrong. Moreau and Jourdan, the crack soldiers of the republic, were to invade Germany while the untried Bonaparte was to hold an Austrian army in Italy. But Jourdan was defeated and Moreau forced to retire, so that Austria was free to move troops southwards over the Brenner against Bonaparte. The old hands failed, the novice triumphed. Bonaparte had been expected to hold one army: he routed five.

His first move was to drive a wedge between the Austrians and their Piedmontese allies, forcing them to retire on Milan and Turin respectively. The Piedmontese were glad to make a separate peace and Bonaparte turned on Milan. At the Bridge of Lodi he forced a passage across the Adda and took Milan, which enthusiastically accepted him as a liberator. The hardest fighting was to come. The great fortress of Mantua was strongly garrisoned. As long as it held out, the French conquest in Italy would be insecure, for it provided a perfect base for a reconquest. Four times attempts to relieve it were made and in four great battles Bonaparte beat them off. Bonaparte's tactics and strategy throughout the campaign were often faulty but he never made the mistake of indecision, which Austrian quality was worth at least an army corps to him. The failure of the fourth attempt to relieve Mantua led to its surrender in January 1797. The Austrian hold on Italy had gone and the Pope was at once forced to make peace with the irreligious republic.

Campo Formio. The month of March was spent in driving the last Austrian army back through the eastern Alps on Vienna. When at length the plain was reached Austria surrendered and preliminary peace terms were signed at Leoben. During the summer Bonaparte treacherously acquired Venice on the pretence of establishing a democratic republic in place of the old oligarchy, but really with the intention of handing it over to Austria as compensation for her lost territory. The Treaty of Campo Formio in October ratified the bargain. Austria gave up her claims to Belgium and Milan and recognised the Rhine as the French frontier. In return she received only Venice.

General Bonaparte, the promising young officer of 1796, was now the most famous man in Europe. His amazing self-confidence, courage and success, together with a willingness to allow plenty of plunder, had won him the affectionate allegiance of his army. They had already become his soldiers rather than the Directory's. With them behind him he had become a politician as well as a soldier. The treaties with Piedmont, with Austria, with the Pope were his work, not the French government's. He treated the Directory with bare courtesy and less obedience. The Directors bitterly resented and feared Bonaparte's attitude but they dared not remove him for three reasons. He was the hero of Paris to which he had sent, among other confiscated art treasures, the famous horses of St Mark from Venice. He was the paymaster of the Directory, to which he sent enormous sums extorted from the princes and peoples of Italy. He became its master, for it was to his sword that the Directors owed their continued power and the opportunities of plunder which it gave.

Barras and Carnot. While Bonaparte was conquering Italy the Directors had been quarrelling. The majority was led by Barras, an ex-noble, whose one aim was to make money and whose only policy was to continue the war and the Jacobin tradition, since only thus could he retain power. The minority, led by Carnot, wanted an honourable peace and a moderate policy at home. Carnot, as a member of the Committee of Public Safety, had shared the responsibility for the Reign of Terror, but his part in it was indirect and now bitterly regretted. But Barras, who had no objection to terror and was soon to practise it, never allowed Carnot to forget his past. One day, for example, they were quarrelling when Carnot raised his hands in protest. 'Do not lift your hands,' Barras exclaimed, 'they will drip with blood.'

Carnot's party was in a minority among the Directors, but it represented a majority of Frenchmen and, after the elections of 1797, a majority of deputies. The only way Barras could retain power was by removing the opposition. He accused them of being royalists and invented a royalist plot. Then he appealed to the army, whose soldiers were still Jacobins and whose generals wanted the war to go on. Bonaparte sent him an officer from Italy who surrounded the Tuileries, where the Assembly sat. Carnot escaped just in time, but many were arrested and transported to Cayenne, 'the dry guillotine', which killed as surely as the knife though less mercifully. This indirect intervention of Bonaparte's is known as the *coup d'état* of Fructidor (August 1797). The pretence of free government was given up and a revolutionary dictatorship re-established. There was, however, a difference between the dictatorship of the Directory and the dictatorship of the Committee of

Public Safety. The Committee had drawn its support from and had obeyed the mob of Paris; the Directory depended on and had to obey the army.

The Egyptian expedition. When Bonaparte returned from Italy, France had only one enemy left: England. An army was being prepared for its invasion and Bonaparte was offered the command. In 1797 the plan might have succeeded, for mutinies at Spithead and the Nore, due to the atrocious conditions of service in the British navy, had temporarily paralysed England. These mutinies gave the French an opportunity which was never repeated. By 1798, however, order had been restored in the fleet. Bonaparte realised that an invasion was now out of the question and devised a scheme for attacking England in another way, which the Directors welcomed because it removed their over-mighty general to a safe distance.

The plan was to conquer Egypt and use it as a base for a further attack on India. Only fifty years before, France under Dupleix had been as powerful as Britain in India. A great part of the country was still free from British rule and even where England ruled her position was none too strong. If the expedition had reached India, Bonaparte might have succeeded. But no sooner had he landed in Egypt in 1798 than Admiral Nelson destroyed the French fleet at the battle of the Nile. Bonaparte was now isolated in Egypt and, although he conquered that country, it was impossible for him to receive reinforcements from France or to go on to India. Instead he turned north against Turkey, but his army was checked at Acre. The French public were told only of his triumphs and believed that his troops had not entered Acre merely because they were afraid of catching the plague. His popular reputation, therefore, stood as high as ever.

At this moment Bonaparte heard that a new coalition had been formed against France, that his Italian conquests had been lost and that France herself was again in danger of invasion. It was a disaster for France, but it might be an opportunity for Bonaparte. Leaving his troops in Egypt he set sail for France, somehow managing to slip unnoticed through the English fleet. On October 9, 1799, he landed in France. On November 9 he overthrew the Directory.

MAP 2. FRANCE AND AUSTRIA, 1796–1809

1. The bulk of the Austrian dominions lay east of the Alps, but some of her richest provinces were north and south of them. Austria was like the man in the inset, whose legs straddled the Alps but whose trunk lay beyond them. The capital, Vienna, was roughly at the man's waist. France could attack Austria by either 'leg'. The German 'leg' was best reached by the Danube, the Italian by the gap between the high Alps and the Mediterranean.

2. Now look at the principal Alpine passes (1–4). The Brenner (3) is the lowest and most important. It allowed Austrian troops to move from Germany to Italy or vice versa. At the Italian end it was guarded by the Quadrilateral fortresses. The Brenner is the key to Napoleon's first Italian campaign (1796–7).

3. The Villach-Klagenfurt gap (4) is followed to-day by the railway from Vienna to Venice. It was used by Napoleon in 1797. Once he had forced it, Austria made peace at Leoben.

4. The St Gotthard (2) enters Switzerland by a high saddle between the Rhone and the Rhine. It is not really suitable for armies, but was used by the Russian Suvoroff in 1799 on his way to attack the French at Zürich.

5. The Great St Bernard (1) is the highest road pass and impracticable for generals—except Napoleon, who used it in 1800 to cut off the Austrians who were besieging Genoa. His victory at Marengo turned the Austrians out of Italy but did not end the war since Austria still had her German 'leg'. Hence the importance of Moreau's victory at Hohenlinden.

6. In 1805 Napoleon had to crush Austria before she could be reinforced by Russia. A lightning concentration of seven armies at Ulm wiped out the Austrian advance guard and gave Napoleon Vienna. He was lucky in crossing the Danube and winning Austerlitz before the main Russian army came up.

7. In 1809 Russia was neutral. Napoleon's difficulty in crossing the Danube this time—it took him six weeks and one drawn battle—showed how lucky he had been in 1805. Properly defended the Danube was almost as big an obstacle as the Alps. Once it was crossed, however, Austria was relatively defenceless, as the victory of Wagram showed.

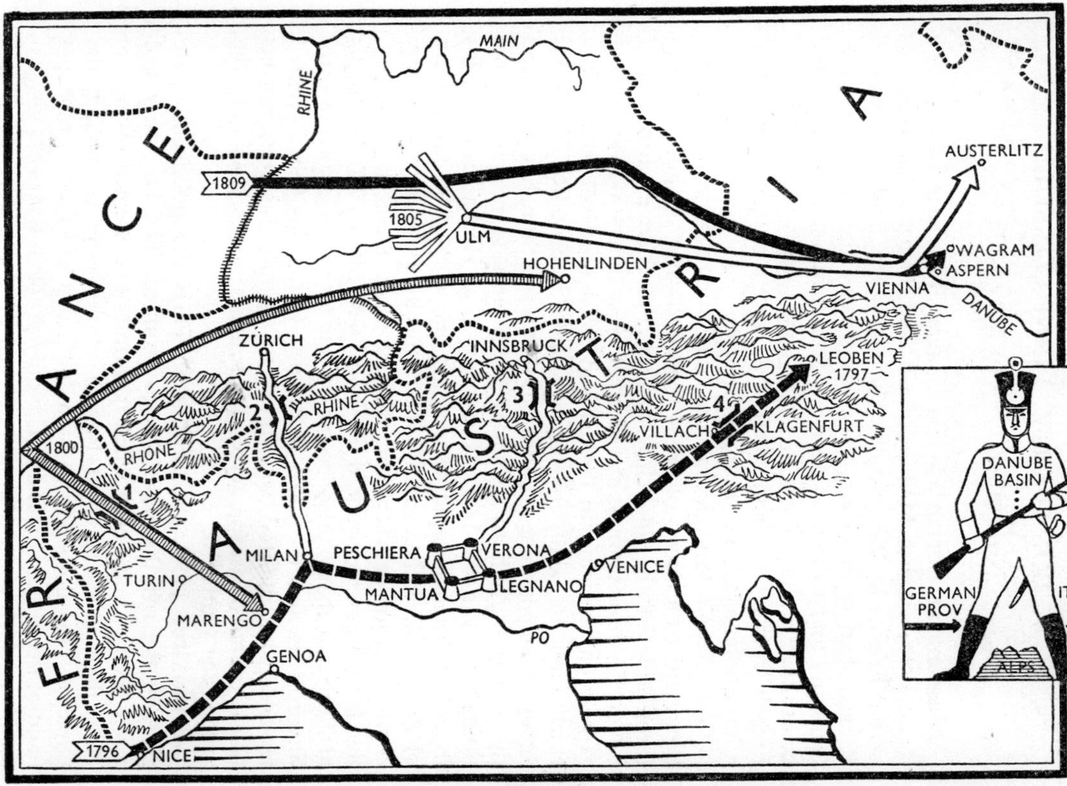

Map 2. FRANCE AND AUSTRIA, 1796–1809

7. BONAPARTE, FIRST CONSUL

November 1799 to May 1803

Masséna at Zürich. While General Bonaparte was slipping through the English Mediterranean fleet, General Masséna was saving France in Switzerland. The allies—England, Austria and Russia—had reconquered Italy, overthrown the French army in Germany and were massing for a final invasion of France. But before they could do this in safety it was necessary to drive the French from Switzerland where they were a danger to the communications of armies invading France.

From every side armies marched on Zürich near which Masséna lay. Across the snows of the St Gotthard came the last and most formidable of the encircling armies, the Russian army under Suvoroff, whose ugliness was as great as his superstition and whose exploits and atrocities in Turkish and Polish wars had brought him equal fame for skill, for courage and for brutality. The odds against Masséna were so enormous that it was thought safe to send the Austrian Archduke Charles northwards to prepare for the invasion of France. Then Masséna struck. He had watched his immediate enemy stupidly spreading one wing of his army north of Zürich, the other south with its back to the lake—the only communication between them through the bottle-neck of the city. Suddenly Masséna pounced on the helpless enemy, who lost all his artillery and many prisoners as he cut his way out of the trap. Suvoroff reached the Lake of Lucerne to find his friends routed beyond recall; he, too, was forced to abandon his guns and his baggage and to escape across far rougher passes than even the arduous St Gotthard.

Brumaire, 1799. Meanwhile in Paris the politicians were plotting the overthrow of the government. The Directory had never been popular even when its armies had seemed invincible. Now that on every front French armies were crumpling, the Directory was doomed. Masséna's victory had saved France; it had not saved the Directory. The only problem was what government or which general should succeed it. Some were planning a return to Jacobinism; others a royalist restoration. The success of either group meant prison, exile or death for the Directors. One at least of their number was determined to save himself in the only possible way by being the first to organise a *coup d'état*. The abbé Sieyès hunted busily for the right soldier for his purpose. He tried General Bernadotte, he approached General Moreau, at last he found General Bonaparte.

In November 1799 (Brumaire in the revolutionary calendar) Sieyès and Bonaparte agreed that parliament should be persuaded to adjourn to St Cloud, where it would be safe from 'Jacobin plots' under the 'protection' of General Bonaparte. The other Directors were to be forced to resign, and parliament was to be bewitched into doing nothing while the abbé Sieyès drew up a perfect new constitution—he had taken a hand in making every constitution since 1789, but in spite of their depressing failure he still believed in himself and his constitutions.

The first stage of the plot was successful: parliament moved to St Cloud, three of the five Directors resigned and the others were detained in the Tuileries. Then plans went wrong. The deputies at St Cloud grew suspicious; General Bonaparte was howled down as he entered their chamber and he and his escort of four grenadiers were rudely hustled out. The *coup d'état* and its leaders were in danger. It was not General Bonaparte who saved the day. It was his brother, Lucien, who harangued the wavering troops and persuaded them against all evidence that the deputies were in the pay of Pitt. It was General Murat who marched his men into the Chamber and ordered them to throw the deputies out of the windows. The Directory had failed to commit the gentlemanly suicide which Sieyès had planned. It was a pity, for Sieyès valued above all things constitutional good manners. He had one consolation: the hitch had been due to Bonaparte's bungling. Plainly there would be no need after all to fear him as a possible rival. Sieyès could govern France in peace. But just a month later, when the new constitution was ready, Sieyès found that he had been pensioned off to his country estate and that General Bonaparte had made himself First Consul and virtual dictator. The people of France ratified the arrangement at a plebiscite by 3,000,000 votes to 1500.

Masséna at Genoa. Bonaparte was First Consul, but France was still in danger. His own army was marooned in Egypt, the army of the Rhine was shattered, the army of Italy was a fragment besieged in Genoa. The Austrians were only waiting for that city to fall in order to invade France. Cut off from the land by the Austrians and from the sea by the English, Genoa under General Masséna starved but would not surrender. There was no sign of the promised relief. Despair ruled the city but still Masséna held on knowing that he alone was saving France from invasion. At last a messenger broke the blockade, rowing a boat along the coast and eventually swimming ashore through a fierce fire from the English ships. Then at last Masséna surrendered, for his work was done: the First Consul had crossed the Alps.

Marengo, Hohenlinden, Lunéville. Bonaparte's passage of the Great St Bernard, the highest of the principal Alpine passes, undertaken in the face of fierce snowstorms, more than a month before the time when it is normally open to traffic, had completely surprised the Austrians and cut off their retreat. If Bonaparte could beat their army in the field—and he never doubted his ability to win a battle—then not only was France saved but Italy reconquered. A defeat, however, would have been as disastrous for him as for the Austrians, since he too had no possible line of orderly retreat. The armies met at Marengo and at first the Austrians won all along the line. Their commander even sent a messenger to Vienna to announce his victory. In the evening, however, General Desaix, whose force had been detached from the main French army, rejoined Bonaparte. With his fresh troops the tide was turned, the Austrians shattered, Italy re-won. Bonaparte's gratitude to Desaix was increased by his death at the moment of victory. There was no one left alive to share Bonaparte's glory.

France had accepted Bonaparte as First Consul because there was a chance that he might bring order out of anarchy at home and peace with honour in Europe. It was a gamble but it was being justified. Marengo established his power in France, but it did not bring peace in Europe. That came only after a winter campaign in Germany, where General Moreau defeated an Austrian army at Hohenlinden. By the Treaty of Lunéville Austria recognised the Rhine as the French frontier and agreed that Italy should be virtually a French protectorate. Russia too left the coalition; but still there was not complete peace. England was still undefeated, still determined to beat France. The First Consul had promised France peace, but it must be peace with honour—on that both he and France were agreed.

The Armed Neutrality. He saw his chance to humble England by organising the neutral powers against her. The nations of the Baltic provided indispensable naval stores—hemp and tar—and were also one of the world's granaries. (For many years the world price of wheat was the Danzig price as the Chicago price is now.) The strict English blockade hit their trade, illegally as they held. England, for instance, considered food as contraband; they limited the term to actual war supplies. It was easy to rouse their feeling and that of the United States against England and, as a result of Bonaparte's diplomacy, the Armed Neutrality was formed. The Baltic powers, including Russia, agreed to defend by arms their view of neutral rights. In this way Bonaparte hoped to weaken English sea power by depriving her of marine stores and English industrial power by depriving her of markets. Thus weakened he believed that she would make peace with France on terms as favourable as those accepted by Austria at Lunéville. The plan might have succeeded but for two unexpected events. Admiral Nelson, disobeying orders, destroyed the Danish fleet and reopened the Baltic to British trade. The Tsar of Russia was murdered and his successor, Alexander, made peace with England. The Armed Neutrality had fallen to pieces.

Treaty of Amiens. Bonaparte's plan to dictate peace to England had failed. Nevertheless peace was made. Bonaparte needed it to complete his reorganisation of France and her conquests; England needed it to restore her commerce. Preliminary terms were signed in October 1801, and six months later by the final Treaty of Amiens England recognised France's European annexations, while France accepted England's conquests of Ceylon from Holland and of Trinidad from Spain. Other conquests were returned. Both in England and in France men went mad at the news of peace. In London the crowd unharnessed the horses from the French representative's carriage and pulled him in triumph through the streets, cheering Bonaparte the while. In France Bonaparte worship grew rapidly. The people showed their gratitude by electing him First Consul for life by 500,000 more votes than he had secured two years before when he was made First Consul for ten years.

The peace of Amiens proved to be only a truce. Early in the summer of 1803 France and England went to war again, a war which lasted without a break until Bonaparte's abdication. This time the struggle was of a different nature. The French armies of the revolution had been missionary armies spreading through Europe the revolutionary faith. The kings of Europe and the English oligarchy had fought them, not because they were French, but because they were revolutionary. The armies of Napoleon Bonaparte were French armies, fighting for the glory of France, not revolutionary armies of the old type. They were opposed because the greater France, for which they fought, could only be achieved at the expense of a lesser Austria and a lesser Britain. Indeed the old cause of hostility had nearly gone. France under the Consulate had become a respectable, conservative nation of whose ideas no European power need be afraid. The new wars were for the conquest of territory, not for the victory of ideas.

Why war was renewed. It is doubtful whether Bonaparte or England wanted war in 1803. It is certain that war was inevitable. Englishmen genuinely believed that it was their birthright to command the Mediterranean and other seas. They were determined to hold what their fathers had won. They did not argue the question on grounds of right and wrong, but were simply determined to defend their possessions which were now threatened by France's annexation of Belgium, by her protectorates over Holland and Italy, by her alliance with Spain, by the example of Bonaparte's Egyptian campaign and by the prospect of a new French empire in America based on Louisiana (the Mississippi valley) which had just been acquired from Spain. England was also beginning to think herself the hereditary and sole manufacturer of Europe. The Treaty of Amiens had been welcomed because it seemed to offer England the return of her 'natural market' for manufactured goods in France. Bonaparte's decision to protect French industry from English competition was another grievance.

France's claims were just as genuinely held as England's. She claimed as hers by natural right the frontiers which Caesar had once assigned to Gaul, frontiers which included people to whom French was as foreign a tongue as it is to Englishmen. She claimed further that these frontiers should be surrounded for her protection by a fence of buffer states whose object was not to benefit their citizens, but to ward off any blows that might be aimed at France. Lastly, France claimed to be the centre of European civilisation and as such to be the greatest power in Europe.

The most sacred prejudices of France and England were so contradictory that war was inevitable. The actual pretexts were relatively unimportant, but illuminating because they illustrate the French determination to control Europe and the English determination to control the Mediterranean.

Switzerland was a French bastion in central Europe from which the communications of any power invading France could be threatened. In 1803, the Helvetian republic, a French revolutionary off-shoot, was in disintegration and Bonaparte intervened to give Switzerland a constitution which, while satisfying the Swiss, left their country under France's control in foreign policy. Switzerland legally was none of England's business; but practically it was, since although England could afford to buy help from central Europe, this help could not be effective while Switzerland was in French hands. The other dispute concerned Malta which Bonaparte had taken from the old crusading Knights of St John. England had soon conquered it from the French; but under the Treaty of Amiens she had to hand it back to the Knights and thus neutralise it. England, however, refused to evacuate the island and thus broke the treaty.

In May 1803 war broke out again. To finance his army Bonaparte promptly sold Louisiana to the United States, thus leaving North America to the English-speaking peoples.

8. THE REORGANISATION OF FRANCE 1800 to 1810

Making the Code Napoléon. In the Tuileries the Council of State is gathering: grave and learned men, famous cabinet ministers, prudent lawyers, discreet civil servants—'the fifty least stupid men in France', as one of them said. They have been debating for many days the laws which shall govern France and they have lost their bearings in a tangle of legal details. It is seven in the morning—the council meets early and sits late for, as Bonaparte told it, 'we must earn the money France is paying us'. The agenda is dull, the councillors listless. Suddenly a roll of drums is heard, the ushers throw open the doors, the councillors rise, the First Consul hurries to his desk, nods to his two fellow consuls, 'the two arms of his easy chair', and with a brisk 'Come, citizens, to work', starts the day's business. The discussion turns on the obedience which a wife owes her husband, always a favourite subject with Bonaparte. One of the councillors asks whether this obedience had been laid down in the old laws of France. 'What a question!' Bonaparte retorts. 'Do you not know that the angel told Eve to obey her husband?' The law of France still repeats the angel's order.

Of course these interventions were the exception, but the 'Code Napoléon', the summary of a Frenchman's rights and duties, contains little of which Bonaparte disapproved and much which but for him would not be there. It is moreover doubtful whether, without his tempestuous driving force, the heavy task of overhauling all the law of France would ever have been accomplished.

In a settled society like that of England we take the laws for granted, accepting them as unthinkingly as we accept the presence of the river Thames. It is difficult therefore for us to understand how greatly the laws shape our lives: how different we should be *if* we lived under different laws, how different the French are *because* they live under different laws. No Frenchman of 1800, when the First Consul set lawyers to work to 'codify' the civil law, could have made that mistake. Not only had there been before the revolution one law for the rich and another for the poor; but since the revolution there had been different laws almost every year. It was Bonaparte's object to sum up the work of the revolution by weeding out the bad laws, keeping the good and making the whole simple to understand.

The Civil Code, dealing with the rights and duties of citizens among themselves, was finished by 1804. At the same time another committee was drawing up a Code of Civil Procedure laying down the way in which the law courts were to work. These codes were followed under the Empire by two criminal codes, dealing with the duty of the citizen to the state and the way in which offenders should be tried and punished, and by the Commercial Code which laid down the rules on which business should be carried on. In his exile at St Helena Napoleon declared that his chief glory was not the forty battles he had won, but his code of laws. Certainly the five codes were the most lasting part of his glory. Not only are they still the basis of French law, but they also spread throughout the whole of the Napoleonic Empire the knowledge of how to preserve by laws the free and equal society which the revolution had taught men to desire.

Social changes and the law. Three illustrations will show how French law differs from English law and how the Napoleonic law is related to the revolutionary and pre-revolutionary laws of France. The first is the law of inheritance. Before the revolution a Frenchman, like an Englishman, could leave his property by will to anyone he liked. The revolution took away this right and in the interests of a more equal distribution of property divided nine-tenths of a man's property equally among his children, leaving him only one-tenth to dispose of as he liked. The civil code raised the proportion he could leave by will to a quarter or a half according to the size of his family. This compromise stopped the dividing up of France into holdings which were too small to support a family, as had been happening during the revolution. It also hindered the growth of large estates and kept France a nation of peasant proprietors.

The second illustration concerns the family. In pre-revolutionary France the power of the family had been great—Mirabeau, as we have seen, had been sent to prison by his father under a 'lettre de cachet'. During the revolution family life, at least in Paris, was shattered by the emancipation of children and wives, and by the easy granting of divorce. The civil code restored much of the father's old power—he could send his son to prison if he paid for his keep—and much of the husband's power—his wife was forbidden to sell, give away or mortgage any of their common property without his written consent. Divorce was again made more difficult. During the revolution it had been obtainable if either husband or wife complained of 'incompatibility of temper'. The conservatives wanted to abolish divorce altogether—the Roman Catholic church has never allowed it—but the civil code continued to allow divorce if both husband and wife wanted it and if the relations agreed. If the family is still the most important thing in French social life, if it plays a larger part in France than in any other European country, this is largely due to the civil code.

The third illustration comes from the criminal law. In pre-revolutionary France an accused man was neither allowed to be defended by lawyers nor to know the case against him, while he was forced to answer the questions of the judge who was preparing the case. When this judge had finished he submitted a report on which the full bench of judges returned a verdict and, if necessary, passed sentence. The revolution in its early admiration of England introduced trial by jury. It allowed the accused to be defended by lawyers and did not force him to give evidence against himself. This system favoured the accused as much as the old system had favoured the prosecution. Napoleon again produced a compromise between the pre-revolutionary and revolutionary systems. The old system of preliminary investigation by a judge sitting in secret was restored; but, in spite of the war-time unpopularity of all things English, a jury was retained for the trial proper. It was, however, an emasculated jury, for the defendant lost his right to object to the jurors, who were chosen by the prefect, the government representative. French law is still less favourable than English law to an accused man.

The new constitution. The civil code laid down the principle of a family despotism under the father; the new constitution established a political despotism under Napoleon Bonaparte. Here again Bonaparte's work is mid-way between the old régime and the revolutionary order, or rather disorder. The traditional government of France was a strong monarchy. This Bonaparte restored. Both as First Consul and as Emperor he ruled alone. He had the advice of an expert Council of State, but he was not bound by their views; still less was his power limited by the feeble array of parliamentary assemblies—a Senate, a Tribunate and a Legislative Body—which acted largely as a sounding board for his voice.

The same unity of command was restored in local government. The revolution had made it an affair of committees. There were elected boards for this, that and everything; the ordinary citizen was bored by too much electioneering, and local government became a wire-puller's paradise. Then the revolution had abolished the old provinces with their cherished traditions and privileges and instead had carved up France into a number of equal sized 'départements'. A man might be proud of Alsace, he could hardly be proud of the Département of the Haut Rhin. Henceforward there were to be no more Gascons, Bretons or Normans but only Frenchmen. Bonaparte kept the new 'départements' but substituted a prefect appointed by himself for a committee elected by the 'département'. In each village or commune he substituted a 'maire' appointed by the prefect for one elected by the villagers. The motto of the revolution had been 'the republic; one and indivisible'. Bonaparte accepted this, but added 'the government of the republic; one and indivisible'.

The revolution preserved. Both Bonaparte and the Bourbons believed in authority, but they used it for different purposes. When Bonaparte was made First Consul his proclamation began: 'Citizens, the Revolution is true to the principles from which it sprang; it is now at an end.' It had been finished, not 'put on the shelf'. His first coins showed himself as Emperor, but still bore the inscription 'République Française'. He insisted that the social reforms of the revolution must be maintained: there must be no question of the peasants losing the land they had taken, every career must remain open to talent—'every private must carry a marshal's baton in his knapsack'—every Frenchman must be treated equally by the law no matter who his father or what his trade. This was the essential French revolution. The aim of the revolution, as Bonaparte said, was not the overthrow of the Bourbons; but incidentally, because of Louis XVI's opposition, it had been necessary to abolish the throne and even to execute the king. For a similar reason—to safeguard the results of the revolution—it was now necessary to appoint a new king under a different title. The Bourbons had ruled by Divine Right, the Bonapartes were to rule by popular vote. In 1801, when the anniversary of the Fall of the Bastille was celebrated, Bonaparte gave a toast 'To the 14th of July! And to the French people, the sovereign ruler of us all!' As a French historian remarks: 'He wished them to be the sovereign people because he felt they were ready to delegate their sovereignty to him.'

Peace with the church and the nobles. Bonaparte's mission was to reconcile all Frenchmen by persuading them to accept the revolution. The government must cease to make war on royalist Frenchmen; outlawed Frenchmen must cease to make war on the revolution. Two great measures pacified France. The first was the Concordat, or agreement with the Pope, made in 1801, which ended the religious quarrel begun ten years before by the Civil Constitution of the Clergy. The church accepted the revolutionary land settlement and recognised the Consulate. In return the salaries of the clergy were paid by the state. The French government was given the right, which had been enjoyed by the old kings, to nominate the bishops. In this way the Bourbons lost their chief support, the blessing of the church.

The second measure was the recall of the émigrés. The great body of royalist refugees, numbering more than 100,000, were pardoned. They disliked the word, but they liked the opportunity. A Frenchman in London wrote: 'Nearly all of them swore that it would be dishonourable to accept such an amnesty, but while the words were still on their lips they were making preparations to leave; those who had set out on the previous day were bitterly reproached by those who were going to follow them on the next.'

Economic and social progress. With a united nation behind him Bonaparte embarked on a policy of expansion. The old roads which had become almost impassable were remade, new roads were built. Canals were dug, harbours drained, Paris rebuilt. Everywhere unemployment vanished and men in their unaccustomed prosperity blessed the name of Bonaparte. Education was encouraged. It, too, was to be 'one and indivisible' in the service of the state, teaching boys to be good Frenchmen and supporters of the government. A national university of France was founded to replace the old separate provincial universities, and secondary schools were established in the big towns. Bonaparte's was the first modern dictatorship based on popular consent. He realised, as his successors have done, that popular consent is won by propaganda.

The new nobility. It is also won by encouragement of good service to the state. In 1802 Bonaparte founded the Legion of Honour, still coveted by Frenchmen. It was given to soldiers and civilians alike, and was generally welcomed since it carried no title and was not hereditary. The Legion of Honour gained support for the government; the creation of a new nobility under the Empire weakened it. The old nobility accepted new titles, but sneered at the man who gave them and at their fellow nobles. Greed, the worst characteristic of Napoleon's servants, was encouraged by title-hunting. Ordinary men complained that the equality for which so many Frenchmen had died was being sacrificed to Napoleon's vanity. They felt that the man they had made ruler to safeguard the revolution was trying to become a king like other kings. They were right; but, unfortunately for Napoleon, his nobles were not quite like other nobles. The first of the new peers, the Duke of Danzig, was married to an ex-washerwoman who used constantly to exclaim: 'When I used to do the washing.'

Napoleon's legacy. It is difficult for Englishmen to regard Napoleon Bonaparte as anything but a soldier; to a Frenchman he is as much statesman as soldier. He is the founder of modern France. His personal despotism disappeared, but the rest of his work lasted. His Concordat governed the relations of church and state until the beginning of this century; local government is still organised as Bonaparte left it; the framework of the educational system is still his; his laws are still the laws of France, and the social equality which he saved from the anarchy of the revolution is still the most precious possession of every Frenchman.

MAP 3. NAPOLEON IN NORTHERN EUROPE

1. Notice how flat the country is except in the south and west, where there were many towns, villages and roads. In the east the population was scattered and communications bad.

2. The great power in the north German plain was Prussia; Austria was the leading state in the mountains and hills of the south. The only easy route from Austria to Prussia was by the gap made in the Erzgebirge by the Elbe as it leaves Bohemia (Austrian territory) for Saxony. This is shown by the most southerly of the three white arrows on the map. Napoleon made Saxony a strong buffer state between Austria and Prussia, giving its ruler the Grand Duchy of Warsaw and including it as a French protectorate in the Confederation of the Rhine.

3. The broken black arrows show the course of the Jena campaign (1806). Its characteristic was Napoleon's speed—the battle of Jena was fought only six days after the French began to advance from the Main. They reached Berlin eleven days after Jena and Lübeck within twenty-four.

4. The black and white arrow shows the Eylau-Friedland campaign of 1807. The cold, the shortage of food and the terrible roads nearly defeated Napoleon, but he was fortunate in securing a good peace from Russia at Tilsit after his victory at Friedland, which followed a drawn battle at Eylau.

5. The heavy black arrow shows the starting-point for the disastrous Russian campaign of 1812. Napoleon was lured as far east as Moscow—one and a third times as far from Tilsit as Tilsit is from Lübeck.

6. The three white arrows show the campaign of 1813 in which Napoleon's aim was by holding Saxony to prevent the concentration of the three allied armies against him. The loss of his cavalry in the Russian campaign was fatal to him. Because he was not able to follow up his victory at Dresden, the key town guarding the Elbe gap, the allies were able to crush him at Leipzig.

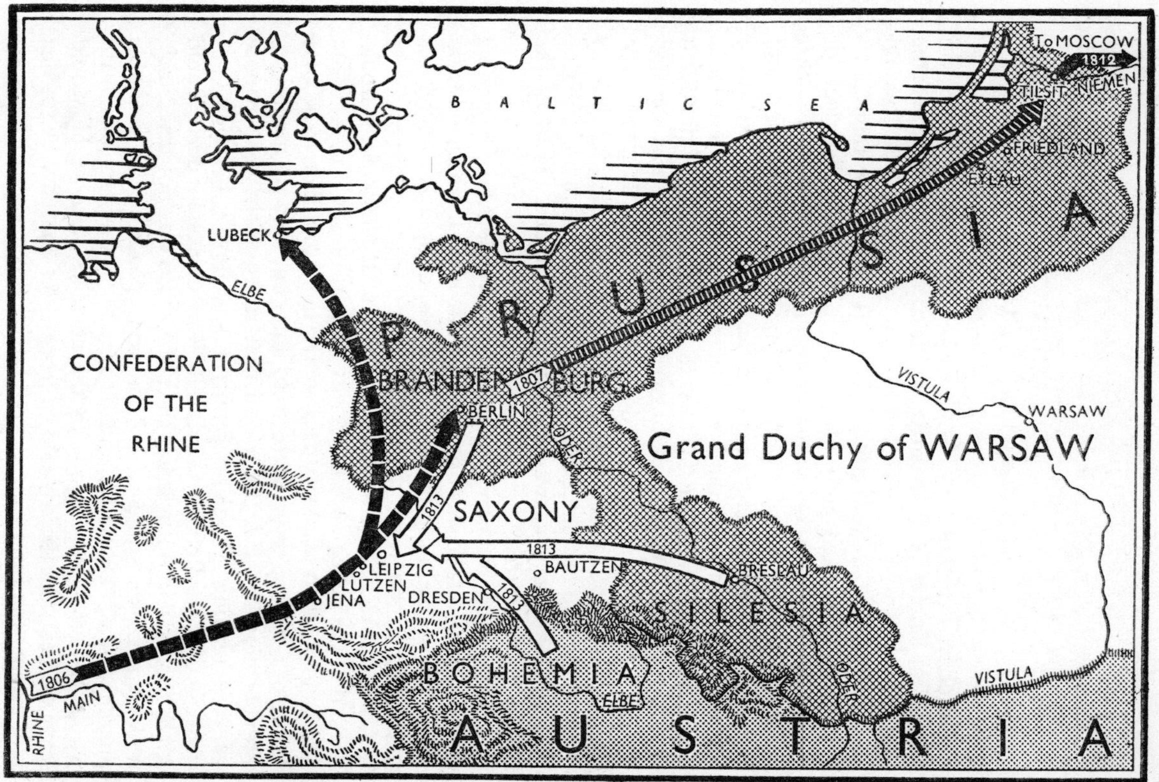

MAP 3. NAPOLEON IN NORTHERN EUROPE

9. NAPOLEON

May 1803 to July 1807

Fouché. Bonaparte was still First Consul when war began again, but before the first battle had been fought he had become Emperor of the French. The story of how this happened would be melodrama if it were not tragedy. It shows how republican opinion was tricked into supporting the re-establishment of monarchy. The first character in the story is Joseph Fouché, Napoleon's Minister of Police, a former priest but 'first and foremost a man who minds his own business and after that other people's'. He was one of the few who dared to contradict Napoleon. The Emperor once taunted him with having voted for the death of Louis XVI. 'Quite true,' answered Fouché, 'it was the first service I had the occasion to do Your Majesty.' It was by the execution of another Bourbon that Fouché finally cleared the way for Napoleon to become Emperor. Having no scruples himself he believed that everyone was equally unscrupulous. Most of the people he met were.

Napoleon disliked him but valued his services. At the beginning of 1803, however, he was not putting quite his usual high value on them. The special ministry of police was shut up and Fouché was unemployed though not disgraced.

A good royalist plot would help Fouché. It would show Bonaparte that the police were inefficient without Fouché's supervision and incidentally it would help Bonaparte in his ambition to become Emperor. France was devoted to him but not to monarchy. France must be made to realise that it would have to choose between making Napoleon Emperor and having the Bourbons back. As long as Bonaparte was only First Consul his death would leave France without a head, since the office was not hereditary, and the royalists would have a chance to profit by the inevitable anarchy. If Bonaparte were made hereditary Emperor, his death would be less advantageous to the royalists because he would leave an heir, and there would be less risk of anarchy. A royalist plot against the First Consul would drive this argument home by showing Frenchmen that only Bonaparte's life stood between them and the Bourbons.

The Cadoudal conspiracy. A royalist plot would be useful, a royalist plot there should be, a royalist plot there was—probably encouraged by Fouché. An émigré, probably in Fouché's pay, encouraged the exiles to think that there was a real chance of a joint royalist-republican rising against Bonaparte, who was to be ambushed and, if necessary, shot. Georges Cadoudal, one of the fiercest of the La Vendée leaders, was sent over with British government money on a British warship to Bilville near Dieppe, where friends helped him to climb the tall cliffs. For five months he lived in Paris undiscovered by the police, then he sent to England for Pichegru, a republican general who had turned royalist, and the Comte d'Artois, Louis XVI's younger brother. The former was to win over General Moreau, the republican victor of Hohenlinden; the latter was to rally royalist support when the rising took place. The Comte d'Artois stayed at home; Pichegru came, but was unsuccessful. Moreau was prepared to consider a rising in his own favour, but would have nothing to do with the Bourbons.

The plot had failed; it was now discovered. Probably Fouché had told Bonaparte about it: at any rate the First Consul began to accuse the official police of blindness. Two minor members of the gang admitted that Cadoudal and Pichegru were in Paris, that Moreau was implicated and that a Bourbon prince was expected. For six weeks the cliffs at Bilville were closely watched, but no prince came. Then Moreau was arrested—an excellent by-product of the plot since Moreau was a possible republican rival to Bonaparte. But Paris was suspicious. Moreau was a popular hero, a good republican, the last man in the world to be mixed up in a Bourbon plot. Where were his fellow conspirators? It looked suspiciously like a 'frame up'. However, the police caught Pichegru and eventually Cadoudal. The government breathed again; the plot was going to work.

d'Enghien's murder. It was still, however, necessary to produce the Bourbon prince. At this stage news reached Paris that the Duc d'Enghien, a cousin of the pretender, was living just across the German frontier, secretly visiting Strassburg and entertaining the republican traitor, Dumouriez. The Duc d'Enghien would do. Fouché advised Bonaparte to kidnap d'Enghien, bring him to

France, 'try' and execute him. The plan was carried out in spite of the fact that Bonaparte and Fouché learnt from d'Enghien's papers that he had never visited Strassburg, that his visitor was not 'General Dumouriez' but a harmless M. Thumery and that d'Enghien was in no way connected with the plot.

Emperor of the French. Still d'Enghien was dead and Fouché delighted: the judicial murder of a Bourbon prince would rally discontented republicans and horrify royalists. In May Bonaparte was proclaimed Emperor, in June Cadoudal was executed and Moreau exiled, in July Fouché was reappointed Minister of Police, in the autumn a plebiscite ratified Bonaparte's election as Emperor and in December the Pope himself came to Paris and hallowed Napoleon, who however insisted on crowning himself. Cadoudal summed up the whole affair from prison: 'We have done more than we expected; we came here to give Paris a king and we have given her an Emperor.'

The naval war and the 'Army of England'. England went to war with France for the second time in the spring of 1803. It was two years before she persuaded Austria and Russia to join her in the Third Coalition. During the interval Napoleon's soldiers waited impatiently for a chance to defeat the inferior British troops, while the English sailors longed for a chance to destroy the weaker French navy. Neither side could reach the other.

Napoleon tried three main schemes to make an invasion of England possible. His first idea was to build a flotilla of auxiliary fighting boats which could protect the crossing of his troops; but the flotilla was expensive, unseaworthy and a drain on the resources of the undermanned and ill-equipped French navy. Its hope lay in a surprise crossing, but experience showed that it would take several tides to embark the French army. After the Spanish fleet had reinforced the French—Spain declared war on England at the end of 1804—the plan was changed. The English fleet was to be lured away to the West Indies on a wild chase of the French fleet (England was as sensitive then about West Indian sugar as she is to-day about Middle Eastern oil). Napoleon would thus get a temporary command of the sea by a stratagem instead of a victory. Villeneuve attempted this and failed; so in the late summer of 1805 Napoleon ordered a third scheme to be tried.

Villeneuve was to obtain a real and lasting command of the sea by a successful battle. On paper this was possible since the French fleet in home waters was as numerous as the British, but in fact it was impossible because the French were poor seamen. As Villeneuve pathetically remarked: 'We only know one manœuvre, to form line, and that is just what the enemy wants us to do.' The enemy got their wish. Months after Napoleon had marched against Austria, after the invasion of England had been abandoned and when a battle had become useless, Villeneuve obeyed Napoleon's orders. His line was duly destroyed at Trafalgar by Nelson's two attacking columns.

Napoleon's marshals. The army, which set off in 1805 to punish Austria for renewing the war, was very different from the ragged troops with which Bonaparte had defeated her in 1796 and from the improvised army with which he had humbled her in 1800. It was the Grand Army which had been trained and drilled for two years in preparation for the invasion of England. It was commanded by the first generals in Europe. Napoleon's marshals, mostly self-made men, soon showed central European governments the advantages and central European peoples the attraction of a 'career open to talent'. Their pre-revolutionary careers illustrate the difference between the French Empire and the Austrian, and show how much the former was the heir of the revolution, how little (except in Napoleon's despotism) of the 'ancien régime'.

Only two of the seven corps commanders, Marmont and Davout, had been officers before the revolution, and Davout had been cashiered for mutiny. Bernadotte, who was to die King of Sweden, had been a regimental sergeant-major; Soult had been a private in the infantry, nursing an ambition to become a baker; Lannes had once been apprenticed to a dyer; Ney, the son of a barrel cooper, had been a cavalry trooper; and Augereau demands a sentence to himself. His father was a Paris mason, his mother a German street-seller and he himself had been in turn a footman, a waiter, a trooper in two French cavalry regiments (he escaped from the second after killing an officer who had struck him on parade), a commercial traveller, a sergeant in the Russian army, a guardsman in the Prussian, from which he deserted at the head of sixty men, fought his way across the frontier to Saxony and became a dancing master. The cavalry was under Murat, the future King of Naples, an innkeeper's son who had been intended for the church, but who had enlisted instead; the Imperial Guard was commanded by

Bessières, an ex-barber. The chief of staff was Berthier, who had been an officer in the American War of Independence. Masséna, who was given a separate command in Italy, had run away to sea as a cabin boy, enlisted in the army and risen to be company sergeant-major before becoming a smuggler.

Ulm, Austerlitz, Pressburg. In 1805 the seven army corps, 'my seven torrents' as Napoleon called them, reached the centre of Germany before the enemy knew they had crossed the frontier. Speed before and speed after the battle was the practice of the Grand Army for 'in the new warfare a battle was the breaking of the crest of a wave, the flood which swept irresistibly after it was the real victory'. The Russians were supporting the Austrians; Napoleon could beat them separately, but combined they would be too strong for him. It was necessary to prevent their junction and to do this Napoleon must place his army behind the first Austrian army on the Upper Danube. The seven corps converged on Ulm on the Danube from positions as remote from one another as Hanover and Boulogne. Soult's men averaged fifteen miles a day for four weeks, including rest days. Perfect staff work and fierce marching won the first stage: no other army but the French could have attempted it; no French army but the Grand Army of 1805 could have carried it out. The two years' training on the Channel coast was justified. The first Austrian army was cut off from its friends and did not know it. It soon found out. There was one fierce skirmish and then the Austrians surrendered.

Napoleon at once pushed on to Vienna and then on again to force his enemies to battle before they were reinforced. His great problem was to cross the Danube by the wooden bridge at Vienna before it could be blown up. How could he do it? Two Gascons, Marshals Murat and Lannes, put on their full-dress uniforms and rode across the bridge. The Austrians were too surprised to fire; instead they sent for their commander. 'What do you want?' he asked. 'What, don't you know about the armistice?' they answered. While they talked, French grenadiers destroyed the mines. Not all the Austrians were as foolish as their general. A sergeant ordered his men to fire; Murat complained 'Is this your famous discipline?' and the general ordered the sergeant's arrest. A gunner started to lay his gun at the marshals. Lannes sat down on the barrel. In this way the bridge was won, and Napoleon hurried forward.

The Austrians and the first of the Russian armies rashly decided to fight at Austerlitz as they already outnumbered Napoleon, instead of waiting for reinforcements. They had forgotten that the French could march. Marshal Davout hurried north from Pressburg and arrived on the eve of the battle: seventy miles in forty-four hours, and a battle to follow. It was the first anniversary of Napoleon's coronation. The allies lost 26,000 men and 180 guns. They also lost the war. Austria made peace by the Treaty of Pressburg ceding Venice and Dalmatia to France, Tyrol to Bavaria and her Swabian lands to Baden and Würtemberg.

From Jena to Friedland. Napoleon had kept Prussia neutral by giving her the English possession of Hanover, a gift she greatly coveted yet feared to take. In 1806 he suggested to England that he should take it away again and exchange it for Sicily. The English peace negotiations fell through and Prussia decided to go to war: after all, she thought, the Prussian army was still the army of Frederick the Great. It was not. The Prussian army was routed on one day at the two battles of Jena and Auerstadt. For three weeks the pursuit continued: speed and bluff between them rounded up bewildered enemy columns and tricked heavily garrisoned fortresses into surrendering to light cavalry. Augereau had the pleasure of capturing his old Prussian regiment, still with the same colonel and sergeant-major, and in tipping the men who had served with him. Finally the pursuit ended for lack of opponents.

There was still Russia to be dealt with. That same winter Napoleon pushed on to Warsaw and then onwards again through sand which turned to mud when rain came. At last he brought the Russians to battle at Eylau. It was an ill day; a blinding snowstorm caused Augereau to lose direction and expose his flank to the Russian artillery. His corps was cut to pieces and only a superb cavalry charge by Murat against unbroken infantry and artillery enabled the French to escape defeat. But for the first time the Grand Army had been badly punished and Europe knew it. Napoleon retired to winter quarters. In the spring he was more lucky. He managed to catch the Russian army in a favourable position at Friedland and won another first-class victory. It was the anniversary of Marengo. Napoleon had beaten the last of his European enemies and the Tsar Alexander made peace at Tilsit.

Map 4. NAPOLEON'S EUROPE

1. France in 1789 is marked black—that and the rest of the shaded areas on the map represent the French Empire at the height of Napoleon's power.

2. The plain grey shading represents additions made to France herself—Belgium and Holland, western and north-western Germany, Danzig, western Italy and the eastern shore of the Adriatic. 'France' therefore included nearly all the great ports of Europe, which was necessary if the Continental System was to have a chance. The Roman numerals mark I Antwerp, II Rotterdam, III Amsterdam, IV Bremen, V Hamburg, VI Lübeck and VII Danzig. But Napoleon failed to secure Heligoland, which was a base for much smuggling.

3. The Arabic numbers represent other states ruled by Napoleon or his relations:

1 is the kingdom of Italy, of which Napoleon's step-son, Eugène de Beauharnais, was viceroy.
2 is Naples, ruled first by his brother Joseph and then by his son-in-law, Murat.
3 is the Grand Duchy of Berg, ruled first by Murat and then by Napoleon's nephew, Louis.
4 is the Kingdom of Westphalia, ruled by his brother, Jerome.

N.B. (i) Spain was ruled by Napoleon's brother Joseph, formerly of Naples, but French control was never thoroughly established.

(ii) Napoleon's brother Louis was made King of Holland but removed for laxity in enforcing the Continental System.

4. The other shaded areas represent states under French 'protection'—most were in Germany and formed part of the Confederation of the Rhine. Notice the Grand Duchy of Warsaw, formed out of Prussia and Austrian Poland, which won Napoleon the support of the Poles.

5. Napoleon failed to secure control of the Mediterranean because he was unable to conquer Gibraltar (A), the Balearic Islands (B), Sardinia (C), Sicily (D), Malta (E), or the Ionian Islands (F) except Corfu. Most of these had English garrisons.

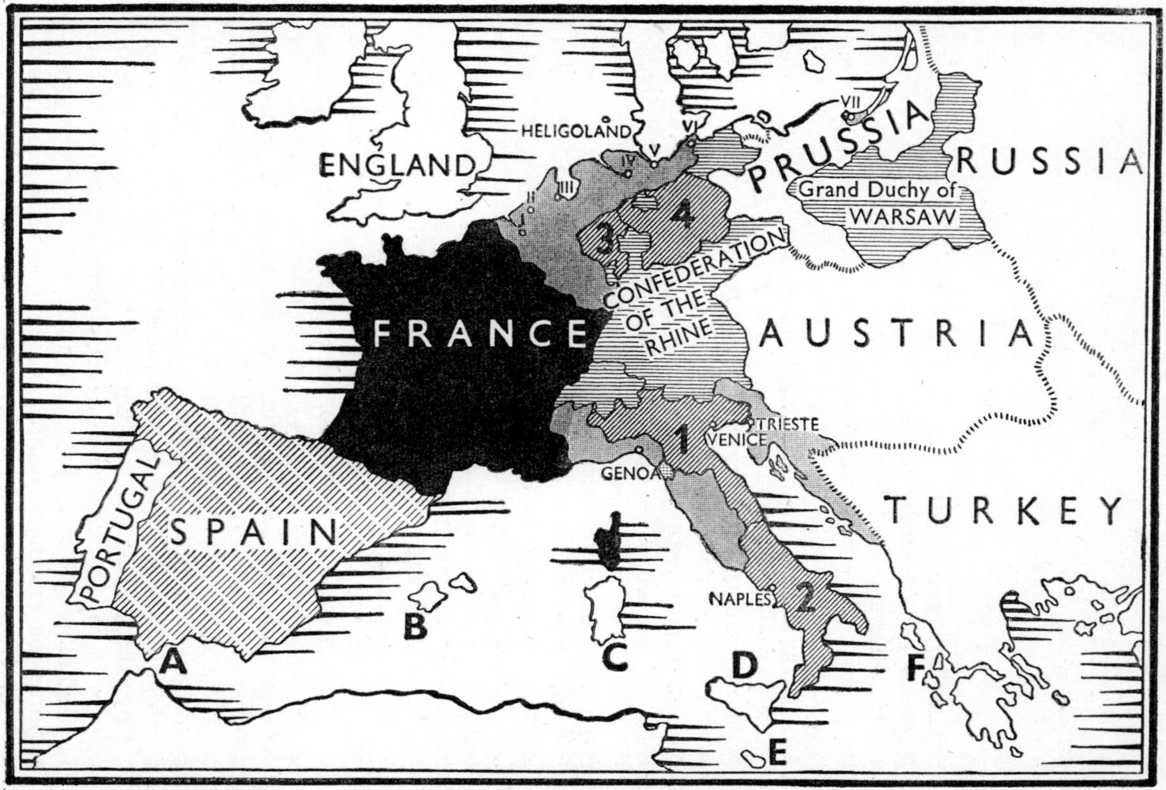

Map 4. NAPOLEON'S EUROPE

10. NAPOLEON'S EUROPE

1805 to 1810

The Treaty of Tilsit. Ten days after the battle of Friedland Napoleon met Alexander at Tilsit. Their conference took place on a barge in the middle of the river Niemen which at the moment was the frontier between the territories they controlled. In less than two years Napoleon had scattered the forces of the Habsburgs of Austria, the Hohenzollerns of Prussia and the Romanoffs of Russia. There was not an army in Europe which could stand up to him. Napoleon had used his repeated victories over Austria, his instantaneous annihilation of Prussia, to depress their power by robbing them of territory. He determined to use his victory over Russia in a different way: he would be magnanimous, Alexander should become his friend, he would even increase Alexander's empire; but in return Alexander must help him against England. For, though Napoleon had destroyed the armies of the Habsburgs and the Hohenzollerns and the Romanoffs, his own navy had been ignominiously destroyed by the English fleet. But there was still one way in which the rich merchant aristocracy which governed England could be humbled.

Before Friedland Napoleon already controlled so much of Europe that a boycott of British goods seemed a possible way of winning the war. If he could persuade Alexander to adopt the so-called 'Continental System' of boycott he felt sure that England would be forced to make peace on his own terms. Alexander was sufficiently fascinated by Napoleon to fall in with the Emperor's plans and to abandon his allies (Prussia, Sweden, England), but only on condition that he was allowed to conquer Finland from Sweden and be entitled to a liberal share of the decaying Turkish Empire which he and Napoleon proposed to divide between them.

Copenhagen bombarded. On these terms the Treaty of Tilsit was drawn up. It contained, however, a secret clause by which, if England refused to make peace, Alexander and Napoleon agreed to force the remaining neutral states to join the Continental System. Denmark was the principal state concerned, for she not only controlled the entrance to the Baltic but also possessed a large navy, which in Napoleon's hands would more than make good the losses of Trafalgar. Denmark was herself well-disposed towards Napoleon—she had unpleasant recollections of Nelson's behaviour at the battle of Copenhagen in 1801. England knew that there was a secret clause in the treaty, but she did not know for certain what was in it although she guessed the main idea. She repeated 'the Nelson touch'. A fleet and an army were sent to 'persuade' Denmark to accept an 'alliance' and allow her fleet to be interned in England. Persuasion failed, but a bombardment of neutral Copenhagen succeeded. Napoleon lost a possible naval reinforcement; Britain lost more—the goodwill of the continent.

The Continental System. This goodwill was vitally important to Napoleon because without it his Continental System could not work. He could bully kings into accepting the System, but he could not bully hundreds of thousands of merchants and consumers into keeping it. What precisely was it? It is contained in four principal measures. The Berlin decree of 1806 forbade any commerce or correspondence with England, made all British subjects liable to arrest and their goods to confiscation, and closed French ports to ships which had previously visited British ports. The Milan Decrees of 1807 proclaimed that any ships which visited English ports should lose their neutrality and become lawful prize.

Later Napoleon came to the conclusion that all the colonial goods in Europe (coffee, cotton, tea and sugar were the most important) were of British origin although they might arrive in neutral ships marked as neutral produce. The Trianon tariff in 1810 put a prohibitive duty on colonial goods which made them impossibly expensive. The Fontainebleau decrees of the same year were designed to stop the widespread smuggling. Long trains of waggons carrying contraband used to pass through Osnabrück, for instance, on their way to central Germany. The decrees ordered the public burning of all English manufactured goods found in Europe and offered rewards to informers. The two measures of 1810 made the System unpopular by making it effective. For the first time the ordinary consumer as well as the merchant was directly hit.

Its effect on England and Europe. How far did the Continental System work? Its object was to produce such widespread bankruptcy in England that the trading classes would force the English government to make peace. It nearly succeeded. The price of raw materials rose rapidly—that of silk, for instance, was quadrupled—and the cost of insurance for cargoes was appalling. In Russia alone British merchants claimed to have lost in 1810 £1,500,000 worth of goods through the seizure of neutral shipping in the Baltic. That winter the bankruptcy rate in England was nearly three times the usual. So great was the distress that big manufacturing towns, such as Leeds and Bolton, petitioned the British government to make peace. It is possible that Napoleon might have succeeded but for one curious gap in his System. His aim was to force England to peace through bankruptcy, not through starvation. He, therefore, allowed the export of grain from the Baltic to England, thinking that the money he received for it—since he would not accept goods—would drain England of gold. He would have had a better chance of starving her by stopping the export of corn since bad harvests made England short of bread.

Ultimately, however, English trade benefited by the System. The Orders in Council, which were England's answer to the boycott, turned the principle of the Milan Decree against Napoleon and laid down that all ships trading with countries belonging to the Continental System were lawful prize. Neutral shipping on the other hand was encouraged to use English ports and to carry English goods to continental destinations. Before long almost the whole of the world's colonial trade came to England and English manufacturers began to find new export markets outside Europe.

The Continental System was a boomerang which finally brought down Napoleon, but its economic effects on Europe were not all bad. French industry was encouraged—the silk trade of Lyons had never been as flourishing—and, if France lost her overseas markets, she gained new markets in the enormous territories which Napoleon controlled. The absence of colonial goods was responsible for many inventions and new processes. When sugar became scarce Napoleon ordered his scientists to find a substitute. By the end of 1810 they had perfected beet sugar and in 1811 Napoleon ordered about 80,000 acres to be planted with sugar beet.

But if France prospered in spite of privations, the rest of Europe only suffered. Napoleon's vassal states had to accept French manufactured goods, but they were not allowed to sell their own in France unless they paid the full duty. They got all the disadvantages and none of the advantages of the Continental System. Napoleon's economic policy was too French and too little international. It was also too ambitious, for he tried to control every aspect of economic life. He thought he could make commerce 'manœuvre like a regiment'. He could not. But Napoleon was not trying to do something which can never be done, but only something which could not then be done. Twentieth-century Europe has many examples of 'planned economies'.

The Grand Empire. A great French historian said that the Continental System was the *raison d'être* of the Grand Empire. That is an exaggeration, since the system of vassal states, of which the Grand Empire was an extension, had been started by the republic before Europe had heard of General Bonaparte. But the final extensions of the Grand Empire—into Spain and Portugal, along the north-west coast of Germany, over the Papal States of Italy—were made to close the gaps in the Continental System. The System could not have existed without the Grand Empire, which put almost the whole coast-line of Europe from Stettin on the Baltic to Ragusa on the Adriatic under the personal control of Napoleon.

The French Empire proper was enormous. It contained Dutchmen, Germans, Belgians, Spaniards, Italians and Jugo-Slavs. All these were ruled from Paris and were represented in the Council of State and the Senate although some of the outlying districts had Governor-Generals with extensive local powers. Thus the Illyrian province, the eastern shore of the Adriatic, profited greatly from the energetic road-making of Napoleon's oldest friend, its governor, Marshal Marmont.

Three other parts of the Empire proper deserve special notice. Holland had originally been a dependent republic, then a dependent kingdom under Napoleon's brother, Louis; but in 1810 it was annexed to France because Louis could not be persuaded to carry out strictly enough the regulations of the Continental System. Two years earlier the Papal States had been seized and the Pope made prisoner because, although he reluctantly agreed to join in the boycott of English goods, he would not promise that all Napoleon's enemies should automatically be his. In 1810 Napoleon's son was made nominal King of Rome, and Rome was 'promoted' to be the second city in the Empire. This was hardly a compensation for being the first city of the church. Napoleon had made peace with the Pope by the Concordat, not because he was a convinced Catholic but because, knowing others were, he thought it good politics. In 1808 he made war on the Pope because, in the interests of his Continental System, he thought it good politics. He had forgotten that millions of people were still good Catholics and would be as much antagonised by the arrest of the Pope as they had been reconciled by the Concordat. This forgetfulness was to prove expensive—particularly in Spain.

The third part of the Empire with which we are concerned is the Rhine provinces. They were almost the earliest conquests of revolutionary France and, although German-speaking, the most loyal. In the old days they had been governed by such a vast number of princes, knights and bishops that they had developed no national life of their own. The French annexation gave them the benefits of the French land reforms and a strong, landowning peasantry was established on the ruins of the feudal system.

The vassal kingdoms. Beyond the Empire proper lay the vassal kingdoms. The first of them in time and prestige was the kingdom of Italy founded in 1805 as the successor to the revolutionary Cisalpine and Italian republics. Its king was Napoleon, who appointed his step-son Eugène de Beauharnais, a decent, modest soldier, as his viceroy. From his capital at Milan Eugène ruled a territory which eventually included Venice and the eastern half of the Papal States. Most of the vassal kingdoms were kept in the family: three brothers and a brother-in-law became kings. They had clamoured so persistently for this promotion that one day Napoleon complained: 'One would think I was depriving them of the heritage of the late king, our father.' They got their way, but they were allowed no independence. Napoleon's word was as much law in the vassal kingdoms as in the French Empire. In the same year in which Louis Bonaparte became King of Holland, Joseph was made King of Naples. From there he was moved in 1808 to the more important but infinitely precarious kingdom of Spain, while Napoleon's brother-in-law, Marshal Murat, was promoted to Naples.

Central European changes. Murat had since 1806 been Grand Duke of Berg, the first of the Napoleonic states to be carved out of

Germany. More important was the kingdom of Westphalia, formed a year later, and given to Napoleon's youngest brother, Jerome. Napoleon's policy in the rest of Germany was to strengthen the medium-sized states such as Baden, Würtemberg, Bavaria and Saxony at the expense first of the tiny principalities, free cities, church states and independent knights' fiefs, and later at the expense of the two big powers of Austria and Prussia. The first step in this direction was taken after the Marengo campaign, when the smallest German states east of the Rhine were dissolved and re-distributed by Bonaparte among the bigger states in compensation for the territory which some of them had lost by the French annexation of Germany west of the Rhine.

The second step followed the Austerlitz campaign, when a great part of the Habsburgs' German lands were distributed among the South German states which were then formed into the Confederation of the Rhine. This took the place of the old 'Holy Roman Empire', the surviving medieval machinery for consultation among the German states under the presidency of Austria. The Confederation of the Rhine was entirely under Napoleon's control and provided conscripts for the French army. The third step was taken after the Jena and Friedland campaigns when Prussia was dissected. Her western provinces went to found the kingdom of Westphalia, and her eastern ones to enrich Saxony by the establishment of the Grand Duchy of Warsaw, which was placed under the King of Saxony.

The Grand Duchy of Warsaw takes us outside Germany into the lost kingdom of Poland. In the eighteenth century Poland had been divided among her three strong neighbours, Austria, Russia and Prussia, though not without great squabbles among themselves. Prussia's desire to get the lion's share of the last remaining fragment had been her reason for retiring from the first coalition in 1795. She was now repaid for her greed, since all her Polish provinces were taken away and formed into this new Grand Duchy, which, in Napoleon's eyes, had the further advantage of being a useful threat with which to keep Russia in order. He had only to suggest the revival of the Polish kingdom to cause a rebellion in Russia's Polish provinces—for the Poles had kept their patriotism though they had lost their independence.

Influence of the Grand Empire. The Grand Empire was

short-lived but extremely influential. The kingdom of Italy simply by its name re-kindled Italian patriotism which had seemed dead. The Grand Duchy of Warsaw kept hope alive in Poland. On the other hand German patriotism was not fostered by Napoleon—it was born in hatred of him—but at least his reconstruction of Germany cleared the ground of much inherited muddle and left the nineteenth century free to build a united German nation. Until Napoleonic customs officials and recruiting sergeants made the Empire hated, it was tolerated and even welcomed. If it was not perfect, it was usually better than the government it replaced. Brigandage was suppressed, roads built, education endowed, torture abolished, enlightened laws introduced—for the 'Code Napoléon' followed the French army. Men learned that government is for the good of the governed as well as of the governors. They discovered a society in which merit not birth was the passport to success. Napoleon gave them social equality; he refused them political liberty. In the old days they had known neither so they were free to welcome the one without mourning the other.

THE BONAPARTES

CHARLES-MARIE BONAPARTE, died 1785 = MARIE-LAETITIA RAMOLINO, died 1836

Children:

- JOSEPH, King of Naples 1806–1808; King of Spain 1808–1813 died 1844
- NAPOLEON I, born 1769 First Consul 1799 Emperor 1804 died 1821
  - = (2) MARIE LOUISE, daughter of Francis II of Austria died 1847
    - NAPOLEON (II) King of Rome 1811–1814 Duke of Reichstadt died 1832
  - = (1) JOSEPHINE divorced 1809 died 1814 = (previously) ALEXANDRE DE BEAUHARNAIS
    - EUGÈNE, Viceroy of Italy died 1824
    - HORTENSE died 1837 = LOUIS
- LUCIEN died 1840
- LOUIS, King of Holland 1806–1810 died 1846 = HORTENSE died 1837
  - (LOUIS) NAPOLEON III born 1808 President 1848–1852 Emperor 1852–1870 died 1873 = EUGÉNIE de MONTIJO died 1920
    - LOUIS NAPOLEON, Prince Imperial killed in action 1879
- CAROLINE died 1839 = JOACHIM MURAT, Grand Duke of Berg 1806 King of Naples 1808 executed 1815
- JEROME, King of Westphalia 1807–1813 died 1860
  - Prince Napoleon died 1891 = CLOTILDE, daughter of Victor Emmanuel II of Piedmont

Map 5. THE DECLINE OF TURKEY: FIRST PHASE

1. In 1789 all the country shaded on the map belonged to the Sultan of Turkey, except the small vertically shaded portion which had recently been reconquered by Austria. By 1914 only the black portion was left.

2. Turkey was the greatest Mahometan power. The possession of the holy cities of Mecca and Medina made the Sultan the Caliph or 'Commander of the Faithful' Mahometans throughout the world.

3. The Barbary States—Algiers, Tunis and Tripoli—recognised the Sultan as overlord, but were virtually independent pirate communities preying on Mediterranean shipping.

4. In 1798 Napoleon conquered Egypt, intending to use it as a stepping stone for the re-establishment of French power in India. The destruction of his fleet at the battle of the Nile stopped this plan. He therefore decided to invade Syria and march on Constantinople. His plans were again defeated by England's successful defence of Acre.

5. Russia was gradually conquering the north shore of the Black Sea, which had been until recently a Turkish lake. Her ambition was to secure Constantinople and access to the Mediterranean. She regarded herself as the protector of the Christians in Turkey. In 1807 Napoleon and Russia reached an agreement at Tilsit by which Russia was to be free to take Constantinople if she could. The arrangement was not carried out.

6. Austria and England became friends of Turkey through fear of Russia and France. Austria was afraid of being shut in by Russia and England feared for the safety of India if Russia conquered Constantinople or France Egypt. Austria's jealousy of Russia was one of the main obstacles to a coalition against Napoleon.

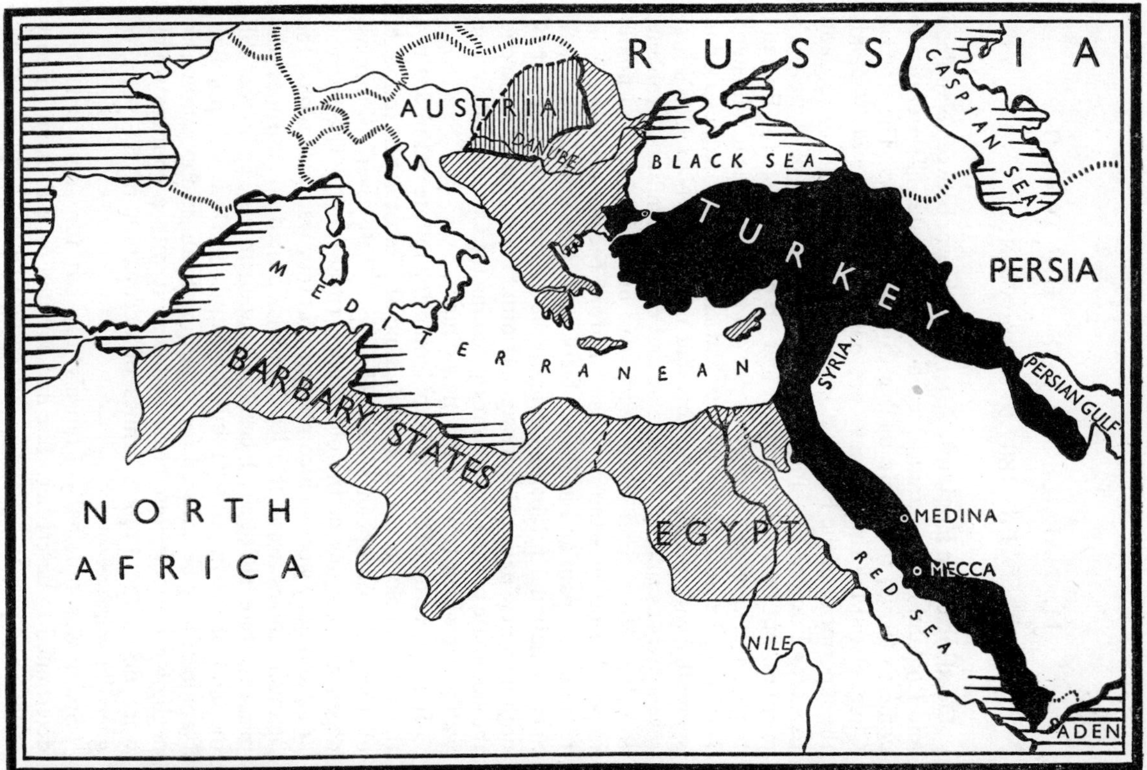

Map 5. THE DECLINE OF TURKEY: FIRST PHASE

11. NAPOLEON: THE TIDE TURNS

July 1807 to January 1813

Deadlock with England. After Tilsit the war between Napoleon and England had apparently reached a stalemate. They refused to come to terms; they were unable to come to blows. The English Channel and the English navy prevented an invasion of England; the superiority of the French army made it impossible for England to attack France. England had been the paymaster of Napoleon's enemies, but by 1807 Napoleon had no enemies left for England to pay. The kings of Europe had proved totally unable to withstand Napoleon because their subjects had either been apathetic or in sympathy with the French. This had made easy Napoleon's first seven years as ruler of France: he had only had military problems to solve. His second seven years were made difficult and finally impossible because the peoples of Europe became actively hostile to him. In Spain and in Russia Napoleon came into conflict with ancient national pride; in Germany he raised up almost out of nothing a new national spirit against him. It was his effort to end the stalemate in the English war by making the Continental System watertight which brought this new danger.

Joseph, King of Spain. Napoleon's first failure was in Spain—the most backward and at that time the most contemptible of all the great European states. The royal family were Bourbons, degenerate cousins of the old French royal family. The Spanish army and navy were so incompetent that they were ridiculous; the Spanish nobles were the least public-spirited in Europe; the peasants were half-starved, cruel and superstitious. Spain was still the land of the Inquisition and the Inquisition was still popular. Napoleon may be excused for regarding Spanish opposition as a negligible obstacle, but pride endowed it with a courage which wore down the French armies of occupation in six years of guerrilla warfare.

Spain was nominally France's ally but for some time had determined to desert her. She had counted on a Prussian victory at Jena and prepared plans for attacking France. Napoleon made up his mind to punish this treachery in a curiously underhand way. Beyond Spain lay Portugal, an even more decadent country, but England's oldest ally and a great market for her goods—and therefore a serious menace to the success of the Continental System. It would be more convenient for Napoleon to deal with Portugal before he broke with Spain. His troops came as allies of Spain to conquer Portugal—which they did without trouble; they remained to conquer Spain. The Spaniards accused their king of selling his country to the French. A rebellion forced him to abdicate and Ferdinand, who hated his father, took his place. This did not at all suit Napoleon's plans. He forced the two kings, father and son, to come to Bayonne in France and there he bullied them both into abdicating. He then decided to make Spain safe for the Continental System by appointing his brother Joseph king. It was a fatal mistake.

Spain would stand any amount of bad government by Spaniards; she would not stand even good government from a foreigner—and that Joseph's government would have been good his record in Naples showed. At once the whole country rose in rebellion. One French army was forced to surrender at Baylen; another was badly beaten by an English expeditionary force in Portugal and agreed to abandon that country in return for a safe passage to France. These two reverses, the Capitulation of Baylen and the Convention of Cintra, were the first outright defeats suffered by French arms since Napoleon had become ruler of France.

The Peninsular War. The detailed history of the Peninsular War may be found in English history books; what concerns us here is its effect on Napoleon's fortunes and on the general course of European history. From the first it was a doubly hopeless struggle for the French. First, it enabled England at last to come to grips with France and on favourable terms. England could only help Austria against Napoleon with money. She could not send her men. She could help Spain and Portugal with both.

The second reason why French defeat was certain was that the French were fighting in difficult country made worse by the fact that every Spaniard was an enemy. French soldiers were ambushed in open country, knifed in dark corners, poisoned in inns. Two hundred cavalry were necessary to escort a single messenger and even then there was no guarantee that the message would get through. Under Spanish conditions the French found they could not live on the country. In Germany everything had been very different. There had been no trouble. 'The Germans are reasonable folk', Marmont used to say, 'and know that soldiers must eat. What one takes from them goes to a useful destination and that consoles them.' The Spaniards would rather die than feed a Frenchman with anything but poisoned meat.

And so in Spain the French army rapidly lost its fighting power both from loss of morale and from the fact that for every soldier in the front line it was necessary to have two guarding communications. The French armies in Spain outnumbered the English, but the French soldiers in the front line were outnumbered by the English. The difficulty of the country with its entire absence of tolerable roads was a greater handicap to the French than to the English. The only French communications were by land and were therefore necessarily long. British sea power, on the other hand, enabled the English base to be moved round the north coast of Spain as the expeditionary force advanced. In this way the English were able to keep their land communications short.

France might have beaten the English or the Spanish alone; she could not beat both together. She ought therefore to have abandoned the Spanish war before it was too late. Spain under English control would not have been a serious danger because the Pyrenees form a splendid natural defence for France. Napoleon, however, refused to give up the struggle because to do so was to sacrifice the completeness of his Continental System. His obstinacy was strengthened by his ignorance of Spanish conditions—the only time he was in Spain was for a few weeks in 1808. The Peninsular War destroyed the French Empire by using up in Spain many hundreds of thousands of men whose presence in Germany would have enabled Napoleon to win the war of 1813. It was even more far-reaching than that. But for his Spanish entanglement there would have been no war in 1812 or 1813. It was only their knowledge of his Spanish commitments which tempted the kings of Europe to try their luck again.

The Wagram campaign. Austria was the first to take advantage of the Spanish war. In 1809 she mobilised her army and went to war. Once again Napoleon swept across Europe and marched in triumph into Vienna. Once again, as before Austerlitz, he was faced with the problem of crossing the Danube. On the north bank the Archduke Charles was waiting for him, hoping that Napoleon would risk a crossing and a battle with his back to the river, his retreat cut off. Napoleon took the risk. The Austrians proved steady fighters and the French found themselves pressed back against the Danube, while Austrian fire-ships destroyed the pontoon bridges which would have enabled them to retreat. The French army came perilously near annihilation, but somehow or other the riverside village of Essling held out although its neighbour Aspern was lost. At last the Austrians withdrew and Napoleon was able to rebuild his bridges and withdraw to the island of Lobau in mid-stream. For six weeks the French remained on their island waiting for reinforcements. When they came Napoleon tried his luck again. Once more he crossed the river and this time his luck returned. A terrific bombardment prepared the way for the victory of Wagram, the last of the overwhelming Napoleonic victories.

The Emperor of Austria hastened to make peace: by the Treaty of Schonbrünn he handed over the rest of his Adriatic coast-line to France and his Polish provinces to the Grand Duchy of Warsaw. Next year his daughter, Marie Louise, married Napoleon. The unfortunate Josephine was divorced to make way for this new wife who might bear Napoleon an heir who would make his power secure.

Outwardly Napoleon had never been more successful; actually the campaign of Wagram brought his end nearer. The apparent contradiction is easily explained. Austria had fought alone, since Prussia was too timid to help her and Russia too intent on acquiring what she had been promised at Tilsit. The only assistance she had received was a singularly feeble attempt by England to invade the Low Countries. The grotesque incompetence of the Walcheren expedition persuaded Europe that the British army was as inefficient as it had been in the early days of the French revolution. Austria, thus left to be defeated in solitude, was determined that this should not happen again. The Wagram campaign, therefore, made it more difficult to build a new coalition against Napoleon. But the campaign had underlined the lesson of Eylau. The drawn battle of Aspern-Essling had shown Europe that a French army, even when commanded by Napoleon himself, was not invincible.

Russia and Spain. The Tsar Alexander of Russia remembered this lesson when in 1811 he determined to risk a war with Napoleon. Five reasons led him to break the Tilsit friendship. Napoleon was unwilling to let him take Constantinople as part of his share of the Turkish Empire which they had agreed to divide at Tilsit. The Continental System was ruining Russian trade. The enlargement of the Grand Duchy of Warsaw made it more than ever a threat to Russia's Polish provinces. His court, the most exclusive in Europe, bitterly resented the French alliance as a betrayal of the principle of aristocracy. His people regarded the French as wicked atheists—as the French ambassador said, it would have needed a change of religion for the Russian people to become friendly with the French.

Russia in fact closely resembled Spain. Napoleon had enjoyed a reasonably sympathetic reception in his German and Italian wars because large sections of the population wanted the social reforms which were the product of the French revolution and which were spread by the Napoleonic armies. There was no such class in Spain or Russia. Those two countries belonged to a previous century in their intellectual development. The French revolution, particularly in its irreligion, was a protest against things which they held dear. When the French invaded Spain and Russia it was not a war of a people against a king, but of people against people and of century against century. It was also in both cases a war in which the countryside itself fought against the French. Both Spain and Russia were inhospitable countries, without roads, without food; the one a land of barren mountains, the other of snowbound plains.

The Moscow campaign. In June 1812 Napoleon was ready to invade Russia. He crossed the Niemen with an army of over half a million men drawn from almost every nation of Europe—Frenchmen, Germans, Poles, Italians, Swiss and Jugo-Slavs. He went in search of the Russian army to bring it to battle, overwhelm it and force a victorious peace as he had so often done before. But he could not find the Russians, who retreated before him, luring him ever further into their thinly populated country, lengthening his already over-long communications.

When at last the two armies met at Borodino the French had only about a quarter of their original army; the other three-quarters were either killed, missing, wounded or on special duty. The remaining quarter were, however, sufficient for their work. After a tremendous struggle the Russians were defeated. Now was the moment for the annihilating pursuit which distinguished Napoleon's great battles. There were fresh troops ready, for the Guard was untouched; but Napoleon remembered how far he was from France and dared not risk the flower of his army. And so the Russians were allowed to draw off in good order without pursuit. The battle might as well never have been fought, had better not have been fought for the French lost 50,000 men and 30,000 horses. Napoleon was growing prematurely old. He was only forty-three; but it was nineteen years since he had looked after the guns at the siege of Toulon and those nineteen years had been spent in almost continuous warfare.

There was nothing for it but to push on to Moscow, the old capital of Russia. If he took it he thought that Alexander would be forced to ask for peace. The French entered Moscow; but Alexander said nothing and his subjects burnt the city about Napoleon's ears. Napoleon and his troops were already unpopular enough in Russia, but their looting of the most sacred Russian shrines in search of plunder made them more than ever hated. The garrisons he had left behind were overpowered, the food convoys were stopped, his stragglers were massacred. It was Spain all over again.

After a month Napoleon decided that he must leave Moscow. At last the retreat began. It was soon reduced to a rout by the hordes of Cossacks who harried the French on every side, and still more by the frost which tortured men used to the vineyards of France and Italy. At the crossing of the river Beresina, which took place in 45 degrees of frost, half the remaining army was killed, but this massacre was in its sad way not the least of Napoleon's exploits—the Russians might have counted on destroying not half but the whole of his army. At last in January 1813 a remnant of 1000 men crossed the Russian frontier—the river Niemen on which Tilsit stands. The last man to leave Russia was Marshal Ney, but for whose daily heroism no French soldier would have returned alive.

12. NAPOLEON'S DOWNFALL

January 1813 to July 1815

Jealousy of the allies. The attempt to extend the Napoleonic empire into Spain and Russia had proved a ghastly failure. It did not follow that the period of Napoleonic domination in central Europe was also over. There were many factors in Napoleon's favour. He had so often beaten the central European powers that only the certainty of success would bring them into the war again. Jealousy as well as timidity was an asset to Napoleon. Austria certainly hated and feared Napoleon, but she hated and feared Russia almost as much. It was not worth risking another Wagram in order to establish the Tsar Alexander as the dictator of central Europe. Napoleon's constant reverses in Spain—the Peninsular War had become a slow-motion version of the Retreat from Moscow—and the complete annihilation of his Grand Army in Russia certainly made it possible for a European coalition to defeat Napoleon; but for some time it looked as if that coalition would not be formed.

Prussia's leadership. It was Prussia which took the decision to try and turn Napoleon out of Germany as he had been turned out of Russia, as he was being turned out of Spain. It was not the King of Prussia, who remained the most timid of all rulers, but the people of Prussia who made this decision. It was Prussia which took the lead in the War of Liberation; it was Prussia which became the spiritual fatherland of all Germans who wanted to see their country free from foreign control. The leadership fell to Prussia and not to Austria because the latter was not even mainly a German power. In 1813 most of its subjects did not speak German. Austria's opposition to Napoleon was dynastic and not national. It was based on the view that the possessions of the Habsburg family were in danger. But by 1813 German patriotism had been born and German patriotism required the purely German leadership which Prussia could give. It was for this reason that ordinary Germans looked to Prussia, which had done less than any of the other great powers against Napoleon; and not to Austria, which had been almost as consistent an enemy of France as England had been and which had suffered considerably more.

German patriotism. This German patriotism was a new thing. The end of the eighteenth century and the beginning of the nineteenth was the golden age of German music, literature and philosophy. It is true that the greatest writers, thinkers and musicians were not political patriots, but they gave Germans something of which to be proud and in many cases indirectly laid the foundations of nationalism. Beethoven indeed dedicated his Eroica Symphony to Napoleon, though he changed the dedication in republican anger when he heard that Napoleon had made himself Emperor. The greatest of all German writers, Goethe, was content to accept the Legion of Honour—he was intensely German in his work, but he believed that one could be happy in Germany without being governed by Germans. The only thing which interested him in government was that it should be efficient and humane. He saw that the French were bringing better government to Germany and on the whole he was grateful to them.

Goethe's great friend Schiller did more to kindle German nationalism through his histories and historical plays—a love of the past is the foundation of nationalism. Schiller, however, took his subjects from all peoples—he wrote about William Tell and Mary, Queen of Scots, and Joan of Arc as well as about German leaders. Similarly the greatest of the philosophers, Immanuel Kant, had nothing himself to say about politics, but his insistence on duty was a moral lesson which men could apply to the state and the country. His pupil, Fichte, turned his teaching to directly patriotic ends and in a great series of 'Addresses to the German Nation' urged his countrymen to throw off Napoleon's rule. Thus German philosophy by its sternly moral teaching made men think of themselves less as individuals and more as members of the community, while German literature by turning men's minds to the past gave them a pride in their ancestors and made them conscious that they were different from other nations. These were the spiritual foundations of the new German nationalism. Its economic foundations were laid by the distress which the Continental System brought to Germany.

German patriotism, then, was a vastly different thing from the patriotism of Spain and Russia. The one was an educated patriotism, born in the universities, spread by the printing press, drawing its inspiration from the past, as all patriotism must, but looking towards the future. It was middle-class patriotism. The other was simple peasant patriotism, based on traditional memories, rooted in unthinking conservatism—religious conservatism, political conservatism, social conservatism: looking to the past and turning its head away from the future.

Reform in Prussia. Since the battle of Jena Prussia, in spite of its king, had been fitting itself for the leadership of the new Germany. A remarkable group of statesmen, many of whom were not Prussians at all, had carried out far-reaching reforms. The greatest of them was Freiherr (Baron) von Stein, a Rhinelander, who first realised that if Napoleon was to be defeated it could only be by adopting those French reforms which made men prefer enlightened foreign rule to tyrannical native rule. He carried through a partial freeing of the serfs—as much as the rich Junkers (landowning squires) beyond the Elbe would allow. Public opinion was prepared for the change by 'Father' Arndt, historian and writer of patriotic songs, who was himself the son of a serf.

Napoleon saw that Stein was preparing Prussia to overthrow the French and forced the king to dismiss him in 1810; but his place was taken by a Hanoverian, Hardenberg, who carried on his work and freed traders from many irritating restrictions. The new patriotism was an intellectual movement and Prussia under Humboldt reformed its education to fit the people to be patriotic. From the preparatory school to the university a new vigour was given to teachers and pupils, and German education established its claim to be the best in the world. Fundamentally these reforms were more important than the army reforms by which Scharnhorst, another Hanoverian, circumvented the restrictions which Napoleon had placed on the Prussian army. After Jena the Prussian army was limited to 42,000 men but Scharnhorst adopted a short service system which passed men quickly through the ranks, thus securing a large trained reserve.

The War of Liberation. Napoleon's complete failure in Russia in 1812 gave Prussia her opportunity. The king's resistance to rash measures was swept aside by the determination of his people and of his generals, who made an alliance with the Tsar behind his back. In March, steady, unimaginative Marshal Blücher,

Marshal Forwards as his men called him, crossed the Elbe. The armies were not unequal in numbers, but the French had Napoleon. The allies on the other hand were filled with the same impetuous patriotism which had carried the early armies of the French revolution to unexpected victories. They were reinforced by irregular Free Corps of volunteers from other parts of Germany, such as von Lützow's Night Riders in which the young poet Körner, whose songs helped to sweep the French across the Rhine, served until his death. The older writers might be disinterested in the struggle; the younger were fierce patriots to a man.

The first stage of the war was short and even, ending with a victory for Napoleon at Bautzen. The hopes of Europe seemed doomed to failure. But the Napoleon of 1813 was not the Napoleon of the great campaigns. Instead of following up his victory and dictating peace, he counted his losses and agreed to an armistice. While it lasted the fourth coalition was completed and Napoleon's fate sealed. His hesitation convinced two cautious men that it was worth while bringing their countries into the struggle.

The Coalition enlarged. The first was Bernadotte, the most jealous of Napoleon's marshals. He had been adopted by the Swedes as their crown prince in 1810 in the hope that he would re-conquer Finland from the Tsar, who had been allowed to take it as part of the Tilsit bargain. In 1812, however, Alexander bought Sweden's neutrality in his war with Napoleon by promising in the Treaty of Abo to allow her to take Norway from Denmark with which she shared a king. Bernadotte now went a step further, made an alliance with Russia and brought his army to Germany to fight Napoleon. More important than Bernadotte was Metternich, the Austrian chancellor. He was as anxious as anyone to overthrow Napoleon's power in Germany and Italy, but he would neither risk intervening before it was safe nor allow Russia to become too powerful. At first he attempted to mediate, offering Napoleon terms which would have left him the Rhine frontier and the control of Italy. This was the first of three offers by which Napoleon would have kept his power in France. He would not, however, surrender his control of Germany and so Metternich brought Austria into the war against him.

Soon after the end of the armistice Napoleon won the last of his great victories at Dresden, but like all his later battles it was indecisive because there was no pursuit to disorganise the enemy. Three months later he was completely beaten at the four days' battle of Leipzig and the French were driven back across the Rhine. Once more Napoleon was offered, and refused, a chance to save his throne. The mutual jealousies of Russia, Prussia and Austria threatened to break up the alliance and Metternich insisted on suggesting peace terms by which Napoleon would have kept the Rhine, the Alps and the Pyrenees as the frontiers of France. Napoleon's refusal was followed by a renewal of the war in the beginning of 1814.

The campaign of 1814. This time the fighting took place in France, which during twenty years of warfare had never seen a battle. The allies were still as jealous of each other as ever; Austria in particular was paralysed by the fear of Prussian and Russian annexations in Germany and Poland. England alone had no interest except the overthrow of Napoleon's empire and she was terrified that this, the main object of all the allies, would be sacrificed to their disagreements. Lord Castlereagh, the English representative at allied headquarters, thought that peace should be made before the coalition broke up and Napoleon recovered the ground he had lost. Accordingly the third offer was made to Napoleon. France, by the terms suggested at the conference of Châtillon-sur-Seine, would have been reduced to her old frontiers as they were before the revolution. The campaign of 1814, however, showed Napoleon at his best as a general. He was doing so well at the time of the conference that he felt he could afford to reject the allies' terms. He was mistaken.

Two events decided his fate. First, Castlereagh managed to get the allies on to better terms and the result of his diplomacy was seen in the Treaty of Chaumont by which they agreed not to make peace separately, but to continue the war until France had been reduced to her pre-revolutionary frontiers. The treaty also provided for periodical meetings between the allies even after the conclusion of peace—the alliance was thus made permanent. Napoleon's repeated refusals to consider generous terms had almost convinced the allies that there could be no peace in Europe while he remained ruler of France and they had therefore agreed to sink their differences.

The second event which brought Napoleon's fall was the inevitable turn in the fortune of war—inevitable because France was weary and disspirited, recruits were few in number and poor in quality. Napoleon's strategy could hold off for a time the inferior allied generals; but it could not take the place of a well-trained and well-equipped army. At last the Prussians defeated Napoleon at the battle of Laon and shortly afterwards the Tsar of Russia and the King of Prussia entered Paris in triumph.

Who shall rule France? The next problem was to decide how France should be governed. Above all the allies wanted to find a government that would last. They had made up their minds that Napoleon must go—he was sent to govern the miniature island principality of Elba—but it was not so easy to find a successor. Some supported a regency in favour of Napoleon's young son. Others suggested that his stepson, the Viceroy of Italy, would do. There was talk of making Bernadotte king, while at one time a republic was considered. The obvious solution was the restoration of the Bourbons, but there were difficulties. It was doubtful whether they would accept the social reforms of the revolution, without which there could be no peace in France; it was still more doubtful whether France would accept them—during the campaign in eastern France the allies had searched in vain for any traces of a Bourbon party. Still, in the end the Bourbons seemed the best solution, or at least the one to which there were fewest objections. And so, surrounded by doubt, Louis XVIII came back to Paris.

The Hundred Days. The doubts were soon justified. In less than a year Napoleon had grown tired of exile and returned to France. For one moment France hesitated; then she turned in enthusiastic welcome to the man who had brought her so much glory. Marshal Ney set out from Paris promising to bring back Napoleon in an iron cage: one hurried scrawl from his old master brought him to Napoleon's side. In less than three weeks from his landing in France Napoleon was back in the Tuileries.

It had been easy to scare the Bourbons off the throne; it proved impossible to frighten the allies into accepting his return. They decided on immediate war and Napoleon had to set to work to raise another army, his last. He marched north to the Low Countries where his last campaign was to be fought. Its story may be read in English histories but, except for the military historian, the details are unimportant. Napoleon might have won the battle of Waterloo; he could not have won the war—for the same reason that he was bound to fail in 1814. The coalition against him had outgrown its early jealousies and was too strong for the remnants of the once great French army.

And so in July 1815 Louis XVIII came back to Paris the second time, King of France not by the will of the French people but by the decision of the allies. Napoleon was hurried off to St Helena to die there of cancer six years later. The greatest chapter in French history was at an end.

PART TWO

METTERNICH'S EUROPE

1815 TO 1852

13. RE-MAKING EUROPE

1814 to 1815

[In this chapter no attempt is made to separate the decisions made at Paris (regarding France) and at Vienna (for the rest of Europe.)]

The double task. In the autumn of 1814, when Napoleon was safely in exile, the Duke of Wellington left his armies and travelled slowly across Europe with his colleague, Lord Castlereagh, to Vienna. There they found half the kings and all the statesmen of Europe trying to piece together as much as they could of the old Europe whose pleasant, formal life had been so rudely disturbed twenty-five years earlier by the sack of the Bastille. The kings had suffered in two ways: the new doctrine of liberty, equality and fraternity had attracted their subjects at the same time as Napoleon's armies had invaded their frontiers. To save their thrones—an endeavour in which they often failed—they had been obliged to do as they were told by Napoleon and sometimes even by their subjects. Thus the Emperor of Austria had had to give his daughter, Marie Louise, in marriage to Napoleon; while the King of Prussia had been forced to abolish serfdom. The statesmen at Vienna, therefore, had two duties to perform. They had to restore and to secure most of the old frontiers of Europe, though it was permissible for the greater states to swallow some of the lesser. They had also, and it was a more difficult task, to make people forget that, although Napoleon had enslaved nations, the French revolution had set men free.

Prince Metternich. The leading members of the Congress were perhaps as nearly equal to this task as statesmen could be. The one who most clearly understood its implications was Prince Metternich, a comparatively young man though he had been Austrian foreign minister for five years. He had no romantic belief in the divine right of kings, but he supported their absolute power, because he saw in it the best guarantee of stability. The French revolution had preached equality, which must be a constant threat to property, and had taught men to believe that the people had a right to choose their government. Even Napoleon had claimed to be Emperor by the will of the French people expressed in a plebiscite. The only alternative was the return of the old dynasties who claimed to rule a country just as an ordinary man might claim to own a house—because it was his property. In this way alone, Metternich thought, could Europe avoid the disturbing theory of popular consent with all its risks of revolution. For similar reasons Metternich was opposed to nationalism. He disliked it because it was associated with the French revolution; because it would have been fatal to his own state, the Austrian Empire, and because it would have meant revolutionary changes at a time when stability was essential.

Metternich raised Austria from a discredited and dismembered state to a position of authority in Europe which even Napoleon had scarcely enjoyed. He brought the most fertile and civilised parts of Italy under Habsburg rule and regained almost all Austria's lost territories except the southern Netherlands, which were joined to Holland to create a strong barrier kingdom between France and Germany. His wishes were respected throughout central Europe and his will dominated politics until 1848.

The principles on which Metternich worked may have been sensible: unfortunately they were not exciting. In a war-weary world it was just possible to acquiesce in them, it was impossible to be enthusiastic. Many people, especially Napoleonic veterans and young people, could not even acquiesce and therefore it was necessary for Metternich to censor the newspapers, to prohibit public meetings, to imprison opponents and occasionally to send out expeditionary forces to reinstate the less pleasant kings in the enjoyment of their property—the right of misgoverning their subjects. But Metternich himself was not cruel or a persecutor. It takes enthusiasm to be a persecutor, and enthusiasm was as foreign to Metternich's nature and as dangerous to his system as it was to that of Sir Robert Walpole, who had secured the same sort of breathing space after Louis XIV's wars that Metternich did after Napoleon's. Metternich indeed had only one great weakness, vanity—though that is a serious enough fault in a statesman. He had, moreover, one insurmountable handicap. He served the Emperor Francis, whose conservatism was so rigid that he preserved even those abuses which made his throne insecure.

Hardenberg and Prussia. Hardenberg, the chief Prussian representative at Vienna, was a man whose interests and whose capacity were both more limited and more virile than Metternich's. He cared relatively little for the general European settlement but a great deal for the fortunes of Prussia. Metternich tried to avoid discontent; Hardenberg to win consent. He was one of the group who a few years earlier under Freiherr von Stein had discovered that the only way to fight the French was to fight them with their own weapons. Not that Hardenberg was a democrat. In the years after 1815 it was his influence which kept Frederick William III, that rarity, a weak-willed Hohenzollern, from granting Prussia a parliament. His reforms were designed to make Prussia strong, not to make Prussians self-governing.

At Vienna Hardenberg's attention was concentrated on gaining as much territory as possible for his country. He surrendered to Russia most of his claims to Poland and received in return large territories in central Germany and the whole of the Rhineland north of Mainz. Germany's rivers are her main roads and Prussia now controlled the Rhine, the Elbe, the Oder and the Vistula, leaving only the Danube to Austria. Three years later in 1818 Prussia used this commanding position, which Hardenberg had won, to introduce a Customs Union (Zollverein) which was gradually joined by most of the German states except Austria.

The German Confederation. Both Prussia and Austria were parts of 'Germany', which was then only a rather ineffective League of German Nations with headquarters at Frankfurt where all the German states sent representatives to a diet or assembly. Before Napoleon the name of this league had been the Holy Roman Empire and the reigning Habsburg had been Holy Roman Emperor. Napoleon had destroyed this organisation in 1806; at the Congress of Vienna it was revived as the German Confederation (Bund) with the Austrian Emperor as president ex officio. The result of the territorial changes of 1815 was to make Prussia a much greater power in Germany, and, by the loss of most of her Polish provinces, a much more exclusively German power than she had been before. Hardenberg, therefore, wanted to give the new Confederation real authority, confident that Prussia would be able largely to determine how that authority would be used. Austria, on the other hand, by her new Italian territories had become even less of a German and more of a cosmopolitan power than she had been before. Metternich, therefore, did not want Habsburg policy to be closely tied to the apron strings of German interests and wished the Confederation to be as nearly a sham as possible. Metternich's policy prevailed. It was the Zollverein, not the Confederation, which was the parent of the modern German Empire.

Talleyrand. M. de Talleyrand, the French representative, was in some ways the most remarkable member of that remarkable group. He could look back to the days before the revolution when he had been Bishop of Autun. Then he had written the official justification of the revolution of August 10, 1792. During the Reign of Terror he had found it prudent to live in America, but with the coming of the Directory he returned to France, resolved, like Charles II of England, 'never to set out on his travels again'. From then on he had always been on the winning side. He had served Napoleon and now he used his influence to procure the restoration of the forgotten but unforgetting Bourbons. He persuaded them to grant France a constitution, though of a very sickly sort, but he could not persuade them to employ him for longer than was necessary to re-establish the dynasty.

Talleyrand, the man, inspires contempt. Talleyrand, the statesman, inspired confidence. Men might mistrust his motives, they respected his judgment. He secured admission to the Congress on the plea that it was not France but Napoleon whom the allies had been fighting. Rarely after a great war has a defeated country received such generous terms of peace. France was not only allowed to preserve her original territory intact but also the early conquests of the revolution. She was asked to pay no indemnity. Even after Napoleon's Hundred Days, when the terms were revised, France was still treated with extraordinary leniency. She lost nothing in Europe that had been hers before the revolution, and the money which she was asked to pay towards the allies' expenses was only 700,000,000 francs.

The moderation of these terms is due partly to Talleyrand's diplomatic skill (although Talleyrand himself had fallen from power before the final treaty was signed), but partly also to the fact that the leaders of the allies, with the exception of the Prussians, were not nationalists. The peace left France relatively free from the bitterness of humiliation because it was a peace drawn up by men who were not only immune from, but actively opposed to, the new nationalism which Napoleon and the French revolution had set loose in Europe. Metternich and his friends dealt with kings, not nations; they saw no more point in punishing than in rewarding a people.

Alexander and the Holy Alliance. Apart from these three practical statesmen stood the Tsar Alexander of Russia, a dreamer whose opinions they despised but whose power they respected. The Vienna settlement increased his already great empire by incorporating in it most of Prussia's former Polish provinces, which Napoleon had given to the Grand Duchy of Warsaw. The greater part of the old Polish kingdom thus fell to Alexander, who promised to rule it as a constitutional king with the help of a parliament, a promise he soon forgot.

Alexander came to Vienna more than half a liberal, filled with high dreams of a Holy Alliance of all Christian kings, princes and governors which would secure peace and justice, religion and piety, by the practice of undefined Christian principles. Opinion was divided: some thought the Tsar mad, others merely a hypocrite. All were impressed with the importance of remaining on good terms with the absolute rule of so great an Empire. His neighbours humoured him by taking the pledge and joining the Holy Alliance. They felt sure that he would never let the application of his principles interfere with the enjoyment of his privileges. They were right. In a few years' time an attempt to assassinate Alexander finally cured him of any liberal feelings. For the rest of his life he was Metternich's faithful disciple.

The Holy Alliance was still-born, a pious formula which converted nobody and deceived only its author; but the Holy Allies soon turned themselves into an effective committee for the government of Europe. England, which had refused to sign the Holy Alliance, joined Russia, Austria and Prussia in the Quadruple Alliance for the preservation of the Vienna settlement.

Lord Castlereagh. This was largely the work of the British representative, Lord Castlereagh, a man whose character reflected both the determination and the timidity with which Napoleon and the French revolution had filled the English governing class. It was he who had overcome the jealousies of the allies, brought them together in the fourth coalition and prevented the danger of a separate peace. He represented England's inflexible determination to overthrow Napoleon. But he was equally responsible for the stern measures which were taken to smother every movement for political and social reform in England. He also represented, therefore, the panic to which the rich men who governed England had been reduced by the French revolution. English radicals hated him more than any other member of the government—and with reason, for, if he helped to free Europe from Napoleon, he hoped to saddle England with the perpetual rule of a selfish upper class.

Castlereagh was as great a conservative as Metternich, with whom indeed he was on excellent terms; but conservatism in England could not help being different from conservatism in Europe. Europe was ruled by absolute kings. Its statesmen could therefore cheerfully accept Alexander's Holy Alliance of kings, and while they laughed at its romanticism, use it for their own purposes. England had chased James II, its last absolute king, into exile. It could hardly, like Austria, denounce all revolutions, since the English constitution was the product of the Glorious Revolution. European conservatives wanted to preserve autocracy, English conservatives wanted to preserve aristocracy. They could go a long way together; they could not go the whole way.

The Quadruple Alliance. Castlereagh's Quadruple Alliance represented the limits of their co-operation. It was, first and foremost, an alliance against any Bonapartist restoration. It was also an alliance against any effort to change by force the frontiers agreed on at Vienna. It provided also for regular meetings of the allies 'for the purpose of consulting upon their common interests, and for the consideration of the measures which...shall be considered the most salutary for the repose and prosperity of Nations, and for the maintenance of the Peace of Europe'. Metternich and his friends thought that any attempt to secure a constitution was a threat to the 'repose and prosperity of Nations' which must be put down by force. Castlereagh, who was after all a constitutional minister, could not accept this as a universal rule, though he was prepared to allow Metternich to suppress some particular constitutions.

Some machinery for international consultation was needed to deal with the economic and political problems raised by the industrial revolution and the rise of a vigorous new middle class. Unfortunately the Concert of Europe set up by the Quadruple Alliance was the wrong machinery for the purpose, since most of the people who were to do the consulting were altogether out of sympathy with the demands of this powerful new class. The English aristocracy might and did make friends with middle-class manufacturers; the European autocrats treated them as enemies. The Concert of Europe was on paper the most useful result of the Vienna settlement, but in practice the most obstructive. It postponed, though it could not prevent, the rise of the middle class to political power. In doing this it discredited the idea of international consultation.

THE HABSBURGS

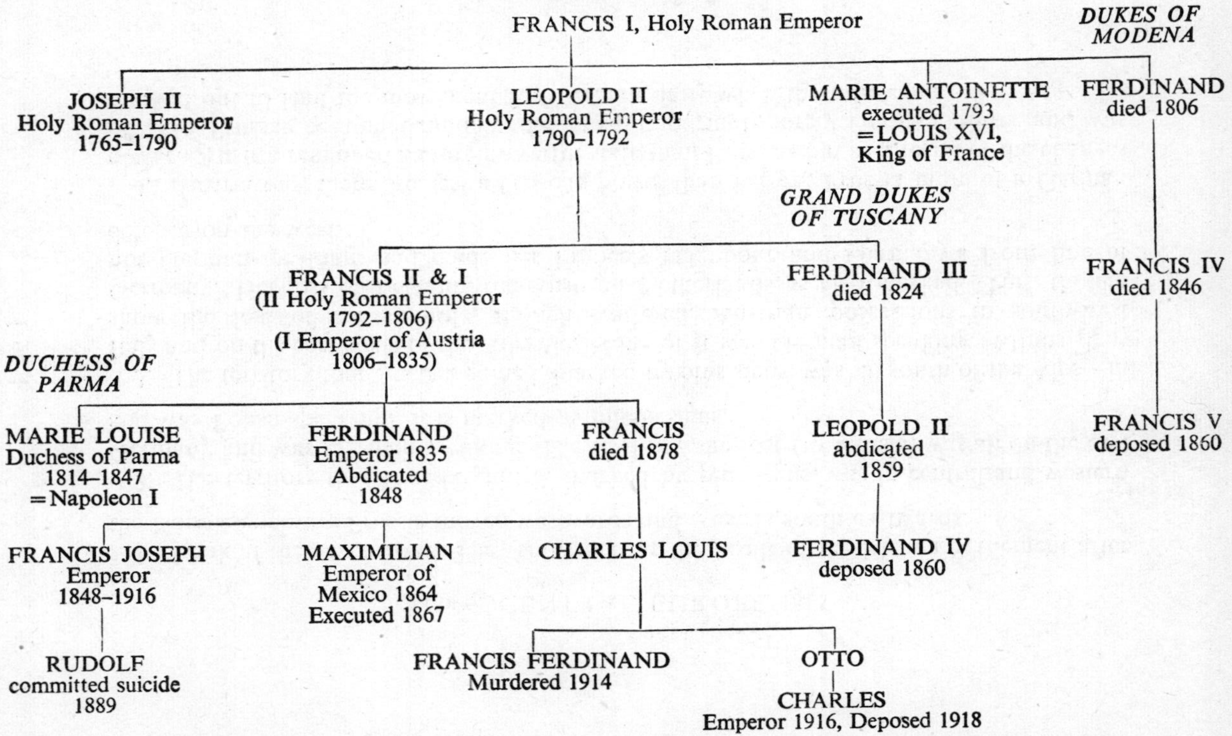

MAP 6. CENTRAL EUROPE, 1815

1. Look at the two arrows. They show that as the result of the Vienna Settlement after the Napoleonic wars Prussia moved westwards and Austria south-eastwards.

2. The territory that Prussia gained, marked by plus signs, was in central and western Germany and was German-speaking. The territory she lost (to Russia) was all on the east and was Polish-speaking. It is marked by minus signs.

3. The territory that Austria gained, marked by plus signs, was all south of the Alps—in Italy and on the east coast of the Adriatic. None of it was German-speaking. Minus signs show the loss of considerable, though scattered, Austrian possessions in south-west Germany. Her loss of Belgium, the Austrian Netherlands, is also marked. This, though not German-speaking, had made her France's neighbour and Germany's front line of defence on the west.

4. Austria was, therefore, less a German power than before; Prussia more of a German power. Austria remained a more powerful state than Prussia; but as a result of the changes of 1815 Prussia became definitely the most important purely German power and was marked out to lead the movement for German unity when the time came.

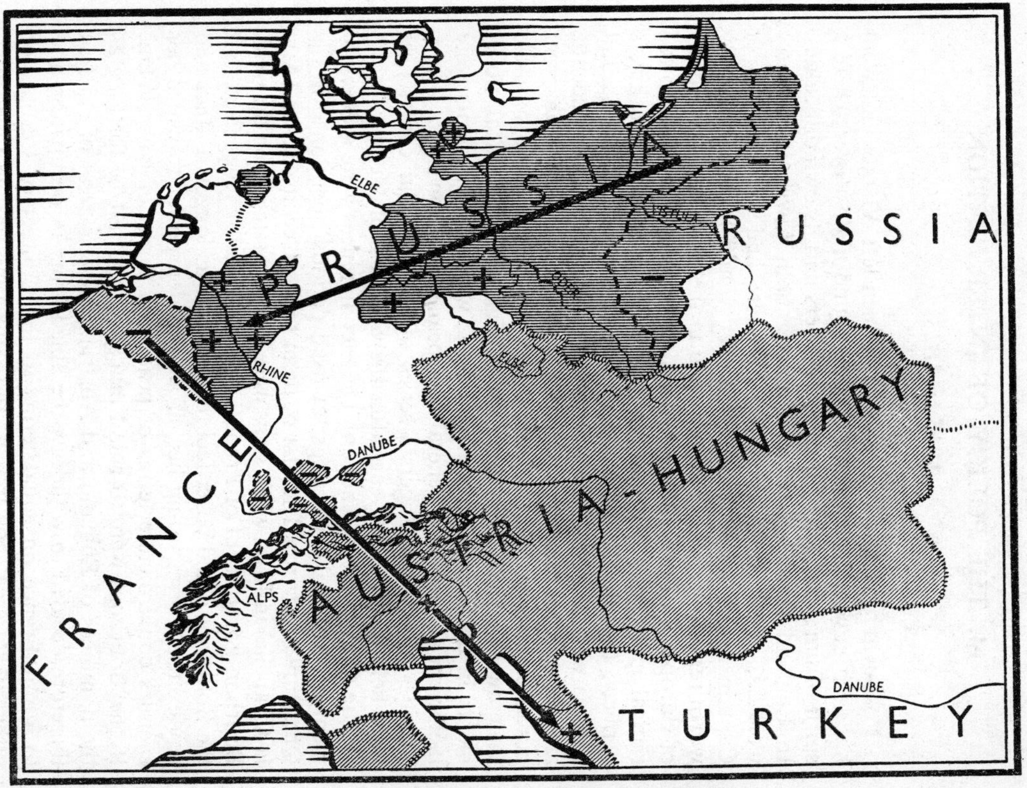

MAP 6. CENTRAL EUROPE, 1815

14. THE POLICY OF INTERVENTION
1815 to 1823

Post-war unrest. When the statesmen left Vienna and went home to govern their countries they found almost everywhere a spirit of unrest. Each of the great powers had its own difficulties. In England there was trouble among both town and country workers, although the Peterloo massacre and the activities of Captain Swing were so mild that they would have been hardly worth reporting if they had occurred in the more backward parts of Europe. In France the parliamentary elections, which showed radical gains, and the murder of the Duc de Berri, the king's nephew, thoroughly upset the returned Bourbons.[1] Austria and Prussia were alarmed by a liberal movement among the students, and by the murder of a minor conservative poet.[2] The great powers, however, had efficient police forces which could suppress their own radicals. They were more frightened by radical movements in the Mediterranean countries, where governments were inefficient and where a general European revolution might begin. To deal with these dangers a series of international conferences were held in the eight years after the Congress of Vienna.

The Congress of Aachen. These conferences were a development of the regular meetings 'at fixed periods' mentioned in the Quadruple Alliance. The first of them took place at Aachen (Aix-la-Chapelle) in 1818, when steps were taken to strengthen the weak Bourbon government of France. It was generally recognised that the Bourbons could never feel really secure as long as France had any excuse for regarding herself as a beaten power to which her enemies could dictate terms of peace and forms of government. If the Bourbons were to rule France it could not be for long as the nominees of England and Austria. The Congress of Aachen therefore decided to withdraw the allied army of occupation and to admit France on equal terms to the Concert of Europe, or rather of its great powers. At the same time the old Quadruple

[1] French internal affairs are discussed in Chapter 18.
[2] German internal affairs are discussed in Chapter 20.

Alliance was kept alive to stop any proposal of a Bonapartist restoration.

Castlereagh's difficulties. From the first England's attitude to the Concert of Europe was half-hearted. On the one hand Castlereagh shared the general fear of revolution. He spoke with horror of the 'organised spirit of insurrection which is systematically propagating itself throughout Europe, and which explodes wherever the hand of the governing power from whatever cause is enfeebled'. On the other hand he drew a distinction between a constitution and a revolution—two things which to Metternich were equally diabolical. He knew that 'the House of Hanover could not well maintain the principles on which the House of Stuart forfeited the throne'. The British parliament would never agree to do what Alexander wanted and promise to go to war against any country which tried to obtain a constitution from its king. Further, parliament was reluctant to commit England in advance to any automatic action. It might agree to co-operate with other European powers in some particular action for a definite end, but it wanted to be free to decide when the moment came. Castlereagh saw his dream of international co-operation fading because the great powers of Europe wanted him to promise too much, while the British wanted him to promise nothing at all except to keep the Bonapartes off the throne of France. The outcome of the Congress of Aachen was that England refused to guarantee the absolute right of kings to govern as badly as they liked, but agreed to attend further conferences if the international situation made them necessary.

The international situation was soon bad enough for another conference. In January 1820 the people of Spain rose in revolt against their Bourbon king, Ferdinand. They were followed at short intervals by the peoples of Portugal, Naples, Piedmont and Greece. In 1820 and 1821 no European ruler could feel safe on his throne.

The Spanish revolution. To understand the Spanish revolution it is necessary to go back to the year 1812, when Ferdinand was Napoleon's prisoner in France, and when a few educated Spaniards were establishing an anti-Bonapartist government at Cadiz. These men had drawn up a constitution on the early French revolutionary model. Ministers were not allowed to sit in parliament, the church was not to be represented and the deputies, after serving in one parliament, were to be ineligible for election to the next. On the other hand, only the relatively prosperous middle class were to be allowed to vote. It was in any case a bad constitution; it was an impossible constitution for Spain—but then any constitution was impossible in Spain, where hardly any one could read or write. What Spain needed was good government, not self-government. The constitution of 1812 was drawn up by the leaders of the small Spanish middle class, who shared the anti-clerical views of eighteenth-century France. The Inquisition not only existed in Spain, but was popular with Spaniards. The liberals abolished it. The peasants, influenced by the priests, resented this, and enthusiastically supported Ferdinand, who, when he was restored to the throne in 1814, made haste to abolish the constitution and to persecute the liberals.

By 1820, however, Ferdinand was as unpopular as the liberals. Spain wanted firm, conservative government and the suppression of liberalism. It did not want the rule of the brutal, vicious and corrupt men with whom Ferdinand surrounded himself. The Spanish colonies in South America were in revolt, and though Ferdinand planned to send an expeditionary force to subdue them, he allowed his corrupt friends to embezzle the money which would have provided them with proper equipment. A mutiny broke out in the camp at Cadiz and the rebellious soldiers were soon joined by the discontented people of the towns. To save his throne, Ferdinand was forced to accept the constitution of 1812.

In the neighbouring state of Portugal, where similar conditions existed, there was a similar revolution. In 1807 the king had fled to the colony of Brazil when French troops invaded the country. He had refused to return, leaving the Englishman, Marshal Beresford, as regent. Once the Spanish revolution had broken out in 1820 Beresford found it impossible to control any longer his ill-paid troops, so, while he went to Brazil to persuade the king to return, the regency was overthrown and a constitution proclaimed. A year later King John did come back to Portugal, a poor, toothless, shuffling king, who, according to the English minister, made a 'sort of grunting noise between each word', and looked half listless, half frightened. However, he duly swore to accept a constitution on the lines of the Spanish one.

Europe beyond the Pyrenees was a world of its own, which

might have been allowed its revolutions without interference from the rest of Europe. England certainly thought so. The Duke of Wellington, who knew more about Spain than most men, announced that 'there is no country in Europe in the affairs of which foreigners can interfere with so little advantage as Spain'. Alexander of Russia, on the other hand, was anxious to march an army into Spain and crush the sinister revolutionaries. Metternich in theory agreed with Alexander, but in practice he sided with England since he was as much afraid of Russian domination in 1820 as he had been in 1813, and was most unwilling to see a Russian army marching through Germany and France. There the matter might have rested but for the spread of revolution to Italy.

Trouble in Naples. A few weeks before the rebellion in Portugal there had been one in Naples, which was misgoverned by another branch of the Bourbon family. The revolutionaries demanded the introduction of the Spanish constitution. The king, another Ferdinand, promised to accept it but secretly appealed to Metternich for help to overthrow the liberals. Naples was as backward and insignificant a country as Spain, and quite as little fitted to govern itself. It might have been left to discover this for itself but for the fact that Naples was in Italy. Metternich knew that Austrian rule in north Italy was profoundly unpopular, and realised that, unless he suppressed the Neapolitan revolution, he would soon have a Lombard revolution on his hands. The educated and vigorous people of north Italy would never be content to accept Austrian despotism if the ignorant and idle Neapolitans were allowed to rule themselves.

Metternich accordingly decided to intervene and accepted the Tsar's suggestion of a conference at Troppau to bless his crusade for autocracy. Alexander was, of course, delighted that something was going to be done at last about those deplorable revolutionaries. Castlereagh's position was as usual ambiguous. He was quite prepared to allow Austria to do whatever she thought necessary, provided that she acted on her own responsibility. He was not prepared to allow her to invade Naples as the representative of the Concert of Europe, because this might be used as a precedent to make England agree to some future intervention somewhere else of which she disapproved. Consequently, when the conference met, England was represented only by an observer.

Troppau and the Neo-Holy Alliance. The conference at Troppau did much more than merely arrange for the suppression of the Neapolitan revolution. It authorised a general guarantee of all kings against all revolutions—the very thing which England had prevented at the Congress of Aachen. This arrangement between the kings has sometimes been called the Neo-Holy Alliance. The first to benefit from the royal trade union was of course the King of Naples. He was invited to attend a second conference at the beginning of 1821 at Laibach, where the final plans were to be made. Naturally his subjects were most unwilling to let him go, but after he had repeated in the most solemn way his oath to respect the constitution he was allowed to leave the country. No sooner was he outside the country than he invited the Austrians to march in and overthrow the constitution. The Neapolitan army was defeated and Ferdinand was able with the support of Austrian troops to start a wholesale persecution of liberals.

But the Italian troubles were not yet over. The day after the Austrian victory in Naples there was a military revolt in Piedmont. The army leaders wanted a constitution for Piedmont and an end of Austrian influence in Italy. The king abdicated in favour of his brother, Charles Felix, whose conservatism was impeccable but who, unfortunately, was not in Piedmont at the time. A young cousin, Charles Albert, was therefore made regent. He accepted a constitution, partly because he wanted to and partly because he could not do anything else since the liberals were in power. Charles Felix, however, was furious and appealed to the Austrians, who lent him an army which beat the Piedmontese army at the battle of Novara—the second time that year that an Italian army had been defeated by an Austrian. Charles Felix returned to Turin to start the inevitable persecution of liberals.

The Carbonari. The first example of intervention had been a triumphant success, thanks largely to the feebleness of the liberals. The movements of 1820–1 in Italy were the work of mysterious secret societies called the Carbonari, or charcoal burners, from their habit of meeting in the secrecy of woods as well as from the elaborate ritual, based on the technical terms of charcoal-burning, which they used. They were formed to fight Napoleon's domination of Italy, they survived to fight the Austrians. The movement soon spread from its original home in Naples to other parts of Italy, and soon there were branches in every Italian state pledged to free Italy from Austrian interference and to set up constitutional government.

It was a romantic movement, well fitted to appeal to sons of prosperous Italian families, and to catch the generous sympathy of freedom-loving foreigners such as Byron, the English poet, and Louis Napoleon, the future Emperor of the French, both of whom were members. But just because it was a romantic movement, appealing to young men of leisure with a taste for secret passwords, it had little backing among the hard-working peasants of the countryside. Its most practical members were the young officers in the various Italian armies, but their efficiency was limited by the fact that they still thought in terms of Piedmont or Naples rather than of Italy. There was no co-ordination among the movements and the Piedmontese liberals allowed the Neapolitans to be defeated before they made their own revolution. The liberals of 1821 failed because they represented a class and not the nation, the separate states and not Italy.

The Congress of Verona. No sooner was Italy quiet than Greece was in arms. In 1821 Greece was still part of the Turkish Empire, but a serious rebellion had broken out. Metternich was in despair, for the sworn enemy of all rebels, the Tsar Alexander, was seriously thinking of helping the Greeks. Of course his help would not have been disinterested: it would have had for its object a greater Russian empire carved out of the Turkish. But this was precisely what Metternich most feared. In this he and Castlereagh were in agreement, and so the last Congress of the series begun at Vienna was called to meet at Verona in 1822 with the object not of encouraging, but of discouraging, intervention.

When the Congress met, however, it found itself so much concerned with Spain that the Greek rebellion dropped into the background. Ferdinand of Spain was now nominally a constitutional king, but he was spending his time organising a civil war against constitutional ministers. Alexander forgot his interest in Greece for the time being in his renewed desire to help Ferdinand by sending him a crusading army pledged to restore bad government in Spain. Once again Metternich's fears of Russian domination proved stronger than his principles. On no account must Russian intervention be allowed. The French also were determined that they would not have a Russian army marching through France, but they were quite prepared, and had indeed decided, to send one of their own. England meant if possible to stop any intervention in Spain, whether by France or Russia. The English representative was instructed to inform the conference 'frankly and peremptorily that to any such interference, come what may, His Majesty will not be a party'.

The outcome of the conference was a compromise. The three eastern powers—Russia, Austria and Prussia, the Neo-Holy Allies—agreed to protest to the Spanish government against the Spanish constitution. They refused to authorise military intervention in Spain either by France or Russia. England flatly refused to join in any protest to Spain. France also refused to join the eastern allies, not through any dislike of intervention, but because she was determined to intervene on her own.

This was the end of the attempt at international government begun at Vienna in 1815. It had failed for two reasons. The first was the refusal of Britain to join in any guarantee of the absolute power of kings. The second was the inability of the kings, and particularly of the Emperors of Austria and Russia, to overcome their jealousy of one another and wholeheartedly to accept the balance of power established at Vienna.

THE BOURBONS

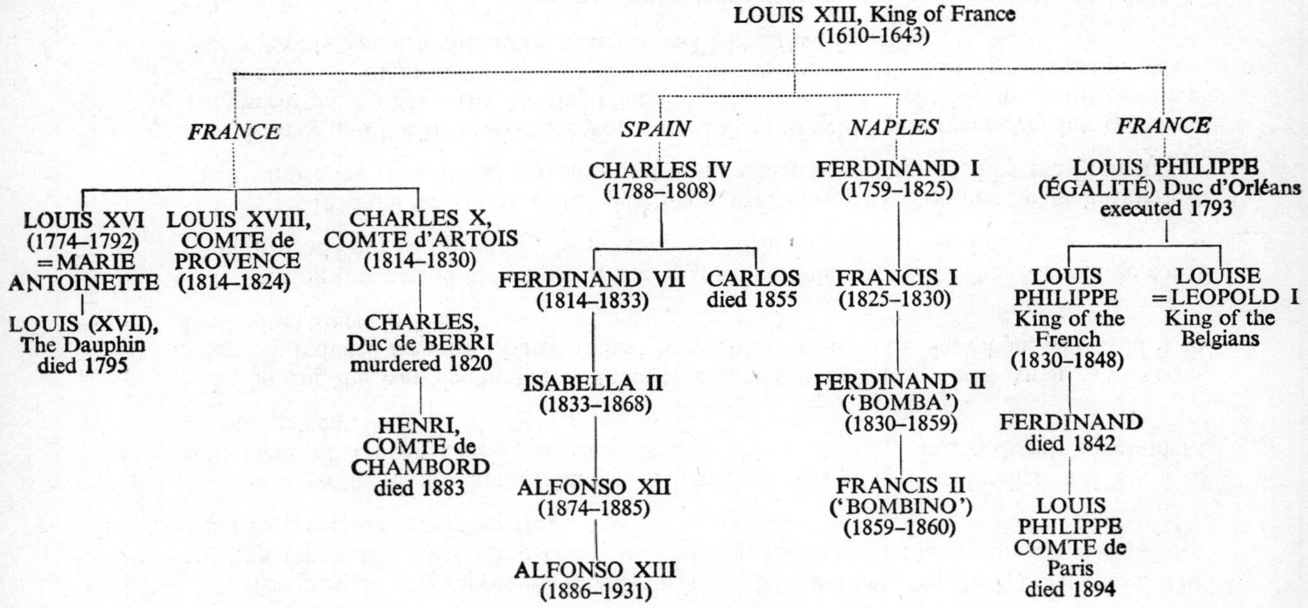

Map 7. CANNING AND THE NEW WORLD

1. The grey shading represents Spanish territory about 1789; the black Portuguese. The shading represents what these countries claimed, but their actual colonisation was confined to the more accessible areas.

2. It is wrong to think that the Spanish empire was altogether derelict in 1789. The settlement of Upper California was then only just starting under direct government encouragement.

3. During the Napoleonic period the area claimed by Spain in North America was considerably reduced because Louisiana had been given by Spain to Napoleon, who sold it to the United States in 1803.

4. During this period also most of the Spanish colonies rebelled. The Portuguese royal family, turned out of Europe by Napoleon, established themselves in Brazil.

5. After 1815 the Bourbons, restored in Spain, tried in vain to reconquer Spanish America. Their failure led to a liberal revolution at home which was suppressed by the French.

6. France and the Holy Alliance would have liked to help Spain reconquer her colonies. They were prevented partly by the United States under President Monroe—then quite a minor power—but more by Canning's threat to use the British fleet against them. After 1826 the only Spanish colony in America was Cuba (lost in 1898).

7. In Portugal Canning was even more successful, using the navy not only to establish a constitutional monarchy in Brazil (replaced by a republic in 1889) but also to drive reactionary French influence out of Portugal itself.

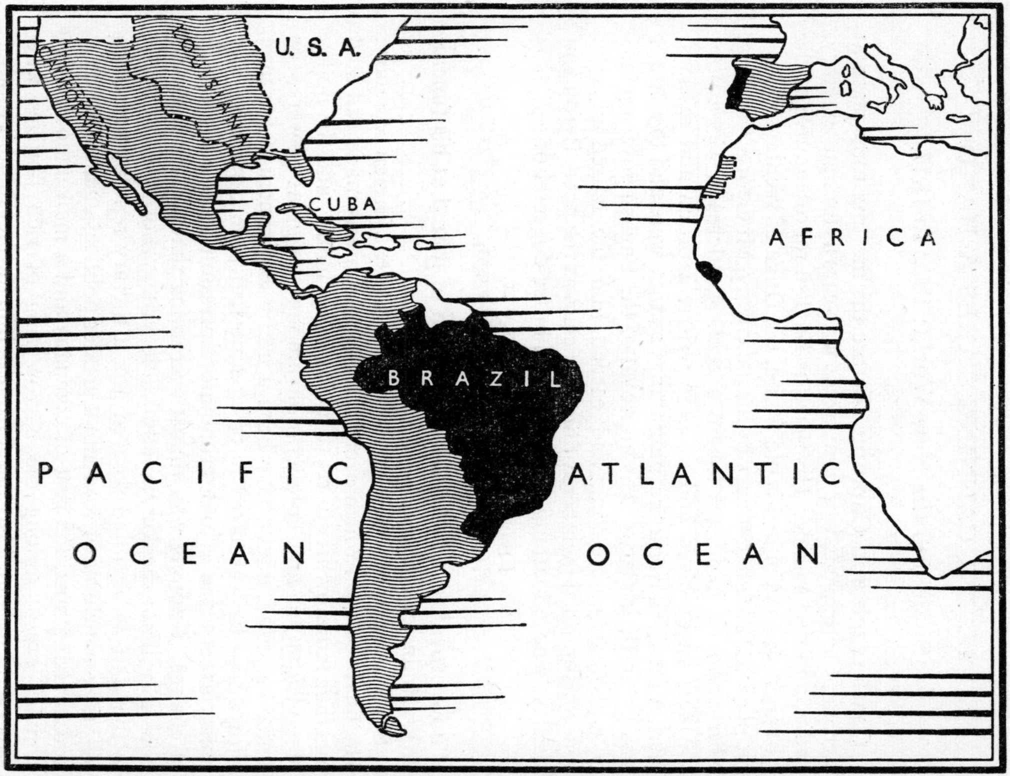

MAP 7. CANNING AND THE NEW WORLD

15. MR CANNING TAKES A HAND

Spain and the New World: 1822 to 1827

Castlereagh and Canning. Before the meeting took place at Verona, Castlereagh, worn out by his ten years as foreign secretary, had committed suicide. His place as English representative at the Congress was taken by his old friend, the Duke of Wellington, and his place as foreign secretary by his old enemy, Mr George Canning, with whom he had once fought a duel. Castlereagh was a poor speaker, an involved and obscure writer, a proud and reserved man, a firm believer in keeping things as they were and the 'lower classes' in their places. His approval of the English government's policy of strike-breaking had earned him the hatred of such generous-minded members of his own class as the poets Byron and Shelley, whose loathing burst into the bitter doggerel of

> I met Murder on the way—
> He had a mask like Castlereagh.

It now provoked a brutal demonstration of hatred at his funeral in Westminster Abbey.

But with all his faults Castlereagh was a great foreign secretary. The final coalition which overthrew Napoleon was made and held together more by his efforts than any other man's; and the policy of consultation and co-operation among the great powers, for which he stood, might have been a useful experiment if the men with whom he was asked to co-operate had been capable of any policy except one of indiscriminate repression of any suggestion of change. Canning was almost everything that Castlereagh was not—a brilliant speaker, an admirable writer whose satirical verse in the *Anti-Jacobin* had secured him a literary reputation before his statesmanship had won him a greater political one, a charming and amusing host, the best of company and a man unboundedly popular except among the narrow ring of rich families who governed England. They never forgave him the fact that his mother was an actress.

The policy of isolation. Yet for all their differences Castlereagh and Canning were agreed in their view of what English foreign policy should be in 1822. Both thought that England must oppose Alexander's desire for universal intervention: in Greece to help the rebels because they were Orthodox Christians and everywhere else to crush the rebels because they were liberals. Consequently Castlereagh's death made little immediate difference to English policy. Canning and Wellington by their non-co-operation at the Congress of Verona were merely carrying out decisions already taken. But although Castlereagh and Canning had reached the same conclusion they did so for very different reasons. Castlereagh went back to English isolation with extreme reluctance, for he really believed in the Concert of Europe and detested liberalism; Canning welcomed the return with enthusiasm because he believed in keeping England free from foreign alliances and detested the tyranny of the Holy Alliance and its friends. At the end of his life Castlereagh had dissociated England from the 'European trade union of kings'. Canning soon made her the protector, wherever possible, of liberal movements. This was the natural continuation of the policy of Castlereagh's last years, but it is most improbable that he would have approved it.

Canning was foreign secretary for the five years from 1822 to 1827, during which time two important affairs occupied most of his attention. He was concerned in safeguarding the revolutions by which the Spanish and Portuguese colonies in South America made themselves independent of Europe, and he was largely responsible for the policy which enabled the Greeks to throw off the rule of the Turks. The two movements developed side by side at the same time but, as they were otherwise unconnected, it is simpler to separate them and to describe in this chapter only Canning's dealings with Spain and her colonies.

The French in Spain. It was beyond Canning's power to stop the French intervention in Spain which was designed to overthrow the recently established parliamentary government. Although it was less than eight years since the battle of Waterloo, England had no army fit to fight a European war. The French were further encouraged by the news that King George IV and some of the cabinet ministers wished them well, and so in April their army crossed the Pyrenees and started on a triumphal march to Madrid. They had come to Spain to restore the king's absolute power but they were soon horrified by the use he made of it. A French garrison kept Ferdinand on the throne but he would not listen to French advice. They wanted him to justify their invasion by treating his people well; he insisted on wholesale massacres which could only disgrace both them and him. It was his duty, he announced, to offer God 'holocausts of piety'. It was at any rate a congenial task. He did not stop the persecution of those who had been in any way concerned with the constitution until there were no more left to execute. Canning had failed to save Spain both from absolute government and from absolute misgovernment.

England, France and Portugal. He was able, however, to stop the westward spread of tyranny. In Portugal there was a stiff struggle between English and French influence. For a long time Portuguese independence had been guaranteed by England; it was now threatened not so much by foreign invasion as by foreign encouragement of rebels. The continental despots were ready to support any movement which would get rid of the constitution introduced in 1821. France in particular was active in this direction. Success would have made France a dangerous rival to England since gratitude, which in politics is usually a 'lively expectation of favours to come', would have made Portugal a faithful ally of France. But this time France was unsuccessful. Canning did not enjoy seeing his political ideals overthrown; he dared not see England's influence eclipsed. British sea power, which was unable to save liberalism in Spain, was well able to protect the constitution of Portugal. Canning saw that it did.

There were two stages in Canning's dealings with Portugal. In the first the actors were, besides Canning himself, the old king, John; his younger son, Miguel, and the French ambassador. John and Miguel were on the bad terms which were customary in the Portuguese royal family. Miguel was an out-and-out absolutist, a man of the same type as Ferdinand of Spain. He was Metternich's favourite. John was a man of much less decided views, though dislike of his son inclined him to a vague liberalism. Both England and, with some hesitation, France supported John against Miguel. Miguel was at first successful and got the constitution suspended, but in 1824 he was sent into exile after he had failed to overthrow his father. There followed a diplomatic struggle between England and France. England tried to keep John loyal to the English alliance, while France hoped to turn him into her good friend and make him a confirmed absolutist. This duel ended in 1825 with a victory for Canning and the recall of the French ambassador.

The second stage of Canning's Portuguese policy started with the death of John in 1826 and the succession of his elder son, Pedro. Pedro had to choose between being Emperor of Brazil or King of Portugal. He chose Brazil, but before abandoning Portugal he gave her a constitution under which she was to be governed by his sister as regent for his daughter. Metternich and the French were furious at the parting gift of a constitution. They did everything in their power to prevent its coming into effect and, if they did not advise Ferdinand of Spain to help Portuguese rebels to prepare an invasion, they certainly did nothing to stop him. Canning now saw his chance to end French and Austrian intrigues in Portugal. He promised the constitutional government every assistance to repel attacks from Spain, even if it should mean war. It did not. The threat was enough.

The independence of Latin America. Meanwhile across the Atlantic the Spanish and Portuguese colonies were trying to get their independence recognised. Brazil, the Portuguese colony, had established a constitutional monarchy of its own with the King of Portugal's elder son as emperor. The mother country was persuaded to recognise the separation in 1826 thanks largely to Canning's persuasion. The Spanish colonies had proclaimed themselves independent republics and destroyed almost all the armies sent against them, though there was still fighting in Peru. Clearly Spain alone could not hope to overthrow the rebels, but she might have succeeded with the help of France, which was playing with a scheme for establishing Bourbon princes in semi-independent South American kingdoms. These princes would in practice have been dependent on France. Absolute government, which Englishmen disliked, and French influence, which they feared and detested, would thus have been supreme on both sides of the Atlantic.

England, however, was in a position to prevent this thanks to her command of the sea. Canning told the French ambassador, the Prince de Polignac, in October 1823 that England would oppose by arms any attempt to reconquer Spanish America with an expeditionary force which was not purely Spanish. France was unwilling to risk a war and Canning insisted on Polignac's putting his promise not to intervene in a signed Memorandum which was circulated to the other European powers.

The Monroe Doctrine. In December James Monroe, the President of the United States, made an announcement of American policy which has come to be known as the Monroe Doctrine. He laid down three principles which were to guide his country's conduct of foreign affairs. First, the United States would not take part in European politics; secondly, she would not tolerate any European intervention in either North or South American affairs, and lastly she would not admit the right of any European country to found new colonies in unsettled parts of America. The Monroe Doctrine is of great importance because it has always remained the inspiration of American foreign policy and for many years the United States, as one of the great powers, has been able to see that it is carried out. But in 1823 the United States was only a minor power and President Monroe would never have dared to send his famous Message to Congress if he had not known that England would make good his boast and guarantee the freedom of America from European intervention.

For three years the Holy Alliance and the French tried to find some way round the Polignac Memorandum and the Monroe Doctrine. The French secretly played with the idea of intervention; Metternich organised a Congress which he hoped would be able to find some means of giving the victorious Spanish Americans something less than the independence they had conquered. Canning refused to attend the Congress and again made it clear to France that any intervention, no matter how disguised, meant war. At the end of 1826 he took the final step of recognising the governments of the Argentine, Mexico and Colombia as independent republics. From that day the rich states of South America have been free to develop in their own way without any European dictation. England gained great commercial advantages from being the first great power to recognise the new republics.

Shortly before his death Canning summed up the results of his western policy in a great speech in the House of Commons:

It was the Spain within the limits of whose Empire the sun never set —it was Spain '*with the Indies*' that excited the jealousies and alarmed the imaginations of our ancestors. But then, Sir, the balance of power! The entry of the French army into Spain disturbed that balance, and we ought to have gone to war to restore it!...Was there no other mode of resistance than by a direct attack upon France—or by a war to be undertaken on the soil of Spain?...I looked another way. I sought materials of compensation in another hemisphere. Contemplating Spain, such as our ancestors had known her, I resolved that if France had Spain, it should not be Spain '*with the Indies*'. I called the New World into existence to redress the balance of the Old.

It was a proud boast but justified.

16. THE EASTERN QUESTION. I

The Independence of Greece: 1821 to 1833

The Eastern Problem. When Canning placed the barrier of British sea-power between the South American republics and the European monarchies he effectively severed the New World from the Old in politics. The new republics, it is true, continued in close economic dependence on Europe and particularly on Great Britain; but they ceased to play any considerable part in the European balance of power. It was otherwise with Canning's intervention in Greece. If his American policy ended a chapter in European history which had begun in the sixteenth century, his Eastern policy opened a chapter which has, perhaps, not yet been finished. The Eastern Question was one of the main preoccupations of the nineteenth century, and the Eastern Question really starts with Canning.[1]

Turkey and Austria. The first element in the problem was the Sultan of Turkey, the 'Sick Man of Europe'. Little more than a hundred years before Canning's time the Turkish Empire had reached to the gates of Vienna, and Mahometan warships had picked up British ships in the English Channel and sold their captive crews into slavery. The Sultan's dominions were magnificent in extent but wretched in condition. They had been won by the sword and they could only be kept by the sword. Turkish governors oppressed their Christian subjects, who never ceased to pray for the day when they might once more be free in their own land.

For many of them this time had already come, since the eighteenth-century Turkish Sultans were a feeble and degenerate race who were as incompetent to defend their dominions as they were unwilling to rule them well. Their policy was one of neglect punctuated by massacre and murder. Their Christian subjects were allowed considerable freedom to order their own affairs, but Turkish toleration never won Christian gratitude because it was liable to turn into persecution at a moment's notice or an individual's whim. The Sultans lost province after province and by

[1] It had, of course, existed in a different form ever since the decline of Turkey began at the end of the seventeenth century.

1800 their European empire was confined to the Balkans. Their Christian subjects were discontented and rebellious, though as yet too weak to win their independence for themselves. They needed help from one or other of the great Christian powers, but this help was not given, because each power cared more for preventing its neighbour from growing more powerful than itself than it did for the liberation of its fellow Christians under Turkish rule. It was the jealousy of the Christians, not the strength of the Mahometans, which now preserved the Turkish Empire.

The second element in the Eastern Question was the Austrian Empire. At first the Habsburgs had gained more than any one else from the weakness of the Turks. Vienna was no longer the frontier town of Christendom: two hundred miles separated it from the nearest Turkish soldier. In another direction the first Turkish frontier post was 800 miles away. In enlarging its boundaries Austria had changed its character. It had originally been a German state. It had become a cosmopolitan empire in which the majority of the people were no longer German, though they were ruled by Germans. Its frontiers contained all the Hungarians, some of the Serbians and some of the Rumanians, besides many other races. The remainder of the Serbians and of the Rumanians were still under Turkish rule. This might have been an asset to the Habsburgs for it would have been natural for these Christians across the frontiers to wish to exchange the harsh rule of the Sultan for the gentle, if despotic, rule of the Austrian Emperor. No doubt they would have wished to do so if they had not seen a more attractive prospect open to them.

Russian ambitions. This brings us to Russia, the third element in the Eastern Question. Russia, not Austria, was the natural protector of the unredeemed peoples of the Balkans—and Russia was Austria's hereditary enemy. Russia was their natural protector because she was akin by race or speech to most of them and by religion to all. For many centuries Europe had been divided between two branches of Christianity: the western or Catholic church (from which the Protestants rebelled) and the eastern or Orthodox church. Austria was the greatest Catholic power; Russia the greatest Orthodox. The Balkan peoples were Orthodox, and, since religious differences were still the most important divisions in eastern Europe, they looked to Russia as their friend. If Russia was allowed to help them to their freedom from the Turks, it would only be a question of time before the Serbians and Rumanians inside the Austrian Empire would want to cut loose from it and join their fellow-countrymen in the new national states under Russian patronage. For this reason alone Austria was bound to oppose the further break-up of the Turkish Empire.

It was not only Austria which was frightened by the prospect of Russian expansion in the Balkans. In the eighteenth century Russia had established herself on the Black Sea, from which she had formerly been cut off by Turkish territory. She was now looking westwards to the Mediterranean, from which she was liable to be cut off in time of war by the Turks at Constantinople. The annexation of Constantinople therefore became the principal aim of Russian foreign policy. When Napoleon and Alexander met at Tilsit in 1807 the division of the Turkish Empire in Europe had been one of the matters discussed, and to buy Russian support for his boycott of England Napoleon was prepared to agree in principle to the Tsar's acquiring Constantinople. Nothing came of the scheme, however, for in practice the French were almost as strongly opposed as the English to Russia's becoming a Mediterranean power. In 1812, however, by the Treaty of Bucharest Russia obtained Bessarabia from Turkey.

France and England. The Russian danger haunted the two western powers throughout the whole nineteenth century and made them much less sympathetic than they otherwise would have been to the sufferings of the Christian peoples of the Balkans. More often than not they bolstered up the decrepit power of the Sultan because they feared that his only possible successor would be the Tsar of all the Russias. This generally pro-Turkish attitude of France and England was the fourth element in the Eastern Question.

There were, however, frequent and serious disagreements between France and England over the future of the Syrian and North African portions of the Turkish Empire. Nelson and Sir Sidney Smith had prevented Napoleon from establishing a great French empire in the Near East, but French ambitions did not die with Napoleon. They were several times in the course of the nineteenth century to lead to strained relations with England, and it was probably only the over-riding fear of Russia which prevented another war between the two countries

Serbians and Greeks. When Canning became foreign minister in 1822 the Turkish Empire in Europe was harassed by revolutionary movements. There were two main centres of rebellion separated by a broad belt of unaffected territory. In the north the Serbians had already virtually secured their independence by a long guerrilla war which lasted from 1807 to 1817. Further east, but still in the northern belt of rebellion, the Rumanians had risen in 1821 under Prince Ypsilanti, who had been an officer in the Russian army, but the Turks suppressed this rising without too much difficulty. The rebels got less Russian help than they had expected and the rising was the work of an unpopular Greek upper class, not of the Rumanian people. It seems strange to find Greeks so far north in Europe, but the first lesson of Balkan politics is that east of the Adriatic nothing is too unlikely to be true. There were Greeks scattered all over the Balkans occupying, despite their Christianity, many official positions which their Moslem masters were too lazy to fill. In fact the Turkish civil service was very largely Greek.

The bulk of the Greeks, however, lived south of the undisturbed Turkish territory and it was here, among the Greek peasantry and seamen, that the most serious rebellion broke out at a moment when the Sultan was fully occupied in trying to suppress his disobedient Mahometan viceroy in Albania. In the Morea, the Peloponnese of the ancients, the Christians rose and massacred the Turks. Their leader boasted that he rode his horse into one captured stronghold over the dead bodies of the Turkish inhabitants which carpeted the ground so well that he never once touched earth until he dismounted in the centre of the town. There was no chivalry on either side: 'Turks impaled Greeks in the historic pass of Thermopylae; Greeks set a line of crucified victims on the roads along which a Turkish army advanced.' A Christian population so brutalised by generations of Turkish misrule might naturally have aroused the sympathy of civilised Europe. It is, however, at first sight surprising that the Greek bandits should have aroused the passionate admiration of the best of their contemporaries. To western Europeans the rebel leaders were not men to be pitied and excused, but heroes to be imitated.

It was a natural misunderstanding, for the brigand leaders of the Greek armies were not the Greeks, mostly from Constantinople, with whom Europe was familiar. Such men as Capodistrias, the Russian foreign minister, or Korais, the Corfu schoolmaster who invented modern Greek, were neither brigands nor barbarians. They were highly educated men, as were many of the Greeks in the civil service of the Turkish Empire. It was men such as these who founded the great secret nationalist society, the Hetairia Philike, who revived the knowledge of classical Greek and who made their fellow-countrymen proud of their nation as well as of their religion. It was not difficult for them to win the sympathy of the rest of Europe. They were helped in this task by two enthusiasms of the time: the passion for liberty which even the horrors of the French revolution had been unable to destroy, and the love of the classics which not even the brutality of schoolmasters had been able to suppress. To a generation which knew its Homer far better than we know our Shakespeare the bloodthirsty brigand Odysseus of Ithaca was the true descendant of his Homeric namesake. The Greeks therefore could count from the beginning on a general European sympathy which neither Spanish and Neapolitan liberals nor Serbian peasants could command.

Lord Byron. In 1823 an Englishman set out on a lonely journey which was to transform this mild feeling of good-will into a passionate concern for the independence of Greece. Lord Byron enjoyed in his lifetime, and still enjoys more than a hundred years after his death, a European reputation which no other Englishman but Shakespeare has equalled. When he exchanged his sordid, pleasure-seeking life in Italy for the dirty, fever-stricken but heroic air of Missolonghi he drew the attention of all Europe to the Greek struggle. He died from illness before he had time to strike a single blow for Greece, but his death was more fruitful for Greece than his life could have been. It was worth many victories. It filled so many men in western Europe with a determination that at all costs Greece must be free that in the long run their timid governments were forced to sink their fears of Russian aggrandisement and to intervene to save the Greek rebels from falling again under Mahometan rule.

But as Byron lay dying at Missolonghi in 1824 all this lay in the future. Canning had just recognised the Greeks as belligerents, but his object had not been to support their cause but to be able to hold them responsible for damage to British shipping. At the time it still seemed to Canning more important to save the Turks from the Russians than the Greeks from the Turks. The Greeks seemed to be quite capable of looking after themselves.

Mehemet Ali's intervention. Within a few months, however, it was clear that they were not. Already in January while Byron was writing the last and finest of his poems on Greece—

> The sword, the banner and the field,
> Glory and Greece, around me see!
> The Spartan, borne upon his shield,
> Was not more free.

—Sultan Mahmud had called in the aid of Mehemet Ali, his troublesome and disobedient viceroy in Egypt. Mehemet Ali and the Greeks were equally dangerous to the prosperity of the Turkish Empire, but Mehemet Ali was a Mahometan and the Greeks were Christian. Since the Sultan had to choose between Christian and Mahometan, he chose the latter. The terms of the bargain were that Mehemet Ali should have Crete in addition to Egypt and that his stepson, Ibrahim, should have the Morea when it was reconquered.

Mehemet Ali's co-operation rapidly and completely changed the balance of power in the Sultan's favour. The Greeks had owed their earlier successes to their superiority at sea—the Turkish navy depended largely on Greek sailors. Mehemet Ali, however, had profited by his experience of the Napoleonic wars to modernise his army and his fleet. At sea he had as marked a superiority over the Greeks as they had previously had over the Turks. For the second time in the Greek War of Independence sea power was decisive. In 1825 Ibrahim landed in the Morea and took Sphacteria, the island which had been the scene of one of the most famous struggles of the Peloponnesian War. Soon the whole Morea was in Ibrahim's hands, and before long Missolonghi on the other side of the Gulf of Corinth had fallen after a heroic defence. In the autumn of 1826 the Turks began the siege of Athens, the last important Greek stronghold. For ten months it held out, but in June 1827 its commander, a Frenchman who had served Napoleon, was forced to surrender the Acropolis. The war would have been over if the great powers of Europe had not at last made up their minds to save the Greeks.

Canning breaks the Holy Alliance. The death of Byron had started the conversion of Canning to this new policy; it was completed by the behaviour of Ibrahim. The re-conquest of the Morea had been accompanied by such burning, looting and massacre of non-combatants that men were ready to believe the rumour that Ibrahim was preparing to kill or sell into slavery all its Christian inhabitants and to replace them by Moslem peasants from Egypt or Syria. Canning decided to act. He first sent the Duke of Wellington to make an agreement with the Russians and then followed it up by the three-sided Treaty of London between England, France and Russia.

The three powers agreed to recommend Turkey to recognise Greece as a semi-independent state, free to govern itself but paying tribute to Turkey. If both sides did not agree to an armistice within fourteen days the allies determined to place their naval forces between the two sides in such a way as to prevent the war from being carried on. They had in fact made up their minds to force the Turks to accept their terms. It was a great victory for Canning. Against the wishes of the king and of the majority of the cabinet, who were opposed to the use of force, he had saved the Greeks from extermination, destroyed the Holy Alliance and prevented the Tsar from making Greece a purely Russian dependency.

The second of these three aims was achieved within Canning's lifetime. Metternich never forgave Russia for its attitude to the Greeks. Although Austria was a Catholic state it thought rebels more dangerous than infidels, and was more ready to tolerate the Mahometanism of the Turks than the liberal and revolutionary tendencies of the Greek leaders. Before his death Alexander had begun to prefer the Russian interest of a weak Turkey to the Holy Alliance interest of the suppression of all revolutions; but Alexander was so fickle in his opinions that Metternich was able to hope for the Tsar's repentance. His successor, Nicholas I, was purely Russian in his outlook and Metternich soon realised that the Holy Alliance was dead. The disagreement on the Eastern Question hardened into a general disagreement, which gave liberalism in Europe new grounds for hope. Metternich summed up this side of Canning's triumph when he wrote four years after his death: 'The ministry of Canning has marked an era in the history of England and of Europe.'

Navarino. The first of Canning's aims—the protection of the Greeks—was not secured for three months after his death. An inter-allied squadron under Admiral Codrington went to the Morea, to Navarino, to watch Ibrahim. Sometimes the English ships were reinforced by Russian and French contingents, sometimes they were alone. But Codrington was one of Nelson's old captains and, even when he was left alone with only his own four English ships, he managed to scare back to harbour the vastly superior Turkish fleet when it tried to slip his blockade in spite of Ibrahim's promise to suspend hostilities. As he himself wrote: 'There is no solid foundation for the belief in either Turkish honour or Turkish bravery.' There was unfortunately plenty of foundation for the belief in Turkish atrocities. Unable to slip the blockade, Ibrahim satisfied himself with burning the town of Navarino, cutting down the fruit trees and killing the women and children.

Codrington and the French and Russian admirals determined to sail past the island of Sphacteria into the bay, where they would be in a stronger position to force Ibrahim to keep his word. On October 20, 1827, the allied fleet of twenty-four ships entered the bay and took up their position parallel to the combined Turkish and Egyptian force of eighty-nine ships. A small dispute soon brought on a general engagement. For a time the fight was fierce and Codrington was at one time the only man unwounded on his own quarter-deck.

> The mizzen-mast fell and missed him by a hair's-breadth. Four bullets pierced his hat or his clothing, and a fifth smashed his watch. His own gunners made the Egyptian flagship a wreck in ten minutes; and in three quarters of an hour they sent the Turkish flagship drifting mastless out of the battle with five-sixths of the crew disabled....His general order ran the next day, 'of the Turkish fleet there remain only one frigate and fifteen smaller vessels in a state ever again to be put to sea.' Even Nelson could not have desired a more complete victory for his old captain....Four hours of battle had done more than seven years of diplomacy...and the freedom of Greece was thunderously proclaimed amid the roar of Codrington's guns. (Temperley.)

Sea power had been as decisive in the third and last stage of the Greek War of Independence as it had been in the first two. In England Codrington was the hero of the day. Even the king, who hated the Greeks as rebels, was forced reluctantly to reward him—'I have sent him a riband', he said, 'but it ought to be a halter.'

The independence of Greece. It was some time, however, before the independence of Greece was recognised by the Sultan, and still longer before she was provided with a permanent government. After the battle of Navarino war broke out between Russia and Turkey. This war was the result of the reversal of Canning's policy by the Duke of Wellington after Canning's death. If Wellington did not actually apologise for Navarino, he at least told Turkey that he hoped that it would not interfere with the friendly relations between two countries which had so often been allies. The Turks took this as a promise of English support and promptly went to war with Russia. They lost the war and in 1829 were forced by the Treaty of Adrianople not only to recognise Greek independence but also the Russian conquests in the Caucasus, and a virtual Russian protectorate over the two Rumanian principalities. Finally in 1833 a king was found in Otto of Bavaria for the new kingdom of Greece which was placed under the joint guarantee of France, Russia and England. Thus, if Russia could not be prevented from profiting from the first stage in the break-up of the Turkish Empire, England and France had at least limited Russia's gains to the north Balkans, where she was a danger to Austria, and kept her away from the Mediterranean, where she would have been a danger to them.

THE ROMANOFFS

CATHERINE II
(The Great)
TSARITSA 1762–1796

PAUL I
Tsar 1796
Murdered 1801

- ALEXANDER I
  Tsar 1801–1825
- NICHOLAS I
  Tsar 1825–1855
  - ALEXANDER II
    Tsar 1855
    Murdered 1881
    - ALEXANDER III
      Tsar 1881–1894
      - NICHOLAS II
        Tsar 1894
        Deposed 1917
        Murdered 1918

MAP 8. THE BALKANS AND THE NEAR EAST, 1817–41

1. The northern group of arrows show Christian rebellions, normally with Russian support, against the Sultan. The southern group show Mahometan rebellions, sometimes with French support against him.

2. Montenegro, a Christian state which Turkey never conquered, is marked 1. Between 1807 and 1817 Serbia (2) won its freedom; in 1822 the two Rumanian principalities (3 and 4) secured home rule from Turkey. In 1829 they became Russian protectorates, which worried Austria and England.

3. Between 1821 and 1833 Greece won its independence—the war started while the Sultan was occupied with a Mahometan revolt in Albania. The decisive battle was Navarino (1827), when the Turkish and Egyptian fleets were destroyed by the European powers. Greece was put under international protection, so that the Russian danger was not increased.

4. The Sultan had called in Mehemet Ali, his viceroy in Egypt, to deal with the Greeks. He took Crete as his reward and, as he had already conquered Mecca, Medina and the Sudan, he became a powerful rival of the Sultan. Mehemet Ali was modernising his country with the help of France, which hoped to make Egypt a French protectorate. This worried England and Russia.

5. In 1831 Mehemet Ali conquered Syria and invaded Turkey proper. England was too busy at home (Reform Bill, 1832) to help the Sultan, who had to buy Russian help by giving her control of the Straits (Treaty of Unkiar Skelessi, 1833). This frightened England and France into defending the Sultan against both Mehemet Ali and Russia. The Treaty of Unkiar-Skelessi was torn up and Mehemet Ali turned out of Turkey, though he was allowed to keep Syria.

6. In 1839 Mehemet Ali again successfully invaded Turkey; but England, Austria and Russia once more turned him out and forced him to surrender everything except Egypt and the Sudan. France's failure to support him destroyed her chance of establishing an Egyptian protectorate.

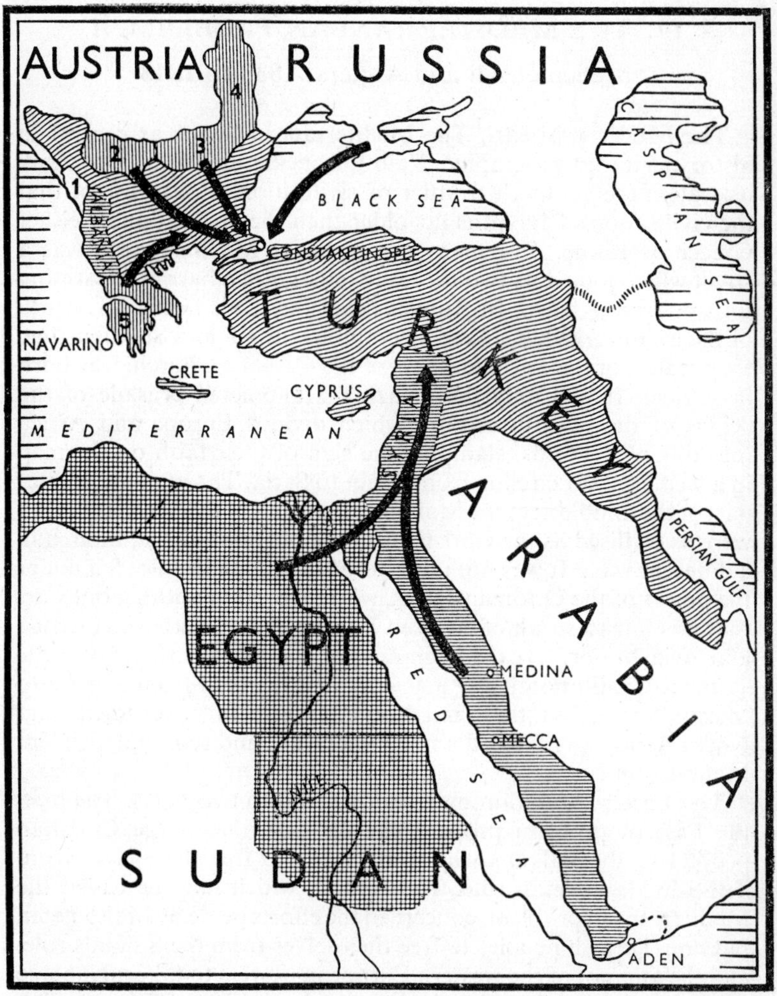

MAP 8. THE BALKANS AND THE NEAR EAST, 1817–41

17. THE MEDITERRANEAN PROBLEM. I

Mehemet Ali and Algiers: 1830 to 1848

The Mediterranean. The Mediterranean Sea is as central in history as it is in geography. Vain Europeans are inclined to think its waters the southern frontier of civilisation, and to forget that the civilisation of Egypt is far older than the civilisation of either Greece or Rome. For many centuries the Mediterranean was a street which joined its two shores together, not a barrier separating Europe from Africa. In Roman times the lands north and south of the Mediterranean shared a common culture, and St Augustine, the greatest of the Latin teachers of the Catholic Church, was born in Africa. It was only the amazing Mahometan crusade of the Arabs of the seventh century which divided Europe and Africa into different worlds, stamping the sign of their faith on the map in a vast crescent circling from Spain to Syria. Through the greater part of the middle ages the Mahometans south of the Mediterranean were as civilised as the Christians of the northern shore, though in a different way. It was not until the Mahometan Empire fell under the power of the Ottoman Turks, who excelled in nothing but war, that it is fair to speak of civilised Europe and of barbarous Africa and Asia Minor. At the beginning of the nineteenth century the Turks were still nominally masters of all the land from Algeria to Mesopotamia, but they had lost their grip in government and North Africa was divided among backward and semi-independent Mahometan rulers.

The Eastern Question may be divided into two parts. The first, the Eastern problem proper, was the question of the Christian peoples of the Balkans who were trying to free themselves from Turkish Mahometan rule. The second, which may be called the Mediterranean problem, concerned the efforts made by Mahometan but non-Turkish peoples to free themselves from the Sultan's rule, and the attempt of the western European powers to take advantage of the movement. The principal parties to the first problem, which is discussed in the series of chapters called 'The Eastern Question', were Austria, Russia, Turkey and Britain. The principal parties to the second problem were England, France and Turkey. At a later stage Italy became interested in it. The two problems frequently overlap, but it is best to treat them separately, and this second problem is described in the two chapters called 'The Mediterranean Problem'. Both problems had this in common: Turkey was losing the power of controlling her empire. Her possessions were coveted by several European powers, any of whom, however, would have preferred the Turks to keep their possessions if the alternative was to see them pass to a European rival.

The gateway to India and Africa. The political link which joined the two Eastern problems was the geographical link which joins the Red Sea to the Mediterranean—the Suez Canal. The canal was not actually opened until 1869, but for two generations before it had existed in the imaginations of English statesmen and French engineers. Even without a canal the possession of Egypt by a hostile power could be made a danger to the British empire in India, as Napoleon had realised. For this reason English policy throughout the nineteenth century was directed to keeping the Russians out of Constantinople and the French out of Egypt, a difficult task since both countries were equally determined to get there. Constantinople in the hands of the Turks was a padlock which confined Russia to the Black Sea; in the hands of the Russians it would have been a key which would have opened the Mediterranean to her fleet and closed the Black Sea to ours. Because the shortest route to India ran through the Mediterranean England regarded the Mediterranean as 'our sea' and was determined to prosecute any trespassers in it. This brought England into frequent conflict with France, who naturally would never tolerate the English view that the purpose of the Mediterranean was mainly to carry English ships to India. Napoleon's expedition to Egypt had attracted the attention of France towards that country and Syria, and the attraction survived the fall of his dynasty.

French ambitions were not limited to Egypt and Syria; due south of Marseilles lay the fine harbour of Algiers, whose pirates had for centuries been a danger to Christian shipping, but which might easily become a great exporting port for the produce of the potentially rich agricultural land around it.

Mehemet Ali. France's opportunity came first in Egypt, then under the rule of that extraordinary adventurer, Mehemet Ali. He had been born in the same year as Napoleon, whose example he imitated, and in one respect outdid, for he founded the dynasty which still reigns in Cairo. His father had been an Albanian peasant and he himself had passed through the stages of tax-collector and tobacco merchant before he joined the Turkish army in Egypt. In 1799 he was rescued from drowning by an English admiral, which may have taught him the value of sea power, but did not teach him gratitude, for in 1808 he entertained the people of Cairo by decorating their city with the heads of British soldiers whom he had routed at Rosetta. In 1806 the Sultan had made him Pasha or Viceroy of Egypt, whose absolute master he soon became after defeating the Mamelukes, the feudal Mahometan nobility of the country.

Mehemet Ali could neither read nor write, but he quickly learnt to use the methods of civilisation both in statecraft and strategy. Napoleon had taken to Egypt scientists and engineers as well as soldiers. Mehemet Ali had been impressed by them and decided to use French experts to modernise the government, to found a medical school, to establish a firm control of Egypt's economic life and to introduce new types of agriculture—olives, cotton, mulberry trees—as well as to organise a fleet and army on European lines. Mehemet Ali had not the least intention of letting Egypt become a French protectorate, but he was quite prepared to get all the benefit he could by being a protégé of France.

He soon set to work to enlarge his territory. In 1818 his stepson Ibrahim conquered the holy cities of Medina and Mecca in Arabia, and between 1820 and 1823 he invaded the Sudan and founded Khartum. Mehemet Ali's next step was to intervene on the Sultan's side in the Greek War of Independence, in return for which he was to have been given Syria, Greece and Crete. All that he actually got was Crete, which he thought a poor reward for his efforts and inadequate compensation for the loss of his ships at Navarino. Since the Sultan would not give him Syria—Greece was no longer his to give—Mehemet Ali determined in 1831 to take it for himself. Soon he was master of Syria and men began to wonder whether Mehemet Ali would dethrone Sultan Mahmoud II and proclaim himself Caliph.

Russia to the rescue. This Mahometan civil war was bound to cause a European crisis. France steadily backed her favourite Mehemet Ali. In England Palmerston, the most determined and outspoken of all foreign secretaries, was in power, but for once he had reluctantly to allow events to take their course because the Reform Bill agitation at home made England too weak to risk trouble in the Near East. This gave Russia her chance. Sultan Mahmoud was so desperate that he was ready to pay any price for Russian help, and so for once the usual diplomatic grouping was reversed and the Tsar appeared as the ally instead of the enemy of the Sultan. The price of his help was the Treaty of Unkiar Skelessi, signed in 1833, by which Turkey secretly pledged herself to open or close the Dardanelles to suit the convenience of Russia. This gave Russia all that she wanted, far more than either England or France could tolerate. The Sultan was assured by them that he need never carry out his promise to Russia, and that they would guarantee him against the consequences of breaking the treaty. But in return he was forced to give Mehemet Ali all Syria.

Second Turco-Egyptian War. This compromise did not last. Mehemet Ali thought that he had been cheated of the results of his victory; the Sultan regarded him as a traitor who had been rewarded when he should have been punished; and England thought that Mehemet Ali's success had made the French too influential in the East. England, therefore, annexed Aden and persuaded the Sultan to deprive Mehemet Ali of some of his powers—for Mehemet Ali was still nominally a subject of the Sultan. As a result a new war broke out between Turkey and Egypt in 1839. Once again Mehemet Ali carried all before him and seemed likely to conquer Constantinople unless the powers stopped him.

Palmerston's triumph. Neither England nor Russia was prepared to see Mehemet Ali lord of Constantinople, for they believed that in fact France was Mehemet Ali's master. They therefore agreed to call a European Congress and dictate terms of peace which would rob Mehemet Ali of his victory. France would be obliged either to abandon Mehemet Ali and fall into line or else to undertake a war against a European coalition, which would certainly have included England and Russia and most probably also Austria. Thiers, the French minister, talked of war and hoped that Mehemet Ali, the 'new Alexander the Great' as he called him, would be strong enough to force the Sultan to surrender before the Congress could intervene. But instead it was Mehemet Ali who surrendered, frightened by the threats of Russia and England backed up by a demonstration from the British fleet. Palmerston had rightly judged that Louis Philippe would play for safety, dismiss Thiers, and desert his Egyptian friends. In 1841 at the Treaty of London, Mehemet Ali was glad to be allowed to keep Egypt, surrendering the rest of his Empire. In the same year the Straits Act closed the Dardanelles to all foreign warships, thus assuring Russia of immunity from naval attack, but depriving her of any outlet for expansion via the Black Sea. Palmerston had played his cards well. Russia had gained nothing and France had lost heavily in prestige through her ineffectual support of Mehemet Ali.

The conquest of Algiers. Egypt was later on in the century again the cause of intense Franco-British rivalry, but for the moment France concentrated on the conquest of Algeria. There she had no immediate European rival, for Spain was now far too distracted by the Carlist Wars to assert her traditional interests across the Straits of Gibraltar. The first step in the French conquest of Algeria had been taken in 1830 when Polignac, Charles X's minister, had seized Algiers itself. Louis Philippe inherited this success, but was at first rather half-hearted about exploiting it. The Algerians found a powerful leader in Abd-el-Kader, a devout Moslem who had made the pilgrimage to Mecca and learned the advantages of Europeanisation in Egypt under Mehemet Ali in 1827. At first he was very successful and it even looked as though the French might be driven into the sea, but Louis Philippe found a brilliant governor-general in Marshal Bugeaud, who gradually wore down Abd-el-Kader's resistance. In the end the opposition collapsed, Abd-el-Kader surrendered and Bugeaud was free to start the work of colonising Algeria with 40,000 French peasant settlers. An attempt to penetrate further west into Morocco which might have brought French territory to the coast-line opposite Gibraltar was checked by a remonstrance from England.

The conquest of Algeria was the beginning of the great French North African Empire, which is quite as remarkable an experiment in colonisation as any of Britain's undertakings, though one conducted on totally different lines. The future complications of the Mediterranean problem were already clear, for now Britain's east-west line of communication via Gibraltar-Malta-Suez-Aden was cut at right angles by France's north-south line from Marseilles to Algiers.

18. TWO BOURBON GHOSTS

France: 1815 to 1830

English and French Restorations. There is a curious similarity between the fortunes of the restored Bourbon monarchy in France after 1815 and those of the restored Stuart monarchy in England after 1660. In both countries the previous king had been executed; in both countries an inefficient republic had given place to a tremendously efficient military dictatorship, that of Cromwell in England and of Napoleon in France. In each case a dispossessed and exiled aristocracy came back thirsting for revenge on the revolutionaries who had killed the king and seized their estates. In both cases they were partially disappointed. French émigrés and English cavaliers both complained that they were neglected while men like Monk, the old Cromwellian, and Talleyrand, the old revolutionary and Bonapartist, were rewarded for their treachery. Neither French nor English exiles regained their estates. The parallel extends to the two kings of the Stuart Restoration, Charles II and James II, and the two kings of the Bourbon Restoration, Louis XVIII and Charles X. Both were brothers and very unlike brothers. Charles II was as unlike James II as Louis XVIII was unlike Charles X. Louis XVIII had some of Charles II's wit and subtlety, and all of his determination to avoid any further travels; both of them were succeeded by obstinate bigots, determined at the risk of almost certain revolution to put the clock back. But here the resemblance ends, for while Charles X would have failed in almost any line of life, James II was a highly efficient naval officer.

Intervention of the allies. The task of the returned Bourbons was, however, vastly harder than that of the restored Stuarts for two reasons. Cromwellianism died a natural death with Oliver Cromwell, for his dreary son 'Tumbledown Dick' had neither the power nor the desire to be a pretender to the throne; but Bonapartism lived long after Napoleon's exile and death and eventually gave France another emperor. Then the Stuarts came back because England wanted them back; the Bourbons came back against the wishes of Frenchmen because France's enemies insisted on their return. So little did France want the Bourbons that Wellington and Metternich, Alexander and Hardenberg had to put them back on the throne twice over. The effect of the Hundred Days was to underline what was already obvious: the fact that Louis XVIII was king neither by the will of the French nor the grace of God, but by the orders of the allies. His subjects could hardly respect a king who owed his position in the first place to the invasion of France in 1814; and the second time to the defeat of a French army at Waterloo in 1815.

The foreigners who restored the French monarchy were not fools; they did all they could to conceal their dictation from the French or at least to make it bearable. This accounts for the generosity of the terms which Talleyrand obtained at Vienna. The Conference was too wise to risk provoking a new Reign of Terror or another Bonapartist restoration by saddling the Bourbons with a humiliating peace which must lower their prestige still further in the eyes of an already contemptuous nation. Thus the first governing factor in the French situation for the next fifteen years (1815–30) was that the security of the restored monarchy was the first care of the European nations, who would have intervened to save it by force if necessary. In fact, however, their tactful consideration in making the original peace terms so lenient and then in withdrawing the army of occupation as early as 1818 were sufficient support.

Bonapartism. Louis XVIII's throne was threatened from three directions: by Bonapartism, by Republicanism and by the die-hard 'ultras' or extreme royalists. The Napoleonic legend had at different times two opposite effects on French public opinion. At first it made France war-weary. For a generation France had been continuously at war, and at the end she had nothing to show for all her sacrifices except another chapter in her already long record of fruitless gallantry. This led for a time to a hatred of imperialism and war, and a sense of relief at the safe and unexciting policy of Louis XVIII. To begin with, then, the new government gained in a negative way from the Bonapartist past. But many thousands of the Emperor's veterans survived. They did not want to fight again themselves, but they did not want their efforts forgotten or decried. The Restoration was a hard period for them. They had made a living by war; they could not make a living in peace. Poverty and unemployment soon made them Bonapartist propagandists. This was something which Louis XVIII could not help, but he might at least have avoided making matters worse by slighting their services to France. The flag they had fought under had been the tricolour, the flag of revolution. Louis XVIII replaced it by the old white flag of the Bourbons, a gesture which pleased the returned royalists but wounded Napoleon's veterans. It suggested that the king was ashamed of them. The court-martial and execution in 1815 of Marshal Ney, 'the bravest of Napoleon's brave', convinced them that they had no rightful place in France under the Bourbons.

The death of Napoleon I in 1821 only made matters worse for the monarchy. In his exile on St Helena Napoleon had re-interpreted the history of his life. He concealed the tyranny of his rule, the pretensions of his court and the dynastic ambition which made him divorce Josephine and marry Marie Louise in an effort to get himself accepted as an equal by the great families of Europe. He posed as the misunderstood liberal, the great friend of freedom, the apostle of equality, the child of the revolution and, therefore, as the natural leader of all those who were dissatisfied with Bourbon France. For the moment the danger was not acute, for Napoleon left no heir fit to lead an insurrection at once; but there were descendants who would before long be ready to take up his cause. Meanwhile the threat remained in suspense; while his memory, freed by the humiliations of his exile from vulgar pomposity, was always there to remind France how painfully the 'do nothing' Bourbons suffered by comparison with the 'little corporal' who had fired her ambition and satisfied her pride.

Republicans and 'Ultras'. The Republican danger was also real, though, like the Bonapartist, more as a threat of what might happen in the future than of what was likely to happen in the present. Neither friends nor foes could forget the revolution. The people of Paris were still probably for the most part Jacobin at heart and Paris was still the greatest single power in France. For the moment, however, the memory of the revolution was chiefly important because of the fear and hatred which it aroused among the extreme royalists. It convinced them that any surrender to the demand of the respectable middle class for a share in political power must lead straight to a new Reign of Terror, just because the great French Revolution had started with a middle-class parliament. The 'ultras', therefore, stood for a complete return to the days before 1789, for the absolute power of the king in politics, for the social privileges of the old nobility and for the complete restoration of the power of the church. Since the 'ultras' were the most loud-spoken friends of the king they frightened away the middle class who might on general grounds and from a love of safety have been ready to support him.

The settlement of 1815. Louis XVIII was fully aware of this danger. He probably shared the social views of the 'ultras', since he was a man who enjoyed life as it had been lived in the good old days for the nobility;[1] but he realised that those days were over, and that the best he could hope for was a compromise which would last. He had the wisdom to set himself against the 'ultras' and to make concessions, though as few as possible, to the spirit of the time. He agreed to grant a charter which set up a French parliament, but few people were allowed to vote for it (there were never more than 100,000 electors under the Restoration), and its powers were strictly limited. It was also made clear that the charter was granted by the king of his own free will and could be abolished by him if he wished. Still, a parliament however powerless was a parliament and neither the 'ultras' nor Metternich were pleased.

Louis took over from the Empire its administrative system of prefects in each département appointed by the central government and of maires in each commune appointed by the prefect; he kept also the Code Napoléon, the masterly reorganisation of French law, and the Concordat with the Pope, which left the church far less political and financial power than it wanted. In all these ways Louis showed his willingness to act as a constitutional sovereign and to retain the best results of the revolutionary period. He failed, however, to prevent the 'ultras' from enjoying their revenge for the Reign of Terror by wholesale massacres of former Republicans and Bonapartists in the south of France during what is known as the 'White Terror'. This short period of lynch law was followed by a series of executions after trials by courts-martial. Marshal Ney was only the best known among many victims.

[1] His favourite minister Decazes, who had been secretary to Napoleon's mother, wanted to become a Gentleman of the Chamber. Louis told him: 'You know the recipe for jugged hare? "First, take a hare." Well, to make a Gentleman of the Chamber, first take a gentleman....'

The Duc de Richelieu. The first and comparatively moderate period of the Restoration lasted until 1820. It is true that Fouché and Talleyrand were dismissed to please the 'ultras', but Talleyrand was soon compensated by being sent as ambassador to London. The chief minister, the Duc de Richelieu, was like his master a sensible man, though an aristocrat. In Napoleon's time he had served the Tsar as Governor of Odessa, but although he found governing France a more exacting occupation—'quels diables de gouverner sont ces peuples civilisés' was his comment—he did it extremely well. He put the finances of the French government into proper order and gave France a prosperity which she had not enjoyed for a hundred years. So well did he govern and so generally contented was France under his rule that the allies agreed in 1818 to withdraw the army of occupation, which was intended to protect the Bourbons from revolution, and to admit France into the Concert of Europe. Under Richelieu, if the monarchy did not inspire enthusiasm it at least secured the assent of most Frenchmen. He had done a good deal to achieve the chief task of the Restoration: to persuade the new France and the old monarchy to live peaceably together.

In 1820 this early period of moderation came to an end. The murder of the Duc de Berri, a prince who was in the line of succession to the throne, gave the ultra-royalists the excuse they needed to get the Duc de Richelieu removed from office, to gain control of the government and to wreck the prospects of a successful restoration. One of the first activities of the new government was the Spanish crusade to help the unspeakable Ferdinand against his subjects.

Charles X. The rule of the 'ultras' lasted ten years, from 1820 to 1830; but at first it was checked by the influence of Louis XVIII. In 1824, however, the tired old king died, glad that the monarchy had lasted his lifetime, convinced that it could not long survive the stupidity of his brother, Charles X. Charles made it clear from the moment of his accession that he was going to be the 'ultras'' idea of a good king and therefore everybody else's idea of a bad one. He insisted on being crowned with all the medieval ceremonies at Reims, where the old kings of France had been crowned. Like the change of flag from the tricolour to the white standard a few years before, this was interpreted and meant to be interpreted as a political declaration. By it Charles meant to proclaim that he believed in the divine right of kings, that he intended to deserve the old claim of the French kings to be the 'eldest son of the Church'. Before the revolution both Louis and Charles had been stupid, pleasure-loving, immoral men. Years of exile had made Louis prudent, but Charles devout.

The religious revival. Charles was not the only man who was converted to a sincere religious belief by the events of the revolutionary period. In fact the period after the Congress of Vienna was marked throughout Europe by a widespread religious revival. It was not only that men's minds had been turned to God by the stupid violence of the wars. The revival had deeper roots. The common-sense, atheist world of the pleasure-loving eighteenth century had been followed by the common-sense, strenuous world of the revolution. Neither left any room for ideals and longings which could not be expressed in terms of material things. That side of human nature, which the last one hundred and fifty years had ignored, was now demanding expression. In England a few years later it took the form of the Oxford Movement which restored to the church much of her lost Catholic inheritance. In Italy and Germany it gave rise to romantic youth movements, often of a strongly liberal character. In France, unfortunately, it took the form of a revival of the political power of the church, which became the close ally of Charles X in opposing every sort of political freedom. There were wise men, great scholars and writers, among these political Catholics of the Restoration, but they were all reactionaries, opposed to every concession to the spirit of change. Both church and state suffered from this identification of religion with conservatism. The church, for instance, gained no benefit from a law which Charles X made punishing with death acts of disrespect to the Host, the consecrated bread in the Communion service; but although the penalty was not enforced, the very large number of 'free-thinking' Frenchmen were outraged by the threat.

There is no doubt that Charles X's inflexible devotion to divine right cost him the support of the middle class, whose leadership in the rapidly expanding industry and commerce of France made them more than ever important. Their leaders were quite prepared to accept the monarchy, for a revolution was plainly bad business; but they wanted a greater share in the government. The first cabinets of Louis XVIII's reign had suited them very well; his later governments and those of Charles X robbed them of the hope that in time they might be given a greater share of power. Towards the end of his reign Charles X made matters even worse by appointing as his chief minister the Prince de Polignac, a man who was ruled by mysterious visions instead of ordinary thoughts.

Polignac and the Ordinances of St Cloud. Polignac did his best to distract attention from rising discontent at home by sending an expedition across the Mediterranean to Algiers to punish its Mahometan rulers for some trifling insult to a French consul. The expedition was a military success, and laid the foundation of the whole French North African empire; but it failed politically. The king and Polignac were determined to be stupid in their policy at home. In 1830 there was a general election in which the moderate liberals won a majority. Charles determined to overthrow the parliament, but he took no measures except on paper to do so. He collected no troops, but he issued on his own authority 'the Ordinances of St Cloud', which dissolved the Chamber of Deputies and declared that he would rule on his own. Paris, however, was determined that if Charles would not rule with a parliament he should not rule at all. Street fighting broke out at once, and the Minister of War, who only heard of the existence of the Ordinances when he read them in his newspaper, was unable to keep control of the capital. From his palace at St Cloud, on a hill above Paris, Charles X saw the forbidden tricolour flying from Notre Dame. That evening he sent a messenger into Paris to find out not the political temperature of the rebellious crowds, but the temperature of the air, for the day had been extremely hot.

The July Revolution. The middle-class liberals were convinced that, much as they might fear a revolution, it was impossible to keep such a king on the throne. Their leaders met at the house of the banker Lafitte and there they arranged to publish a proclamation in favour of Louis Philippe, Duke of Orleans, which had been written by a young journalist called Adolphe Thiers.

The moderates had acted because they feared that otherwise a Jacobin republic would be established by Jacobins. It remained to persuade the Jacobin mob of Paris to accept Louis Philippe as king. Two arguments were successful. First, the other European powers would have intervened against a second republic. Then

Lafayette, the seventy-three-year old hero of the American War of Independence as well as of the French Revolution, agreed to support Louis Philippe. Thus the July Revolution was a compromise. The desire of crown, church and nobility for a new despotism on the old pattern had been finally defeated, but the swing of opinion to the opposite extreme was arrested in mid-air, and the solid middle-class property holders set up a new style of monarchy of their own pattern, though modelled to a certain extent on the example of England.

19. MIDDLE-CLASS MONARCHY
France: 1830 to 1848

Louis Philippe. Louis Philippe, the new King of the French, was a cousin of the king he had displaced. His family, the Orléans branch of the Bourbons, had been on the side of the great French Revolution, and his father had abandoned his title, taken the surname of Egalité and voted for the death of Louis XVI—though this had not saved him from being executed during the Reign of Terror. Louis Philippe himself had fought in the French revolutionary army at the battle of Jemappes in 1794 and he was therefore an ideal compromise king for the situation of 1830—a great aristocrat, whom the well-to-do could trust to protect their lives and property from a second reign of terror, and a sincere liberal who was opposed to the old system of divine right and of noble and clerical privilege.

Louis Philippe's monarchy, the July Monarchy as it is known in French history, was a constitutional monarchy. Louis Philippe's title was King of the French: king, that is, by the will of the French people, and not, like his predecessors, King of France by divine right. In the same way Napoleon had called himself Emperor of the French, not Emperor of France. In Louis Philippe's time the king reigned, but he ruled only through ministers responsible to parliament, although he could and did try to influence their foreign policy. The parliament, however, was elected only by the well-to-do upper middle class of France; the smaller tradesmen and the poor peasants and workmen had no more say in the government than they had had before.

Louis Philippe did all he could to make himself popular and democratic. Sometimes he only made himself ridiculous, and cartoonists were never tired of drawing the tubby king with the umbrella, which became in a way the symbol of his respectable but exceedingly dull reign. He was intelligent and good-humoured—he once helped to finish a caricature of himself which a small boy was scribbling on the wall of his palace—but rather too impulsive and talkative. On the whole, however, he must be placed high among nineteenth-century kings both for ability and for genuine willingness to rule in a constiutional way. The character of Louis Philippe does not explain why his promising experiment in constitutional government ended after eighteen years with the exile of the king.

Thiers and Guizot. Nor is the explanation to be found in any lack of ability among his ministers. France has seldom had two such distinguished statesmen as Adolphe Thiers and François Guizot. Thiers was the journalist who had drawn up the first proclamation in favour of Louis Philippe. Before that, in the time of Louis XVIII, he had published a history of the French Revolution, but he had now been bitten with the desire to make history instead of writing it. 'I would willingly', he said, 'give ten good histories for one successful parliamentary session or one good military campaign.' He was, however, handicapped, like the monarch he served, by a rather doubtful social position. The Habsburgs and the other old royal families of Europe could never quite bring themselves to accept Louis Philippe as an equal; the old aristocracy could never quite tolerate Thiers. When at last he did become chief minister, old Talleyrand summed up the situation in the untranslatable remark: 'M. Thiers n'est pas parvenu: il est arrivé.' Louis Philippe and Thiers were certainly no less presentable socially than the rest of the French rich middle class, whose servants they were; but there is no jealousy like that of one social climber for another, and the July Monarchy suffered from the fact that the 'parvenu' French middle class always rather looked down on its king and his ministers.

Like Thiers, Guizot was middle class by birth and in his earlier life a professional historian. His father, a well-to-do Protestant lawyer, had been executed during the Reign of Terror when his son was only seven. Guizot was generous, prudent, proud and intensely reserved—his memoirs contain hardly a single allusion to his private life. Two of his sayings disclose a self-assurance which only his great ability can excuse. 'Thucydides and Machiavelli wrote and published contemporary history. Why should not I do the same?' Again, he once turned on the opposition in parliament: 'You may pile up your abuse as high as you like. It will never reach the height of my disdain.'

Guizot was a determined friend to liberty as he understood it. But his idea of liberty was strictly limited and only partially suited to the time in which he lived. He believed that every man should be free to carry on his own business and to live his own life without interference from the government and without interfering with others. For this reason he objected both to the absolute power of kings and to the privileges of nobles and churchmen. For this reason he believed in a parliament. But he did not believe that every man should have a vote, a share in political power. The choosing of governments was just as much a skilled art as the actual work of government itself, and both should be reserved to people with the capacity to do it well. In practice this meant the rich middle class. Guizot's views were the result of a serious study of history: he thought they supplied the only government which could preserve the good and prevent the evils of the French Revolution. Other men, Thiers among them, held the same views; but more because they suited their own interests than because they were best for society as a whole. It is significant that the first proclamation in favour of Louis Philippe was drafted at one banker's house, and that another banker was one of his first principal ministers.

The industrial revolution. The political ideas of the July Monarchy were the main reason for its failure, because they prevented it from dealing with the growing social problem. France was just beginning to experience the great economic changes which we call the 'industrial revolution'. They came much more slowly in France than in England, and they had affected only a minority of Frenchmen by 1848; but they were already puzzling men's minds and troubling their consciences. The July Monarchy was founded in the year in which the Liverpool and Manchester railway was opened; before long France built her first railway—from Lyons to St Etienne (1832). By 1841 over 350 miles of railway were working in France. The next year great trunk lines radiating from Paris were planned. The land, embankments, tunnels and bridges were to be provided by the state; the track, rolling stock and stations by private companies. Some progress had been made by 1848 and, although the original bargain was not always carried out, the principle of a partnership between the state and private enterprise was maintained.

The July Monarchy had thus shown that it was prepared in certain cases, where its own interests were concerned, to direct and control private enterprise; but Guizot refused to apply this power of the state to help the poor. The only exceptions he made in their interest was to provide for a great extension of elementary education, which doubled the number of people who could read, and to limit the employment in industry of young children. But in general he believed that all government interference in industry was positively harmful. He held that the poor were poor because they were feckless, lazy or dissolute; self-help would raise them, if they were industrious and prudent, to positions of ease and comfort. His enemies said that his only advice to the poor was 'Get rich'. This was plainly advice which only a few could follow. Distress was widespread and terrible. In 1840 out of every 10,000 men called up for military service from the industrial districts 9000 were unfit; from the countryside only 4000. In the same year the average wage of the French workman was probably no more than 1*s*. 4*d*. or 1*s*. 7*d*. a day. It is hard to save money and become independent, as Guizot recommended, out of such an income.

The government did nothing to help the poor either by making laws to prevent their being exploited or by allowing them to vote for parliament and thus indirectly to make the laws themselves. Generous-minded Frenchmen, however, took up their cause and, in doing so, attacked the whole basis of the July Monarchy. There was in France an important group of socialist writers led by Louis Blanc, and another group of novelists, men like Victor Hugo and Balzac, who also drew attention to the sufferings of the poor. These novelists belonged to what is called the Romantic Movement. The older 'classical' literature had been concerned with certain fixed rules of composition and 'literary subjects'. The romantics claimed freedom from fixed rules of style and freedom to describe all kinds of men and women, including those who were below the dignity of 'classical' authors. Like Dickens in England, they wrote about the poor, and, like him, they were filled with sympathy and indignation.

The July Monarchy tried to make itself a fairly exact copy of English life, which at that time was dominated by the utilitarian thinkers and economists. England and France preserved a monarchy because it was useful, not because it was sacred or romantic. Under utilitarian influence they introduced many sensible reforms, mostly of a material kind. They sought the greatest happiness of the greatest number, but they sought it by ways which in the main guaranteed only the greatest possible happiness for the new, rich manufacturers. This unintentional limitation of happiness to the already happy few was due to the belief that state interference in industry must always be bad, because in the old days it had usually been so. Both in England and in France 'utilitarian monarchy' failed in two ways—it failed to solve or to lighten the social problem, the problem of poverty; and it failed to satisfy the emotional side of man's nature, which demands something more than usefulness as a test of worth.

Belgian independence. The July Monarchy was not much more successful in its foreign policy than in its handling of the social problem. The first crisis of Louis Philippe's reign, which was satisfactorily solved, took the form of a successful struggle for the independence of Belgium. The Congress of Vienna had set up the kingdom of the Netherlands, which united the old Austrian Netherlands (Belgium) to Holland. The new state was intended to act as a 'buffer-state' or policeman set to stop French expansion northwards. This union of Belgium and Holland was from other points of view a highly reasonable experiment which deserved better fortune. By nature not only the whole of the Netherlands, but the industrial north-west corner of France and the entire basin of the Rhine should be one single economic unit. The union of Holland and Belgium was at any rate a step in the right direction. There were several reasons why the union might have been expected to succeed. Though the two countries carry on to a certain extent the same occupations yet they are largely complementary. The Dutch are an agricultural and seafaring community; the Belgians a manufacturing and mining one, so that both would benefit from pooling their resources. But differences of race, language, history and religion, four characteristics which make up 'nationality', combined to make the union unpopular.

The Dutch are a purely Germanic people and the majority are Protestants of a Calvinist type, though there is and always has been a large Roman Catholic minority. The Belgians are almost equally divided between Flemings and Walloons. Both are strongly Roman Catholic and since the days of Philip II of Spain's persecution there have been hardly any Protestants in Belgium. The Flemings are a Germanic people whose language is virtually the same as

Dutch, while the Walloons are French in race and speech. If nationality was purely a matter of race one might expect the Flemings to gravitate towards Holland and the Walloons towards France, but in fact there is to-day and was a hundred years ago a clearly marked Belgian nationalism based on identity of religion and history. At the beginning of the nineteenth century nationalism was a rising force and consequently, although for many generations the Belgians had accepted passively the alien and absolute rule first of Spain and then of Austria, they now objected strongly to sharing freedom with their next-door neighbours. The King of the Netherlands undoubtedly was rather tactless and unimaginative, but he was not a tyrant. On the contrary, one Belgian grievance, that of the Archbishop of Ghent, was that he allowed religious toleration. Belgian priests, who hated toleration of every sort, and Belgian liberals, who would tolerate everything except Roman Catholicism, made strange but effective allies.

Their opportunity came in August 1830, when Brussels, fired by the example of Paris the previous month, rebelled against the Dutch government. At once there was a first-class European crisis. Metternich and the reactionary powers would have liked to intervene to put down the rebellion, but were troubled by a rebellion in Poland. England and France were determined to stop intervention, but did not see eye to eye. England sympathised with the Belgian struggle for independence as she had previously sympathised with the Greeks; but she was as determined to prevent Belgium becoming a province or protectorate of France as she had been to prevent Greece becoming a dependency of Russia. Louis Philippe on the other hand would have liked to see one of his family become King of the Belgians, for this success would have established the reputation of his new dynasty.

The throne was first offered by the Belgians to a younger son of Louis Philippe, but his father had the wisdom to refuse it rather than risk a war in which he might have found England and Austria against him. Next year the crisis was solved by the choice of Leopold of Saxe-Coburg as first King of the Belgians. Leopold made a wise and successful ruler. In his time Belgium flourished —it was the first European state to complete its main railway system—and the Saxe-Coburg family, which had previously ruled only a few square miles of Germany, soon became one of the leading European royal families. The final settlement came in 1839 with the Treaty of London by which the great powers recognised Belgium as an independent state and undertook to respect its neutrality. This was the famous 'scrap of paper' which seventy-five years later was the official ground for the British ultimatum to Germany in August 1914.

Thiers and Palmerston. Unfortunately for Louis Philippe, Casimir-Périer, the cautious and courageous banker who had guided France safely through the Belgian crisis, died not long after. The second crisis of the reign was largely due to the blunders of Thiers, who was a match for Palmerston, his English rival, in audacity and bluff but who lacked Palmerston's means, if the worst came to the worst, to make good his threats by war. This crisis arose in 1840 over the ambitions of Mehemet Ali, the ruler of Egypt.[1] Thiers was completely outwitted at every move in the diplomatic game by Palmerston. All he achieved was to frighten Europe into the belief that France probably still cherished Napoleon's ambitions in Germany as well as in Egypt. The German reaction was shown by the great popularity of a new patriotic song 'The Watch on the Rhine'. Thiers was forced to resign and was succeeded by Guizot.

Guizot and the English 'Entente'. It was fortunate that the beginning of Guizot's ministry roughly coincided with the defeat of Palmerston in England. It was difficult to make friends with Palmerston; it was easy to do so with his successor, Aberdeen. He and Guizot understood each other well and for six years there was an Anglo-French entente. The visits which Queen Victoria paid Louis Philippe were a sign that England fully accepted his position—and England was certainly the most powerful country in Europe. Towards the end of Guizot's ministry, however, the third crisis arose. This time the trouble came from Spain and once again Louis Philippe, whose strength lay in friendship with England, was tempted to revive traditional French ambitions which England would never tolerate.

The Spanish marriages. The Spanish problem became acute in 1833 when the unpleasant Ferdinand died. A long and barbarous civil war followed between his brother, Don Carlos, and the supporters of his infant daughter, Isabella. It was a war without

[1] Described in Chapter 17.

quarter—one of the generals on his death had declared that it was unnecessary for him to forgive his enemies for he had left none of them alive to need it. Officially Isabella's followers were liberals and therefore had the support of England, though their liberalism was anything but tolerant or democratic or enlightened. The Carlists were the favourite party of the church and of reaction and they were supported by Metternich. Louis Philippe openly supported Isabella but he was suspected of secretly helping Don Carlos. In spite of this, however, Don Carlos was driven into exile by 1839. Immediately there was talk of Isabella's marriage, although she was only ten at the time.

Once again, as in the Belgian crisis of 1830, there was competition between the Coburgs and the family of Louis Philippe. England had gone to war before to prevent a French king of Spain and would have done so again. Reluctantly, therefore, Louis Philippe gave up the idea of providing a husband for the Queen of Spain on condition that England ceased to push a Coburg candidate. The objections, indeed, to a Coburg were almost as strong as to a Bourbon. The King of the Belgians was a Coburg, the Queen of England and her husband were both Coburgs, and a few years before Louis Philippe had rather reluctantly allowed a Coburg prince to marry the young queen of Portugal. It was possible, therefore, if a little fanciful, to speak of a Coburg encirclement of France. Eventually a plan was arranged between Guizot and Aberdeen, the English foreign minister, by which Queen Isabella was to marry a Spanish prince and, when they had children but not before, her younger sister was to marry a French prince.

All might have gone well if Palmerston had not come back into power in England in 1846. He soon took steps which Guizot, rightly or wrongly, interpreted as a revival of the Coburg courtship. Guizot acted at once. He arranged for the marriage of Isabella to a Spanish prince who, it was thought, could never have children. At the same time he hurried through the marriage of the French prince to Isabella's younger sister, whose children might thus hope to inherit the throne. It was smart work, but stupid. It finally lost Louis Philippe the friendship of England; he had already lost the friendship of his own working people. As the year 1848 approached, the July Monarchy was left without friends.

20. SONGS BEFORE SUNRISE

Central Europe: 1815 to 1848

Youth Movements. For more than a generation after Waterloo Prince Metternich in Vienna reached an oppressive hand northwards to suppress the German liberals, southwards to crush the Italian liberals. In either country a dozen governments shared his views and carried out his policy faithfully, carefully and at times drastically. Not only the balance of power, but all power was in his hands. Against him were pitted only a small, prosperous but timid middle class and a handful of students. Yet Mazzini, the great Italian liberal prophet, was right when he warned the older generation: 'You know not the secret of the power hidden in these youthful hearts, nor the magic influence exercised on the masses by the voice of youth.' Their elders expected students to study, to be seen and not heard; they themselves thought otherwise. The universities developed a new life as the apostles of liberalism and nationalism. In spite of Metternich they made themselves heard and their views respected. What were these views?

Liberalism and Nationalism. Liberalism believes that men should be as free as possible. First, that men should be free from arbitrary imprisonment, that they should not be imprisoned just because the government disliked them but only for breaking some definite law. Secondly, that men should be free to express their opinions in conversation, in public speeches, in debates and in newspapers without fear of punishment or censorship. Thirdly, that men should be free to worship God as they liked or not at all. Fourthly, that men should be free to earn their living as they liked, to choose their own profession or trade, and that no class or race should be favoured because of its birth. Fifthly, that men should be free to govern themselves, to choose between a king and a republic, and to elect a parliament which alone should make the laws and supervise the government. All political parties in England to-day are liberal within the meaning of this definition, but in 1830 in central Europe no government was liberal. In Metternich's opinion to be a liberal was to be a traitor.

It was even clearer that to be a nationalist in central Europe was to be a traitor. It is hard to define nationalism because, though liberalism has barely changed in a century, the nationalism of to-day is vastly different from the nationalism of 1830. A hundred years ago nationalists were simply men who claimed that all men of the same nation ought to be allowed to form one state: that, for instance, the innumerable little German or Italian states should be merged into one and that a mixed state like that of the Habsburgs, which governed parts of many nations, should be broken up into its separate parts. The history of nationalism in the nineteenth century, then, is the story of how subject or divided races secured their national unity. The essence of this early nationalism in Belgium, in Germany or in Italy was a fight for freedom against despots who were usually foreign. Therefore throughout the period nationalism went hand in hand with liberalism. Both were revolts against tyranny. They fought the same enemies but for different reasons.

The joint victory of nationalism and liberalism has, however, been followed in many countries by a violent quarrel between them. This has happened because the old nationalism, in trying to unite all the fragments of a divided nation, developed an intense, almost religious national patriotism. The work of nationalism has been done but national patriotism remains and must find an outlet. It finds it in two ways. First, it comes into conflict with another national patriotism and hardens into a hatred of foreigners. Italy, for instance, can no longer grow by gathering more and more separated Italians into the nation state, because they are all there already; but grow it must, and so it seeks to grow at the expense of another nation.

Secondly, national patriotism to-day makes new and domineering claims on the individual citizen. Since patriotism has become almost a religion, the nation has become almost a god. It demands the complete and unhesitating service of its worshippers. Instead of Italy existing for the Italians, the Italian now exists only for Italy. In the last century the nation and those who governed it were often opposed to one another because a nation-state could only be created at the expense of the existing governments. Therefore nationalists fought against governments for the liberty of individuals. In this century those who govern a nation are not only members of it but claim to be the incarnation of it. Therefore the nationalist no longer sees any reason for the liberty of the individual. Instead he demands unquestioning obedience and has crushed liberalism to get it. This is why in Germany and Italy the long battle of nationalism and liberalism for freedom has ended in Fascist dictatorship.

The German revival. The older German nationalism, or liberal nationalism as we may call it, was born in the War of Liberation against Napoleon. It was a movement inspired by the universities, where many professors and almost all students were liberal nationalists. It found its expression equally in Professor Fichte's *Addresses to the German Nation* and in Lützow's Night Riders, a body of volunteer irregulars who fought side by side with the Prussian army in the struggle against Napoleon. German nationalism at this time gained a richer knowledge of past glories and learned things of which it could be proud from collections of old German fairy stories made by men like the Brothers Grimm, from collections of folk songs, like *Des Knaben Wunderhorn* (Youth's Magic Horn), and from the work of scholars who studied the great days of the Holy Roman Empire, when in the middle ages Germany had led the world in politics. Philosophers like Hegel preached a form of worship of the state which was ridiculous when applied to the existing petty kings and grand dukes of Germany, but which was an inspiration if applied to the new all-German state for which men were working. Musicians like Schubert gave German poetry a still wider appeal by setting it to music of incomparable beauty.

The students applied these lessons to current politics. In 1815 an organisation was founded which was to play a large part in politics. It was called the Burschenschaft and its greatest strength lay among the undergraduates of the University of Jena. But the Burschenschaft was not confined to one university. In England a student goes to one university and stays there. In Germany he moves about from university to university. Wherever a member of the Burschenschaft went he took with him the gaudy uniform of the association and its colours of black, red and yellow—the colours of the old Holy Roman Empire. The Burschenschaft was a social organisation, providing the beer-drinking and duelling which are the traditional amusements of German students, but its members were also filled with a passionate desire for German unity, and it was this which made them hated by Metternich. The Wartburg, a steep wooded hill with a famous castle not very far from Jena, was a sacred spot in German history, for there Martin Luther had translated the Bible. In 1817, on the fourth anniversary of the battle of Leipzig and the three hundredth of the Reformation, students from Jena University lit a fire on the Wartburg in which they burnt various books and certain symbols associated with Metternich's hated principles. The Wartburg flames were a beacon which rallied German youth and warned Metternich that the European settlement he had so carefully contrived might at any moment go up in flames.

The Carlsbad Decrees. At first he could do little, for the Tsar Alexander of Russia and the second most important German king, Frederick William III of Prussia, were inclined to sympathise with the students, but Metternich's chance came in 1819. In that year an undergraduate of Jena stabbed to death a reactionary writer called Kotzebue. The effects of this murder were great, immediate and unintended. It made the Tsar give up his vague liberalism and become a convinced disciple of Metternich. It had the same effect on Frederick William III. Metternich seized his opportunity and secured the acceptance of the Carlsbad Decrees (1819) by the whole German Confederation. Under these decrees the students' societies were forbidden, the universities brought under stricter government control, a far-reaching censorship was established and a central office for spying on patriotic or liberal agitators set up at Mainz. The few liberal-minded German rulers were forced to fall into line. The Carlsbad Decrees did not destroy the German liberal national movement, but they drove it underground. When Metternich went to Carlsbad for the conference he announced his programme: 'By the help of God I hope to defeat the German revolution just as I vanquished the conqueror of the world.' He succeeded at any rate in postponing it for nearly thirty years. The European revolutions and rebellions of 1830 hardly touched Germany, and the few minor disturbances which took place were made the excuse for a stricter enforcement of the Carlsbad Decrees.

The Zollverein. Prussia and Austria shared the same political ideas until 1848 and, thanks to this, Metternich was able to get his own way in suppressing liberal nationalism throughout Germany.

But he could not prevent Prussia growing in power at the expense of Austria. All the German states at this time charged high tariff duties on imports, which crippled the growth of commerce, and were especially ruinous to Prussia with her scattered possessions. In 1818 Prussia adopted the principle of free trade among her various provinces—until then, as in France before the revolution, there had been customs to be paid at the various provincial frontiers. Prussia soon began to make treaties with her neighbours for the mutual lowering of customs dues. By 1834 Bavaria, Saxony and most of the other German states except Austria had joined her in the Zollverein (Customs Union). Trade was completely free between its various members, while a tariff wall cut them off from non-members. These duties were used to protect German industry against English competition largely as a result of the economic teaching of Friedrich List, who believed in German unity and also in the importance of making an economically self-contained state, complete with manufactures as well as agriculture.

The Zollverein was brought about partly by the stranglehold which Prussia had gained from her geographical position over nearly all north-south trade. But partly it was purchased by the network of roads which Prussia built for her poorer neighbours who agreed to join the Zollverein. In railway development, too, Germany took the lead in continental Europe. By 1848 her main trunk lines were nearly finished, France's only half begun. Above all she had a much better educated middle class than any other country, a class fitted to take the lead in the great industrial progress of those days. It would be wrong then to ascribe Germany's great economic progress entirely to the Zollverein.

The economic effects of the Zollverein have sometimes been exaggerated, but its political results can hardly be over-estimated. Austria remained outside. Against her opposition, Prussia had taken the lead towards the economic unification of Germany, and therefore indirectly towards her political unification. Economically and politically Prussia had stolen a march on Austria. Germany had grown accustomed to Prussian leadership and to the exclusion of Austria.

Italian revolutions. Metternich's difficulty in Germany was that Austria was only one among a number of German states, some of which, such as Prussia, were intensely jealous of Austrian leadership. In Italy his difficulty was that Austria was a German, not an Italian, state. Italian nationalism could never accept German rule and in 1821 it had given Austria some anxious moments, which were described in Chapter 14. At that time, however, the Italian leaders had got no farther than a hatred of foreign rule in Italy. They would have been content if the Austrians had been expelled; they had still to learn that this could only happen if the various Italian states acted together.

The July Revolution of 1830, which set Louis Philippe on the throne of France, hardly ruffled the surface of German life, but it caused a great upheaval in Italy. Revolution started in the most progressive parts of the States of the Church, as the Pope's dominions were called, that is to say, on the far side of the Apennines from Rome. There could be no question that the Papal government was both abominably inefficient and monstrously tyrannical, since even Metternich condemned it. At the same time he defended it by sending an army which easily put down the revolution in the Papal States and in the neighbouring duchies of Parma and Modena to which it had spread. Next year the Austrian army was withdrawn, but no sooner had it gone than the revolution started once more. Again the Austrians returned. This time the French also sent troops to the help of the Pope, much to the annoyance of Metternich, who regarded all Italy as Austria's own preserve. For six years the rival defenders of the Pope remained on guard, the Austrians in Bologna glaring angrily at the French in Ancona. The unsuccessful revolutions of 1830 were the last effort of the Carbonari, for a young Italian refugee, who had taken part in the rising, founded in 1831 a new body called 'Young Italy' which soon took their place.

Mazzini's 'Young Italy'. Giuseppe Mazzini was only twenty-six when he founded 'Young Italy' and most of his supporters were almost equally young. The liberal national movement in Italy as in Germany was a youth movement. 'Young Italy' had a simple, clear-cut programme. It stood for the expulsion of the Austrians, for the union of Italy and for a republic. It was, therefore, almost as much the enemy of the native Italian kings as of the Austrians. It gave Italians a purpose beyond their own salvation, for united Italy was in Mazzini's teaching to become the moral leader of the world. Its members were missionaries as well as politicians. 'Young

Italy' knew what it wanted, but it did not know how to obtain it. It suffered from being led by refugees—it was founded at Marseilles and organised after 1837 from London, where Mazzini made his home and became a regular frequenter of the British Museum reading-room. Exiles are almost bound to lose touch with the day to day politics of their country. They may be an inspiration to an oppressed people, they are rarely prudent guides. Mazzini's 'Young Italy' was an amateurish organisation, riddled with police spies. Its plans were usually known to the Austrian and Italian authorities, and it was unlikely ever to be able to organise a successful revolution. But it could and did make Italian patriots. The members of 'Young Italy' were feeble conspirators, but powerful propagandists. And so if Mazzini did not make Italy, he did make the 'Risorgimento', the awakening of the Italian people to their destiny as a great nation. He gave Italy a purpose and a will.

Mazzini's great mistake was his quarrel with Piedmont, the only Italian state which prided itself on its independence of Austria. Piedmont, therefore, was the best starting-point for an anti-Austrian movement and, after the death of the reactionary Charles Felix in 1831, its king was at least half a liberal. At first Mazzini was delighted with the new king, Charles Albert, and wrote him an enthusiastic letter. The new king, however, was not only liberal but prudent and even timid. He thought it safest to warn Mazzini to keep out of Piedmont on pain of arrest. At once Mazzini's militant republicanism revived, and he became involved in a plot against Charles Albert's life and a futile invasion of his country. Thus 'Young Italy', by becoming the enemy of Piedmont, ruined its chance of freeing Italy.

A liberal Pope. Another group of patriots did not make Mazzini's mistake of neglecting Piedmont. Their leader was Vincenzo Gioberti, a Piedmontese priest, who looked for a free Italy under the presidency of the Pope. The Austrians were to be repelled by the help of Piedmont, and the various Italian states formed into a liberal federation: 'On Turin and Rome hangs the fate of Italy' was how Gioberti put his programme in 1843. But in 1843 Gioberti's programme seemed even more fantastically impossible than Mazzini's, for, as Metternich said, 'A liberal Pope is a moral impossibility'. Yet in 1846 the moral impossibility succeeded to the Papacy to the alarm and horror of Austria.

Pio Nono (Pope Pius IX) was not a particularly clever man, and his liberalism amounted to little more than a mild good-nature which was troubled by the bad government of the States of the Church. He began his reign by granting an amnesty to political offenders and before long he went further and granted a liberal constitution to the States of the Church. Short of declaring himself a Mahometan he could scarcely have taken any step in more striking contrast to the accepted character of Papal policy. In Catholic Italy the astonishing news that the Holy Father himself was actually encouraging the new notions of free speech and individual freedom gave liberalism a new respectability. Pio Nono's election was followed by a wild outburst of Italian optimism. A free Italy seemed at last within reach. Mazzini wrote enthusiastically to the Pope, and Charles Albert announced that if God should permit a war for the freedom of Italy, he would lead the Italian armies.

But if Metternich was wrong in thinking that a liberal Pope was impossible, he was not badly wrong, for Pio Nono soon discovered that a liberal Pope was in an impossibly difficult position. The States of the Church, which stretched right across Italy, were an obstacle to Italian unity which could only be overcome in one of two ways—either of them equally objectionable to non-Italian Catholics. The first way was for Pio Nono to become the head of an Italian federation, as Gioberti suggested. In this case the Pope might find himself engaged in a war against another Catholic nation. The second was to surrender the States of the Church, but neither the Pope nor any responsible Catholic was prepared for this drastic step in 1846.

Map 9. THE AUSTRIANS IN ITALY, 1815–59

1. The black shading represents the Italian provinces of the Austrian Empire—mostly in the fertile valley of the Po. To the south Parma, Modena and Tuscany were ruled by relations of the Habsburg emperor.

2. Nearly all the rest of Italy was under Austrian influence. Austrian troops helped the King of Naples to put down a revolution in 1821, and in 1830 they similarly helped the Pope at Bologna. Piedmont alone was relatively free from Austrian influence, but even there an Austrian army was used against the liberals in 1821.

3. The military basis of Austrian power was the Quadrilateral, the four fortresses of Mantua, Verona, Legnano and Peschiera, which guarded the junction of the road south from the Brenner Pass with the road south-west from Vienna.

4. With the exception of Venice, acquired after the French Revolution, most of the Austrian possessions in Italy had been in Habsburg hands for generations.

5. The French were profoundly jealous of the Austrians and set themselves up as rival defenders of the Pope against liberal movements. For some time a French garrison in Ancona and an Austrian in Bologna watched each other suspiciously while they defended the Pope against his subjects.

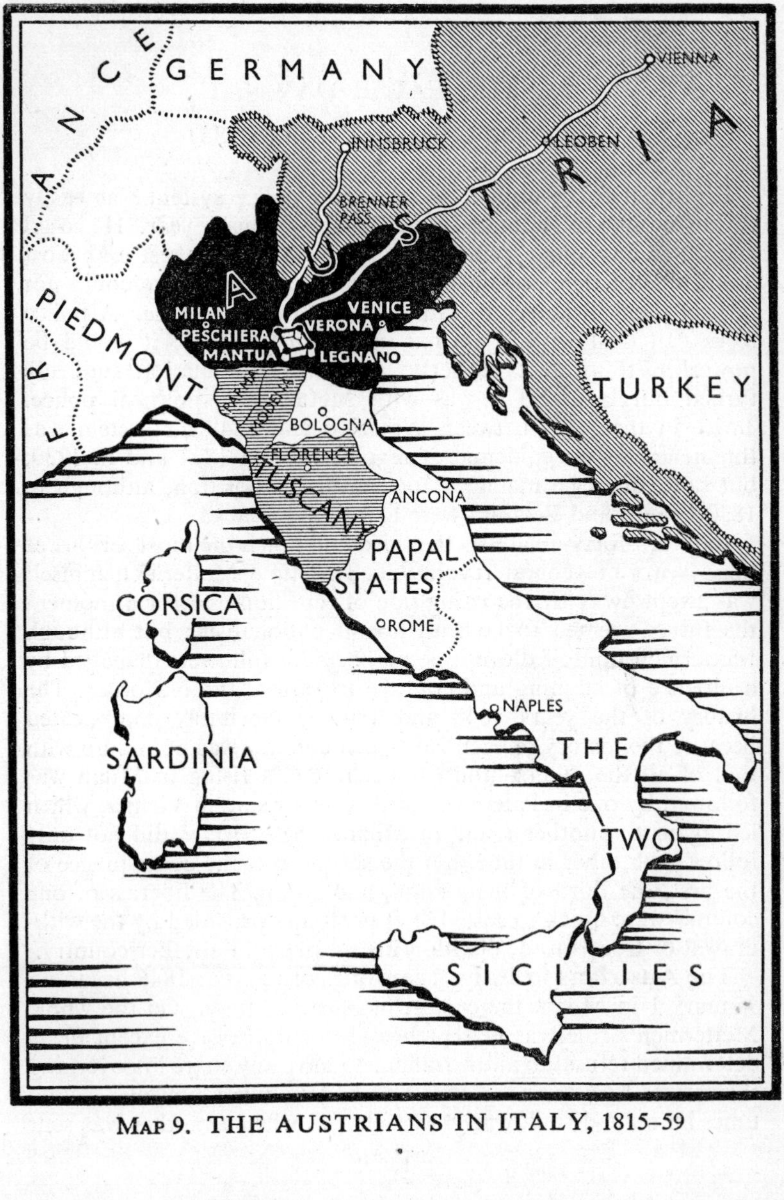

Map 9. THE AUSTRIANS IN ITALY, 1815–59

21. THE FALSE DAWN. I

Italy and Germany: 1848 to 1851

Eighteen-forty-eight. Metternich's police system was easily strong enough to suppress liberals in any ordinary year. He could guarantee any single prince or king, even the weakest and most despicable of them, against his subjects' anger, but he could not protect them if danger threatened all at the same time. A police force can deal with a normal number of thieves; it would be powerless to protect property if half the population suddenly turned burglars. So it was with Metternich's political police. Twice in the years between Waterloo and 1848 his system was threatened by an epidemic of revolutions—in 1821 and in 1830, but both times he managed to recover his position, although in 1830 France and Belgium were lost to autocracy.

Eighteen-forty-eight was the third and much the most serious of these years of general revolution. This time Metternich himself was swept away by the rising tide of rebellion and for a moment the future seemed to be with liberal nationalism; but although Metternich himself did not come back, his followers managed by a mixture of cunning and courage to restore the old order. The history of the years 1848 and 1849 is inevitably complicated because the history of each European country was bound up with that of all the others—thus, for example, a rising in Milan was followed by one in Palermo and this by a rising in Vienna, which led again to another rising in Milan. These risings did not only follow each other in time, but the second was the consequence of the first, the third of the second, and so on. The liberals of one country were quick to seize the opportunity provided by the withdrawal of troops to deal with a liberal rising in another country.

The Austrians in Italy. The risings of the year 1848 began on January 1 in Milan, the capital of Austrian Italy. On the whole Metternich's rule was neither brutal nor oppressive except in its determined refusal to allow Italians to have any share in governing themselves. Austrian red tape and bad manners, however, sometimes amounted to downright cruelty, as when a mother was sent a bill for the rope with which her son had been hanged. Incidents such as these filled Italians with anger, which they could only express in curious roundabout ways. A scientific congress, for instance, was called to discuss potato disease, but every Italian knew that 'potato' was slang for an Austrian soldier. By 1847, then, Austrian Italy was ready for rebellion. The occasion was characteristic. The people of Milan decided to boycott Austrian tobacco. The authorities made a free issue to the troops, and Austrian soldiers paraded the streets smoking immense cigars and puffing the smoke in the faces of the citizens. This was atrocious manners, but hardly an atrocity. It was, however, sufficient to cause a riot, which was suppressed by the Austrian troops at the cost of several lives.

The next news came from the south. Ferdinand II, King of Naples and Sicily, had tried a few years before to conciliate the army by allowing private soldiers to wear moustaches and by creating fourteen new field-marshals, with St Ignatius Loyola, who died in 1556, as an honorary colleague. In 1848 he had to make more drastic concessions. The people of Palermo rose and expelled the Neapolitan troops from Sicily, so that Ferdinand felt he must grant a liberal constitution to save his throne. In February the Grand Duke of Tuscany gave his subjects a constitution, and about the beginning of March Charles Albert of Piedmont and Pio Nono did the same for their states. Then came the almost incredible news that Vienna itself was in revolt and Metternich in exile (he went to live at Brighton). At once the whole of Lombardy and Venetia rebelled against the Habsburgs. After fierce street fighting the Austrian troops abandoned Milan. Venice broke free at a blow, and Manin, a brave and capable Jew, who was descended on one side from the last Doge of Venice, proclaimed her a republic.

Charles Albert's position. But unity and independence cost more than successful riots. Italy needed a soldier-statesman with a trained army. She turned to Charles Albert, King of Piedmont, the one possible leader in the war which must follow if Lombardy and Venetia were to keep the freedom they had taken while Austria was occupied elsewhere. Impatient liberals called Charles Albert 'King Wobble' because he was slow to throw in his lot with them. He had at least four good reasons for hesitation. First, it was well known to be useless to count on Naples for any help.

The Neapolitans had secured a constitution for themselves and that was all that interested them. Secondly, Pio Nono was only prepared to be a liberal as long as it did not involve him in war with Austria. He could not afford as head of the Roman Catholic church to risk a quarrel with the greatest Roman Catholic power. Every intelligent observer knew that this must be his attitude, and he himself made it clear in an address to his cardinals immediately after the beginning of the war. Thirdly, Charles Albert, as King of Piedmont, could hardly be very enthusiastic about risking his throne in order to oblige Mazzini, whose object was to found a republic. Lastly, no help whatever could be got from outside Italy. Great Britain and France both advised Charles Albert to remain neutral. He knew that if he declared war he would have to do nearly all the fighting himself. In spite of all these strong arguments for neutrality Charles Albert decided to fight. His main reason was that to do anything else was to lose the leadership of the liberal national cause for ever to the republicans, and to risk a revolution in Piedmont.

Radetzky. At first it looked as if his decision had been wise. The Grand Duke of Tuscany joined him in the war, and Parma and Modena (whose unpopular rulers had escaped from Italy) voted for union with Piedmont. The important thing, however, was a victory in the war, but this Charles Albert could not secure, for the Austrian troops could not be dislodged from the great stronghold of the Quadrilateral—the four fortresses of Mantua, Verona, Peschiera and Legnano, which guard Austria's easiest approach to Italy over the Brenner Pass. Worse followed. Charles Albert not only failed to win battles, he began to lose them. Radetzky, the Austrian commander in Italy, was a remarkable man. He had been chief of staff at the battle of Leipzig, but that was thirty-five years before. Now at the unusual age of eighty-two he started to win battles with the help of the Austrian 'white-coats', a fighting force which has hardly had justice done to them by English historians. In July he left the Quadrilateral and defeated Charles Albert at Custozza. It was the beginning of the end of the false dawn of Italian freedom.

The Roman Republic. The next stage in the collapse of the Italian movement came with a crisis in the Papal States that winter. In November Pio Nono's prime minister, Rossi, was murdered. Soon afterwards the Pope fled from Rome to Gaeta just across the Neapolitan frontier. From there he appealed to Europe for help against his subjects, who in February 1849 proclaimed Rome a republic. At first sight the substitution of a republic for the rule of the Pope might appear to have been a victory for the Italian liberal national cause. Actually it was a disaster. Italian nationalism had always to reckon with Austrian hostility; the Pope's appeal exposed it to the enmity of all Catholic countries. Louis Napoleon, who had just been elected president of the French Republic, sent an expeditionary force under General Oudinot to destroy the Roman Republic. At first the French invasion met with an unexpected check. Mazzini, who was the political head of the republic, was assisted in the defence by the second great hero of the Risorgimento, Giuseppe Garibaldi.

Garibaldi. While still a boy Garibaldi had become devoted to the cause of Italian freedom, and, like Mazzini, he had had to go into exile. His active nature led him to the 'pampas' of South America instead of to the British Museum reading-room. In 1848 he had come back to Italy, bringing with him a long experience of the guerrilla tactics which he had learned in one of the many unknown wars of South America. His courage was an inspiration, his ingenuity in improvised warfare a godsend to the untrained troops of the Roman Republic. For a time he repulsed both Oudinot and Ferdinand II of Naples, whose forces he twice routed; but in July the French troops, very much on their mettle after a humiliating reverse, took the city. The fall of Rome was inevitable, but Garibaldi's stubborn defence of an impossible position became a heroic memory, which made Italy determined not to be content until Rome became the capital of a united country.

Garibaldi refused to surrender, and before the French entered Rome he left with a few faithful followers. He did not encourage them to hope for safety, but told them plainly: 'I am going out from Rome. Let those who wish to continue the war against the stranger come with me. I offer neither pay, nor quarters, nor provisions; I offer hunger, thirst, forced marches, battles and death. Let him who loves his country in his heart and not with his lips only, follow me.' Garibaldi's retreat from the French armies round Rome through the Austrian armies by the Adriatic was as desperate as he had expected. He started with a column, he ended it all but alone with his wife Anita, the remarkable woman he had loved since he first caught sight of her through a telescope from a ship on the South American coast. The retreat reflects equal glory on the courage of Garibaldi and on the loyalty of the Italian peasants who refused to betray him to the Austrians. Its memory, like that of the defence of Rome, reminded Italians that they not only had a glorious history but men alive who were worthy of it.

Novara and the Austrian re-conquest. Meanwhile Radetzky continued his re-conquest of Austrian Italy. After Custozza Charles Albert's position was desperate, but in the spring of 1849 he made a final effort, entrusting the command to a Polish adventurer. Radetzky won another battle at Novara, where over a quarter of a century before Austria had overthrown the army which Charles Albert as Regent of Piedmont had collected against her. After this second battle of Novara Charles Albert abdicated in favour of his son, Victor Emmanuel II. In August the Austrians starved Venice into submission, after a gallant defence under Manin.

By the end of 1849 or the beginning of 1850 Austrian Italy had been completely recovered; the minor tyrants of central Italy had returned; Pio Nono had come back to Rome, leaving his liberal past behind like a bad dream; and Ferdinand had made himself master once more in Naples and Sicily, earning the nickname of 'Bomba' by the way in which he bombarded his subjects into submission. The defeat of Italian liberal nationalism was complete, but there were two things which gave promise of a brighter future. The first was the heroism of many individual Italians; the second the fact that in 1848 as in 1831 the French had intervened in Italy as much out of jealousy of the Austrians as out of love of the Pope. In 1831 and in 1848 their intervention was against the liberals and nationalists of Italy. In 1859 they were to intervene again—still out of jealousy of Austria—but this time in defence of Italian liberals.

Frederick William IV. The Italian revolutions of 1848 at least threw up one outstanding character in Garibaldi; the German revolutions produced a multitude of amiable theorists but nobody with the quality of leadership. The most significant person in some ways was Frederick William IV of Prussia. He was a patron of artists and writers and shared the romantic love of the middle ages which was characteristic of the German liberal national movement. When he came to the throne in 1840 the liberals had been enthusiastic, but they soon found that he would not grant them the constitution for which they longed. Frederick William IV was in fact, when it came to the point, an orthodox believer in the divine right of kings. He would not, if he could help it, allow a constitution, 'a piece of paper with writing on it, to come between our Lord God in Heaven and this land, like a second Providence, to rule us with its paragraphs and supplant the old sacred loyalty'. And so when 1848 came Prussia was not the first but one of the last of the German states to adopt constitutional government.

The first of the German revolutions was in Baden at the beginning of March. It was not a revolution against the Grand Duke, who was himself a liberal, but a revolution almost led by him against the Carlsbad Decrees and Metternich's tyranny, which up till now had prevented him from giving his people trial by jury, religious freedom, freedom of speech and a parliament to which the ministers should be responsible. Most of the other German rulers quickly followed his example, though it took several days of street fighting before Frederick William made up his mind to give way. When he did give way he did it thoroughly. He at once promised Prussia representative government and undertook to work for a 'free and new-born German nation'. He did everything he could to convince Germany that he was a good liberal. He need not have troubled, for nowhere did absolutism recover more easily than in Germany.

The Frankfurt parliament. In Germany as in Italy the future of the new constitutions depended on the ability of the various states to unite before Austria could recover and crush them. At least they made a better attempt to secure unity than the Italians did. In May, at the demand of the victorious liberals and with the consent of the kings and princes, a National Convention met at Frankfurt to draw up a new liberal constitution for the whole of Germany to take the place of the old, loose Confederation of rulers.

The position of Austria. This raised the delicate question of Austria. The liberals of Germany would of course never have considered union with Metternich's Austria, but Metternich had gone and a constitutional Austria was being born. Austria was the traditional leader of Germany; and Austrians proper are as unmistakably Germans as are Bavarians or Prussians. One party therefore favoured the inclusion of Austria and a constitutional

Habsburg prince as Emperor of Germany. But 'Austria' was very much more than a purely German state. The Habsburg emperor ruled over a large number of non-German races such as Hungarians, Poles, Czechs and Croats. If Austria, therefore, were to lead or even to be included in the proposed new German empire, it was plain that this empire would be only partially German and therefore unsatisfactory from a nationalist point of view. The members of the Frankfurt parliament had to choose between a state that would be unsatisfactory because it would leave several million Germans outside, and one which would be unsatisfactory because it would bring in many million non-Germans.

After infinite hesitation the Frankfurt parliament chose the former course and offered the crown of Germany to Frederick William IV of Prussia. They had drawn up a constitution providing for two houses of parliament: one to represent the various states of Germany and the other to represent Germany as a whole. It was a good scheme, but it was only a scheme. The Frankfurt parliament was hardly a parliament at all in our sense of the word; it was really only a conference whose members had no power to bind their governments. They could only ask the kings to accept their recommendations; but by March 1849, when the new constitution was ready, the kings had plucked up courage to refuse, for while the five hundred deputies had been in committee, three vigorous Austrian generals had been in action.

Before Frederick William IV was offered the crown of Germany he had quarrelled with his own liberal government and dissolved the new Prussian parliament. It might have seemed a waste of time, therefore, for the Frankfurt parliament to offer him the position of constitutional Emperor of Germany. That is not quite true. Germany could only be united round either Austria or Prussia, and so, if union was to come at all, one or other of them had to be asked. This much at least could be said for the choice of Prussia—the Habsburg emperor was perhaps even more unlikely than the King of Prussia to have anything to do with a liberal constitution. Prussia, moreover, had shown by establishing the Zollverein that she was extremely ambitious to be the greatest German power. Frederick William was tempted, but he soon overcame the temptation and declined the crown on three grounds: first, because it might lead him into war with Austria; secondly, because it would have meant that Prussia might lose her separate individuality; and thirdly, because he would have had to give up being a king by divine right and become a king by the will of the Frankfurt parliament, 'the serf of the revolution' as he described it. With Frederick William's refusal the liberal revolutions in Germany were over.

The Prussian Union and the Convention of Olmütz. The problem of Prussian and Austrian rivalry remained unsolved. Frederick William had refused to become emperor, but he had not given up the hope of playing a much bigger part in Germany than his predecessors had done. He founded a Union of German states under his presidency in opposition to the old Confederation under Austrian leadership. Soon an opportunity arose for a trial of strength between the two bodies. Once again Austria proved her ability to regain lost ground. The ruler of Hesse-Cassel quarrelled with his subjects. He appealed to Austria, they to Prussia. Austrian troops invaded Hesse; the Prussian army remained quiet at Frederick William's orders. Before the end of 1850 Prussia had agreed to the Convention of Olmütz by which the new Prussian Union was dissolved and the old German Confederation revived with Austria at its head. Germany was back in the same position in which it had been before the first revolution of 1848. The only consolation for German nationalists was that Austria was refused admission to the Zollverein.

22. THE FALSE DAWN. II

The Danubian Lands: 1848 to 1849

Metternich's failure. Neither in Italy nor in Germany could the liberals have been even temporarily successful if the Austrian Empire had been, as usual, mounting guard over absolutism throughout Western Europe. In March 1848, however, Vienna itself rose in revolt. Metternich realised that he had lost his grip and prudently he preferred flight and a comfortable exile at Brighton to the honour of martyrdom, with which the mob would have been glad to provide him. Once Metternich had gone, the whole central part of the Habsburg Empire, the lands of the Danube basin, fell rapidly to pieces because he had failed in spite of all his efforts to persuade the Habsburgs to bring their system of government up to date. They had refused to make even those reforms which would have made their own authority more secure: 'J'ai gouverné Europe quelquefois, l'Autriche jamais' was Metternich's comment on his life work. It was a failure which nearly proved fatal to the Empire he had served, and would have been completely so but for the magnificent loyalty of the army and the bitter jealousies of the subject races for one another.

The peoples of the Danube. The revolution in the Danube basin differed in several ways from the revolutions in Italy and Germany. In both these latter cases the revolution tried to unite a people who were split up into different states. In the Danube basin its object was home rule for peoples united against their will into one state. In the one case the aim was union, in the other separation. In Italy and Germany there was an old and rich national culture which was seeking political expression. In many of the Danubian lands a national culture was only just beginning to be formed by Slavonic peoples who were still in a stage of primitive simplicity. In Italy and Germany the revolutionaries were largely drawn from a rich and enterprising middle class. Except in Vienna this middle-class element was negligible in the Danubian revolutions. In Hungary the movement was the revolt of a proud people under the leadership, at least for a great part of the time, of their own nobles. Among the Croatians and Serbians it was hardly even a feudal movement. It was rather a peasant movement to a considerable extent under religious leadership, whereas in Italy and Germany on the whole the church was antagonistic to the revolutions.

The revolutions in the different states of Germany and Italy were not only simultaneous but like-minded. The Danubian revolutionaries, though subjects of one state, were usually at loggerheads. The only common factor in all these three main revolutionary centres was antagonism to the House of Habsburg. But even here there is a difference. In Italy the Habsburgs had no friends, in Germany few. In the Danubian basin, however, there was a numerous aristocracy whose natural loyalty, except in Hungary proper, was not to the country they lived in, but to the Emperor they lived under. It was an aristocracy with long traditions of service in the army; and the loyalty of the officer class, not to a country but to an emperor, was reflected in a similar loyalty among the other ranks. This personal allegiance to the House of Habsburg was almost the only unifying factor in the Danube basin, for, although economically the whole area ought to be one, in fact it was split up into two divisions by a tariff wall between Austria and Hungary.

In the Danube valley in 1848 there were three principal races—German, Hungarian and Slav—but only two main political divisions—Austria and Hungary. The different Slav races, therefore, were divided between Austria and Hungary: the Czechs of Bohemia living in Austria, and the Serbians and Croatians in Hungary. In Hungary there was more political freedom than in Austria. In Austria there was no parliament at all, but although the Habsburgs had tried for several centuries to suppress the Hungarian parliament, they had only succeeded in severely limiting its powers. In 1848 the Hungarian parliament had rather less influence than the English parliament before the Civil War. But if there was some political freedom in Hungary, there was no national freedom for anyone except Hungarians, and therefore the Slavs on the whole supported the Habsburgs against Hungary. It was the less of two evils; for, while they suffered national neglect from the Habsburgs, they experienced national persecution from the Hungarians. It was the support of the Slavs, and particularly of the Croatians, which enabled the Habsburgs to re-establish their power, not only in the Danube basin but also in Italy, for Radetzky's army was made up largely of loyal Croat regiments.

Kossuth and Jellacic. The revolutions of 1848 in the Danube basin turn largely on two men, Louis Kossuth, the Hungarian, and Joseph Jellacic, the Croatian. Kossuth came of a respectable Hungarian family and was originally a lawyer by profession. He soon became devoted to the Hungarian national cause and found in journalism and oratory a more congenial outlet. His was a quick, imaginative nature extremely sensitive to the glories of Hungarian history but, unfortunately for his country, entirely unconscious of the claims of other peoples. When the Croatians protested against the restrictions which were being placed on their language Kossuth coldly told them 'I know no Croatian nationality'. It was a fatal ignorance. His great enemy, Jellacic, came of an old soldier family which was equally devoted to the simple Croatian people and to the person of the Emperor. His early life was spent as his father's had been in service on the border between Croatia and Turkey where frontier raids were an almost everyday occurrence. It was a wild but rich life, and it was said of him that his first book was the Iliad, his first plaything a sword. By 1848 he was the recognised leader of his people in the struggle against Hungary.

It was on the thirteenth of March that Metternich fled from Vienna. Four days later a deputation, of which Kossuth was a member, forced the Emperor to give Hungary full home rule. Ferdinand had to promise that in future Austria and Hungary should be entirely independent except that he should be king of both. In Hungary he was to rule as a purely constitutional king whose ministers were to be responsible only to the parliament. Just five days later Joseph Jellacic was appointed 'Ban' or Governor of Croatia by the Emperor without consulting the Hungarians. He had chosen the man who was to save his throne, but at first this was not apparent. In May there was more trouble in Vienna—this time the Viennese wanted to secure a constitution as well as the dismissal of a minister—and Ferdinand himself had to flee to Innsbruck.

The troubles in Prague. In June there was trouble in Prague, the capital of Bohemia, where a conference of the various Slav races was held to demand a free constitution and home rule for the various Slavonic peoples. It was not a very serious matter, for on the one hand all the Slavs were loyal to the Habsburgs—it was on this occasion that one of their leaders coined the phrase 'If Austria did not exist it would be necessary to invent her'—and on the other hand the Czechs were not at that time a fighting people like their kinsmen the Croatians. It was possible for the Habsburgs to feel thoroughly safe as far as their Slav subjects were concerned, since just at this moment Jellacic had gone to Innsbruck to assure the Emperor that he and the Croats would be loyal even if the courtiers themselves proved faithless. 'These gentlemen', he told the Emperor, 'may live, if they wish, when the Empire has fallen; but I—I cannot.' The affair at Prague was indeed so little serious that, even in those desperate days when the Emperor was a refugee, it was possible to send General Windischgrätz to suppress the Czechs at once. His success was the first sign that the tide was turning.

In July Radetzky's victory at Custozza made it possible to regain lost ground in Hungary. At the beginning of August Ferdinand took away all the concessions he had made in March, and in September Jellacic invaded Hungary to see that the constitution was in fact abolished as Ferdinand had ordered. Up to this time Kossuth had been the most extreme revolutionary member of a government in which moderate men were in the majority. From now on he became the undisputed dictator of Hungary. His reign, however, would have been short if yet another rising in Vienna had not caused Jellacic to turn away from Budapest to help the Emperor, who was once again a refugee from his capital. Jellacic joined forces with Windischgrätz and together they defeated a Hungarian force sent to relieve the rebels in Vienna. Vienna fell the next day.

Schwarzenberg changes emperors. It was now obvious to everyone that the Emperor Ferdinand was no longer fit to rule—indeed it was obvious to most people that he never had been fit. Prince Schwarzenberg, Metternich's successor as chief minister, determined at the end of the year to force the Emperor's abdication in spite of the Hungarian war, and so on December 2, 1848, the young Archduke Franz Josef started a reign which persisted until 1916. Schwarzenberg was a man of ruthless determination who believed, so he said, that 'You can do anything with bayonets except sit on them'. He proposed to demonstrate this in Hungary, where Kossuth refused to recognise the new sovereign. Austria and Hungary came to open war.

In April 1849 Kossuth proclaimed Hungary an independent republic. He hoped by this move to secure the support of England and France; but even had they been willing to help him there was no easy way for them to do so, since Jellacic's loyal troops cut off Hungary from the sea. Kossuth's proclamation had in fact just the opposite effect, since instead of bringing about intervention in Hungary's favour, it led to Russian intervention against her.

Russian intervention. The Tsar Nicholas I had two good reasons for helping Franz Josef. He believed that reaction like revolution should ignore national frontiers. He was also particularly troubled about possible repercussions in Russian Poland, especially since Kossuth was employing Polish officers. In 1830, after crushing a serious revolt, he had taken away the limited self-government given Poland by the Congress of Vienna. If Hungary established her independence his good work in Poland might be undone, and therefore Nicholas was glad to lend Austria his army. His help was needed because recently the war had gone against the Austrians, thanks to the brilliant courage of the Hungarians and the inspired leadership of their general. Hungary, however, could not fight both Russia and Austria, and in August she surrendered to the Russians at Vilpagos. Nicholas handed Hungary over to Franz Josef and Franz Josef handed it over for punishment to General Haynau, whose brutality brought him the nickname of General Hyena and secured him a ducking from a crowd of infuriated brewers' draymen when he visited England soon after.

The Habsburg triumph. And so the revolutions of 1848 ended in the Danube basin, as in Germany and Italy, in almost unrelieved failure. The only reform salvaged from the wreck was the emancipation of the serfs, which had accompanied the political revolutions. Prince Schwarzenberg imposed a policy of complete unification under the absolute control of the Emperor at Vienna. The Hungarian Parliament and virtually all other local privileges disappeared. Kossuth in exile was not more disappointed than Jellacic in retirement. There was punishment for the rebels—fourteen Hungarian generals were executed on one day; there was no reward for the loyalists. The Croats had saved the Habsburgs, but the Habsburgs apparently were not grateful.

23. THE FALSE DAWN. III

The Revolutions in France: 1848 to 1852

La France s'ennuie—LAMARTINE

A self-contained revolution. In 1848 France was bored. The eighteen years of Louis Philippe's reign had been dull and France was ready for a change. The other European countries made revolutions in 1848 because they were excited, the French rebelled because they were bored. That is the first thing which distinguishes the French revolutions of 1848 to 1852 from those of the rest of Europe. There were deeper differences. In Italy, in Germany, in the Danubian basin the revolutions were more nationalist than liberal. In France the revolution was not nationalist at all, for France had long been a nation-state: it started as a purely liberal reform movement and ended as a liberal revolution complicated by a social revolution.

In Germany and in Italy the revolutions were almost entirely the work of the middle class; in France they were the joint work of the middle class and the working class, who joined together for a moment because they both wanted changes, but who soon separated because they disagreed as to what changes were wanted. In Germany there were traces of this double revolution—a working-class social revolution side by side with a middle-class liberal revolution—but they were faint and unimportant traces. Except for railways France in 1848 was at a more advanced stage of economic development than Germany and at a much more advanced stage than Italy. The French revolution of 1848, therefore, belongs to a later historical period than the other European revolutions, although, if we leave out the January riots in Milan, it started a month earlier. There is one other important difference. The revolutions of central Europe were all closely related to one another because they were all revolutions against Metternich's system. Both in their early successes and in their subsequent failure it is impossible to understand what happened in one country without considering what happened in another. But the French revolutions of these years were self-contained. The failure of the

Hungarian revolution was directly connected with the failure of the Italian revolution; the rise of Louis Napoleon was entirely unaffected by events outside France.

The history of France from 1848 to 1852 falls into six separate periods: the fall of the monarchy, the period of liberal and socialist co-operation, the period of a purely liberal republic, the period of Louis Napoleon as republican president, the period of Louis Napoleon as prince-president or dictator, and lastly the establishment of the Second Empire. In 1830, when Charles X was going much too fast, and in the wrong direction, the French had applied Louis Philippe as a kind of human brake. They now wanted a more active government and they wanted to take a greater part in running it themselves. Under Louis Philippe only the upper middle class were allowed to vote; the rest of the middle class could see no reason why they should be excluded, and Guizot's enemies, led by Thiers, encouraged them to demand the vote in the belief that, if they got it, they would vote against Guizot.

Reform banquets. This was the atmosphere of the 'Reform banquets' which the liberals organised up and down France in 1847 as a kind of superior political demonstration. They were highly respectable occasions, and only the most moderate and elevating sentiments were expressed over the wine. But in February 1848 the government prohibited an opposition banquet. There was a procession of protest against the ministry, but nothing else. The government, however, thought it wise to call out the National Guard in case of further trouble. To its alarm the National Guard, the mainstay of the July Monarchy, made it clear that Guizot must go. Guizot went and Thiers took his place. All might now have been well, for the middle class were satisfied; they wanted reform, not revolution. Luck, however, was against Louis Philippe. The next evening another political procession accidentally met a body of troops. Someone fired a shot, the troops assumed that they had been ordered to fire, a volley followed and sixteen members of the procession were killed. The news spread and Paris arose. By the morning the streets had been barricaded and, quite unexpectedly, a revolution had broken out.

Louis Philippe abdicates. Louis Philippe found that he could not rely on the middle-class National Guard and he therefore abdicated and went to join the strangely contrasted group of political exiles who had taken refuge in England. He might have saved his throne if he had called on the army, which was loyal, but he had not the heart to face a civil war without the support of the middle class he had served so faithfully. The July Monarchy had fallen. What was to take its place?

Liberals and Socialists. From the beginning there was friction between the liberals, whose 'banquets' had been the unintended starting-point of the revolution, and the socialists, who put up the barricades which turned an ordinary cabinet crisis into a means of overthrowing the monarchy. The liberals wanted to protect private property, to defend capitalism, and to keep on friendly terms with the rest of Europe. The socialists wanted work or maintenance for the unemployed and an economic as well as a political revolution. France was without a government and for the moment the two rival groups co-operated to produce one. A republic was declared and a provisional government set up in which the two leading men were Lamartine, a liberal who had written a romantic history of the Girondins, and Louis Blanc, a teacher and journalist whose vanity and ambition were combined with a real care for the poor. He thought he had discovered a cure for unemployment by a system of co-operative workshops which, run on capital found by the government, would painlessly drive capitalism out of existence.

Louis Blanc's national workshops. At first the radical or socialist wing of the coalition had things its own way, which meant that the working class of Paris dictated to the rest of France. They had declared France a republic without consulting the people and they went on to guarantee work to all citizens. National workshops were set up to carry out this promise by providing labour and wages for the unemployed. It was a makeshift expedient and not a planned socialist experiment, for the men were given unproductive and senseless tasks like digging ditches and then filling them up again. It was really a case of mass bribery by the frightened middle class.

Three things broke up the uneasy coalition between middle-class liberals and working-class socialists. First, the socialists themselves were in a difficult position: they knew that their power rested only on a dictatorship of the Paris proletariat, but they could hardly resist the demand for universal suffrage, the right of all grown-up men to vote, which by calling in the provinces would destroy their power. Secondly, all France went through the revolution of 1848 with its eyes on 1789. Frenchmen were determined to avoid a second Reign of Terror which they believed might happen if Paris was allowed to dictate to the whole country. Thirdly, the coming of railways had made it much more possible than it had ever been before to call in France to suppress Paris. The Second Republic in its most dangerous crisis in June 1848 was able to bring troops and volunteers into Paris from the provinces by train.

The history of the liberal-socialist break is soon told. In April elections were held for the Assembly which was to draw up a constitution for the Second Republic. Every man was allowed to vote and 84 per cent did. The socialists were heavily defeated and realised more clearly even than before that they could only get their way if they set up a dictatorship. In May they made an attempt to do so and proclaimed a government of their own at the Hôtel de Ville. The National Guard easily suppressed the rising and Louis Blanc went into exile.

The 'June Days'. Meanwhile the economic crisis grew worse. In the provinces men threatened to leave their work and go to Paris to the national workshops unless they got higher wages. By the end of June over 100,000 men were employed in the workshops—an increase of 13,000 in a week. The liberal government decided to close the workshops on June 21. They knew that this would probably mean a rising in Paris. They were not mistaken. For four days the Paris mob, swollen by a multitude of unemployed from the provinces, fought the government army under General Cavaignac, a veteran of the African war. It was a stern struggle, for the narrow, winding streets of old Paris made excellent defensive positions. By June 26 Cavaignac's regular troops and the National Guard had crushed the revolt with the help of artillery. No quarter had been asked or given, and the total losses were at least five thousand. Among the dead was the Archbishop of Paris, who had generously attempted the hopeless task of mediation.

The 'June Days' entirely changed the political situation. France was no longer bored; she was frightened. If the July Monarchy had been too dull, the Second Republic was far too exciting. Middle-class citizens never forgot the 'June Days' nor forgave the socialists who had caused them. The Paris working class remembered them just as clearly and never forgave the liberal leaders who had crushed their rising so ruthlessly. The peasantry, who were still the majority of Frenchmen, distrusted the Second Republic, under whose rule the 'June Days' had taken place, almost as much as they distrusted the Paris workers. A restored monarchy was impossible, since the two kings in exile and their followers in France were almost as jealous of each other as they were of the republic. A new solution to the old problem of liberty and order was necessary; a new man offered it to France.

Louis Napoleon. The representative of the Bonapartist tradition in 1848 was Prince Louis Napoleon, the Emperor's forty-year-old nephew. For ten years he had been little more than a romantic joke to European politicians. He had troubled Louis Philippe by two attempts to seize the throne: after the second, on which he took with him a tame eagle as a Napoleonic mascot, Louis Philippe decided to kill his prospects with ridicule. The prince was sent to prison, from which he afterwards escaped to England; the eagle was sent to the zoo. The Napoleonic legend was taken over by Louis Philippe, who had Napoleon's body brought home from St Helena and buried with great ceremony at the Invalides in Paris. After the first revolution of 1848 Louis Napoleon returned to France and was elected to the National Assembly. The republican leaders mildly patronised him—one of them called him 'Our excellent young man'—and were mildly surprised when he became a candidate for the presidency of the republic. They were astonished and slightly alarmed when he was elected president in December 1848 with five and a half million votes compared with the one and a half million given to Cavaignac, who was second, and with Lamartine's eighteen thousand.

Louis Napoleon's power was based on the extreme political skill with which he made use of every favourable circumstance. The glory of his name was reflected for all the army to see in the bold Napoleonic phrasing of his proclamations, and in his habit of wearing military uniform instead of civilian dress. The devout conservative peasantry were won over by his care for religious observances and by the sending of an army to help the Pope destroy the Roman republic. The provincial towns were flattered by the frequent visits he paid them, and the Paris working class, if they did not become Bonapartists, at least preferred Louis

Napoleon, whose interest in the social problem was both well known and known to be genuine, to the liberals of the Second Republic who treated poverty as if it were a crime.

The Republic undermined. These were the groups of people who had made him president; with their help he set himself to undermine the republic over which he presided. He worked in secret—outwardly he was as good a republican as any: 'I shall regard as enemies of my country', he piously exclaimed, 'any who shall attempt by illegal means to change what France herself has established.'

The liberal leaders played into his hands. There were minor socialist disturbances in Paris in June 1849. They were not dangerous—one of the 'leaders' was found following the crowd: 'Je suis leur chef', he told the police, 'il faut que je les suive.' The liberals, however, were so frightened by the socialist menace that they passed laws which gave the president emergency powers and allowed him to censor the press. Then, having made the president more powerful, they made themselves more unpopular. They took the right to vote away from large sections of the working class. The liberals were allowed one further folly. The constitution which gave precisely equal power to the president and to the parliament was clearly unworkable. Deadlocks were bound to occur. Louis Napoleon proposed to revise the constitution and submit the result to a plebiscite. Parliament refused. Louis Napoleon decided to act.

The Second Empire. Three years after he had been elected president, on the anniversary of Austerlitz, he carried out a *coup d'état*. Royalist and republican leaders were arrested, printing presses seized, the powder of the National Guard damped. The machinery of resistance was dislocated. There was no resistance. On December 2, 1851, Louis Napoleon became dictator of France. He kept the title of president, but extended his term of office ten years. Parliament lost most of its powers, though it was once again to be elected by universal suffrage. The president reserved for himself the right to make war or peace and to appoint all government officials. The ministers were to be responsible to him, not to parliament; and he was to appoint the members of the upper house, who were to decide whether the laws proposed by parliament were constitutional. It was a revival of the first

Napoleon's consulate and like the consulate it was submitted to a plebiscite for approval. It was intended, as the consulate had been, to prepare the way for an empire. A year later, in December 1852, Louis Napoleon proclaimed himself Napoleon III and the Second Empire was duly ratified at a plebiscite by nearly eight million votes to a quarter of a million.

PART THREE

CHARACTERISTICS

24. THE TWO WORLDS

Religion and Science

Special creation and evolution. In 1859 an Englishman, Charles Darwin, published a book, the *Origin of Species*, which did more than any other single thing in nineteenth-century history to change man's opinion of himself. The political revolutions of the nineteenth century, important as they were, are insignificant compared with the revolution in thought of which Darwin's book was the greatest single cause.

For more than a generation before 1859 scientists had been trying to explain on new lines how the various forms of life in the world came into existence. They were no longer satisfied with the old idea that men and animals were just, so to speak, 'made different'. In the past people had realised that there were obviously great mental differences between savages and civilised men and physical differences between the various races, between pygmies and giants or between negroes and Chinese. Nobody doubted, however, that all these races were related to one another—within the human family there were admittedly plenty of variations brought about by evolution. But between man and the animals there was thought to be a complete barrier. Some apes might look very like human beings, but it had not occurred to people that apes and humans might be relatives. This belief that each type or species of animal was completely unrelated to any other type or species is called the doctrine of 'special creation'. In the eighteenth century and earlier this had been the accepted view.

Charles Darwin. This was gradually being replaced during the first half of the nineteenth century by a belief in the 'evolution of species', a belief that all living beings were related to one another and that man himself had developed from the other animals. Charles Darwin's grandfather, Erasmus Darwin, had been one of the pioneers of this new evolutionary theory and there had been plenty of others. It is important to realise that Charles Darwin was not the solitary discoverer of a great new truth, but one of a number of scientific men working along similar lines. In his case the inspiration came from a voyage, which he made from 1831 to

1836 as a naturalist attached to the sailing-ship *Beagle*. During the voyage he noticed the large variety of similar but slightly different fossil animals found in South America. He sought for a reason to explain these slight changes and he found it in his theory of 'evolution by natural selection'. Roughly he meant by natural selection that slight changes occur by chance, and that some of these changes are useful in that struggle for existence which confronts all animals, hampered as they are by an inadequate supply of food. The fortunate animals survive, the unfortunate ones die out. The preservation of those chance changes which favour survival, and their gradual accumulation, leads in time to the development of new 'species', new animals altogether. In this way man came into being. The earlier evolutionary thinkers had believed that behind the evolution of species lay purpose and will; Darwin taught that the mechanism of evolution was chance.

Darwin was not even the only man to arrive at 'natural selection' as the explanation of evolution—A. R. Wallace had come to the same conclusion in 1858—but he was the first to make the ordinary, general public listen to the new doctrines. A technical subject of interest only to scholars suddenly became a matter of acute general controversy in newspapers, at home, in the new railway carriages —wherever men read or talked. Some held that to say man was descended from other animals was to degrade and insult him; others held that evolution had given man new ground for hope. The higher forms of life had steadily evolved from the lower and thus the theory of biological evolution supported the general liberal view of the inevitability of progress towards better and better things.

The first result of the new teaching was to shake profoundly the position of the Christian church. If any theory of evolution was true, the accounts of the creation of man, given in the first chapter of Genesis, could not be literally true. If the theory of evolution by natural selection was true, there seemed to be no room left in the universe for God. Science appeared to have given a rational explanation of the universe without leaving any room for the action of God. Other sciences besides biology helped to weaken the hold of the church on the minds of men. Geology made it impossible to accept any longer the old belief that the world had been made in seven days.

The higher criticism. Even more important, perhaps, was the growth of the new science of textual criticism. Ordinary people in the first half of the nineteenth century took the Bible as it stood as the inspired word of God. Science had made it difficult to accept the account there given of the origin of the world; history made it impossible any longer to believe in the literal truth of all the events recorded in the later books of the Bible. Men began to study the origins first of the Old Testament and then of the New. It was discovered that the Bible as we have it was a comparatively recent compilation; that, for instance, the first five books, known as the Books of Moses, were made up of several different accounts written at different times with different motives and joined together by a later editor. The same process was later discovered to have occurred in the case of the New Testament, where several unsuspected sources were found to have been employed by the compilers of the first three gospels—these sources, which may be regarded as the work of earlier 'lost' evangelists, have been to a certain extent recovered by a careful analysis of the existing gospels. The great home of textual criticism was Germany, where most of the important discoveries were made in the course of the nineteenth century. It is easy to understand their disastrous effect on the religious life of simple people who had believed that every line of the Bible was the directly inspired Word of God.

First, Adam fell; then Noah's Ark was drowned,
And Samson under close inspection bound;
For Daniel's blood the Critick Lions roar'd,
And trembling hands threw Jonah overboard.

. . . .

Twelve Prophets our unlearn'd forefathers knew,
We are scarce satisfy'd with twenty-two:
A single Psalmist was enough for them,
Our List of Authors rivals A & M:
They were content Mark, Matthew, Luke & John
Should bless th' Old-fashion'd Beds they lay upon:
But we for ev'ry one of theirs, have two,
And trust the Watchfulness of blessed Q.[1] (R. A. Knox.)

[1] Q (German *Quelle* = Source) is the name given by critics to one of the lost sources of the gospels.

The religious revival checked. The great scientific discoveries of the nineteenth century mark a turning-point in religious history. The years after the French Revolution had seen a great religious revival throughout Europe. The anarchy of the revolutionary days had made men long for order, and the brand-new society which the revolutionaries had preached had led imaginative men to turn back longingly to the romantic glories of the past. Both the conservatism and the romanticism of the post-revolutionary period had sent men back to the church. Both tendencies were combined in the Oxford Movement in England and in the liberal catholicism of men like Lamennais in France.

The scientific revolution of the nineteenth century stopped this return to religion, at least for the time being. Men could hardly fail to contrast the growing certainty of science with the apparently growing uncertainty of religion. In every branch of the physical sciences progress was rapid and undisputed—experiment proved the truth of the scientist's theories, and ordinary men, who saw their streets lighted with gas and their stage coaches replaced by railway trains, were suitably impressed. They inclined to believe that everything the scientists said was true, whether it had anything to do with science or not. And, since many of the best known scientists were men with no use for the church, the cause of organised religion suffered. These men who could work scientific miracles did not believe in miracles—that was reason enough for many people to desert Christianity.

Tightening of Roman Catholic dogma. The effects of this scientific revolution were different in Roman Catholic and in Protestant Europe. In Roman Catholic Europe it led to a tightening of dogma, in Protestant Europe to a weakening. The hopeful liberal Catholic movement was ruthlessly crushed by Pius IX, the man who, when he first became head of the church, had been known as 'the liberal Pope'. Politics had much to do with his decision, but it was caused as well by a horror of the new doctrines which were beginning to be taught. Three important measures mark the progress of this movement in the Roman Catholic church. In 1854 belief in the Immaculate Conception of the Blessed Virgin Mary[1] was made compulsory for Roman Catholics. Pius IX's

[1] The belief that the Virgin Mary, like Our Lord, was born without sin.

reply to those who asked for some relaxation in what they must believe was to add another dogma to the list. Then in 1864 he issued a Syllabus in which he condemned almost all modern developments and declared that the Pope 'neither can be nor ought to be reconciled with progress, liberalism nor modern civilisation'. A few years later he called a general council of the Roman Catholic church to meet in the Vatican, his palace in Rome. He persuaded this council, against the wishes of the best Roman Catholics of the time, to proclaim in 1870 yet another dogma—that of the infallibility of the Pope. From this time the decision as to what is or is not the teaching of the church, the decision as to what is necessary to salvation, rested under God with the Pope alone.

The Roman Catholic reaction to the scientific revolution, then, was to launch a direct and impetuous counter-offensive. It has been successful in putting heart into the followers of religion, but it has repulsed waverers. In Roman Catholic countries there is a clear-cut division between religion and agnosticism, the attitude of those who do not know what to believe (agnostic = in Greek 'one who does not know'). The Roman Catholic church has retained more power and vigour than the Protestant churches, but its influence, though more intense, is less extensive. Its only effect at all on a large section of the population is to make them bitterly anti-religious.

Protestant orthodoxy relaxed. The most immediate effect of the scientific revolution was to make it impossible any longer to believe in the literal accuracy of every statement of the Bible. This was a particular blow to Protestants who had often been taught that 'the Bible only is the religion of Protestants'. Their first reaction was like that of the Roman Catholics. Those who taught the new doctrines were wrong and wicked. Gradually, however, the greater number of Protestants came to accept the new learning and even to find that it enriched their faith. But the old doctrines were not altered: it was left to individuals to interpret them as they liked. This meant another weakening of the hold of the church as an organisation over its members. The Protestant churches were already weaker than the Roman Catholic church; they became in many instances little more than voluntary societies to which one might but need not belong. Everywhere in Protestant countries there was a marked decrease in church attendance. At the same time the slackening of the demands of orthodoxy made it possible for the church to influence those who remained outside it. In Protestant countries there is a blurred instead of a harsh frontier between those who believe and those who do not believe. Christianity has retained more influence in Protestant than in Roman Catholic countries over those who are not members of any particular church.

25. THE MACHINE AGE

The industrial revolution in England and Europe. In a history of England one expects to find a chapter on the industrial revolution side by side with a chapter on Waterloo, or even earlier in the book. In a history of Europe the industrial revolution has virtually no place before 1848 and very little before 1870. Indeed the easiest mistake to make in reading about nineteenth-century Europe is to picture the continent as being like nineteenth-century England, but from most points of view Europe, at least until 1870, was more like England in the eighteenth century. The coming of machinery to Europe is not only a more recent story but also a less interesting one than its introduction to England. It was in Great Britain that the key discoveries were made—it was here that the first looms were driven by power, the first blast furnaces employed, the first railways laid. In England the industrial revolution was accompanied by all the excitement of invention; on the continent it was largely a leisurely process of adaptation. When railways were first started in Belgium, for instance, the work fell largely to the great Anglo-Belgian firm of Cockerill's at Liège; in Austria the first engine drivers were brought over from England and paid large salaries. It was not until the coming of the aeroplane and the internal combustion engine that England began to lose her monopoly of invention. For this reason no attempt is made here to describe the process of invention which re-made the world.

There is one other important difference between the course of the industrial revolution in England and on the continent. In England the building of railways was one of the later stages; in many European countries it was almost the first sign of industrialisation. In the eighteen-sixties it was possible to travel between many of the European capitals by train, but the views from the windows were mainly of a countryside untouched by the changes which had already given England its Black Country.

Increasing numbers. There are four aspects of the industrial revolution in Europe which it is essential to grasp in order to understand its general history. The most obvious of these is the rapid growth of population. Between the outbreak of the French Revolution and the Franco-Prussian War (1870) the population nearly doubled. This was only an accelerated continuation of the steadily increasing growth which had been going on since the time of Cromwell in England. The process has been continued to the present day, though the recent fall in the birth rate makes it improbable that it can go on much longer. The figures can best be shown graphically.

1650	1780	1870	1930
100,000,000	150,000,000	293,000,000	475,000,000

At the same time Europe is only part of the world, and the great increase in the number of Europeans has been off-set by an equally great increase in the inhabitants of other continents. Europe has now, as it had in the time of Napoleon, about a quarter of the total population of the world. The constancy of this proportion is due partly to the ordinary increase of numbers in Asia and Africa, but more to the peopling of the New World in the nineteenth century which is described in Chapter 27. In all countries the increase of population has been caused more by a fall in the death rate, particularly among babies, than by an increase in the birth rate. The population has in fact often increased, and is now increasing in spite of a fall in the birth rate. Fewer children die now because of the better health services and the increased standard of living, which are themselves part of the industrial revolution.

From farms to factories. The second vital aspect of the European industrial revolution is the slow change in the balance between agriculture and industry. In 1800 Europe was overwhelmingly a continent of farmers. In Italy, where there had always been many towns, there was quite a large city population; but even there agriculture was the only really large form of employment. Elsewhere towns were few and small. England alone was beginning slowly to become a nation of townsmen and even here rather less than a quarter of the population was urban. In France nearly 90 per cent of the population lived in villages in 1800. As late as 1871 just over two-thirds of the population of France and just under two-thirds of the population of Germany were still villagers. Of course some of these people were engaged in industry and not in farming—in Saxony, for instance, then the most thickly populated part of Germany, the great textile industry was almost entirely carried on by villagers in their own homes. The overwhelming majority of country people everywhere, however, were farmers. It was not until 1900 that 60 per cent of Germans lived in towns—a figure which England had reached thirty years before.

It is not surprising that with its steadily growing population Europe became a continent of more extensive and more productive farming. The yield of the grain crops increased at an even faster rate throughout the century than did the amount of land under the plough. This progress was due to the abandonment of the old open-field system, to a better rotation of crops and to the use of fertilisers. Once again it is easier to understand the process if it is shown by a diagram:

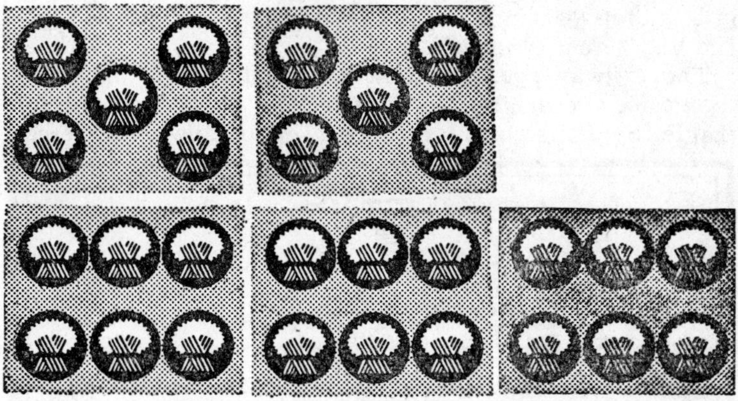

The top line represents the number of acres ploughed and the number of bushels of grain harvested in Europe in 1820; the second line shows the number of acres and of bushels in 1880. The acreage had increased by 50 per cent and the yield by 80 per cent.

The gradual change from country to town which was going on during the century was most marked in the growth of enormous capital cities and in the development of quite new large industrial towns. Here also it is easier to use a diagram to illustrate the growth of three of Europe's capitals:

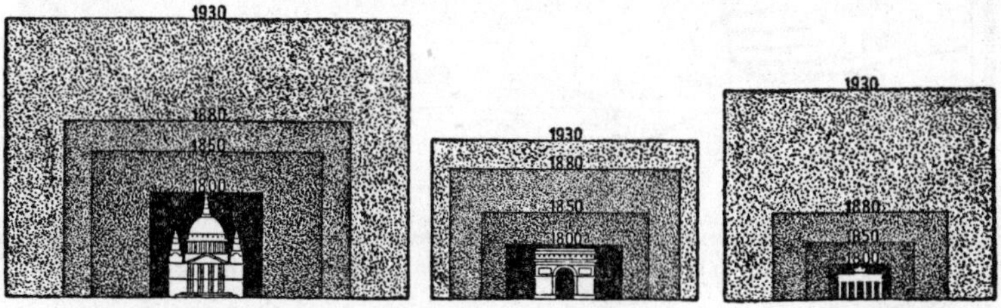

	LONDON	PARIS	BERLIN
1930	8,202,000	2,871,000	4,013,000
1880	3,830,000	2,269,000	1,122,000
1850	2,363,000	1,053,000	419,000
1800	959,000	500,000	172,000

Even more remarkable was the birth of the new industrial towns. Thus Essen on the Ruhr had nearly half a million citizens in 1920 instead of its 9000 in 1850. By 1920 it was as big a town as Vienna had been seventy years before.

The railway age. The most important changes industrially took place after the Franco-Prussian War, but some time before that Europe had secured her main network of railway lines, except

MAP 10. FRENCH RAILWAYS, 1846

in outlying countries, while the telegraph already connected most large cities. The great age of railway construction in Europe did not start until 1848. A railway map of France in 1846 shows how little had been done there up to that time. The only country, except England, which in 1848 had anything like a reasonable railway system was Belgium. There were few engineering difficulties to overcome and, as Belgium had always been densely populated, it was not surprising that she was a pioneer in continental railways.

Germany came a long way after Belgium in railway development, but a good way ahead of any other country. At first the German railways were planned to run east and west to link up the great navigable rivers, which were the main traffic routes. Before long, however, Germany and Austria had completed a through line from the Mediterranean to the North Sea, and in 1851 England made use of this route for the Indian mails. Up to this time the mails had come overland from Marseilles via Paris; now for a few years they were sent from Trieste via Hamburg, reaching London forty hours before they would have reached Paris on the old route.

Napoleon III's reign was the great period of railway construction in France. Before his abdication in 1870 the main network was complete and both flanks of the Pyrenees had been turned. The age of Alpine engineering had been begun with the driving of the great Mont Cenis tunnel. Other countries followed more slowly. In Russia, for instance, hardly any railways had been built before 1857 except the line between St Petersburg and Moscow, which the Tsar insisted should follow a completely straight route. By 1870 the track had been carried on to serve the Black Sea ports, but the rest of the country was still largely without railways.

Iron, coal and Germany. The fourth vital aspect of the industrial revolution in Europe was the change it brought in the balance of power. First England and later the new German empire were the gainers; the other great powers lost relatively in importance. Up to 1870 England was economically by far the most important country in the world. She long retained that position, but after 1870 the United States and Germany became serious rivals. Prussian policy since the formation of the Zollverein had been designed to unify Germany economically. In any case this policy would have made her a great power; but without the coal and iron in which she was or became rich, she could never have become the greatest power on the continent. Germany had her own coal in adequate quantities. In the few years immediately after 1870 her coal production was already twice that of France, although only a quarter of Great Britain's. In 1930, in spite of the loss of rich coal-fields under the Treaty of Versailles, she was producing nearly three times as much coal as France and more than half as much as Great Britain.

Germany was, however, relatively short of iron, the other essential for a first class industrial power. The Franco-Prussian

War was decisive because as a result the rich but undeveloped Lorraine iron field passed from France to Germany. Its loss condemned France to remain a nation of farmers; its acquisition secured for Germany the first place in Europe. On the basis of Lorraine iron Germany built up her immense economic and

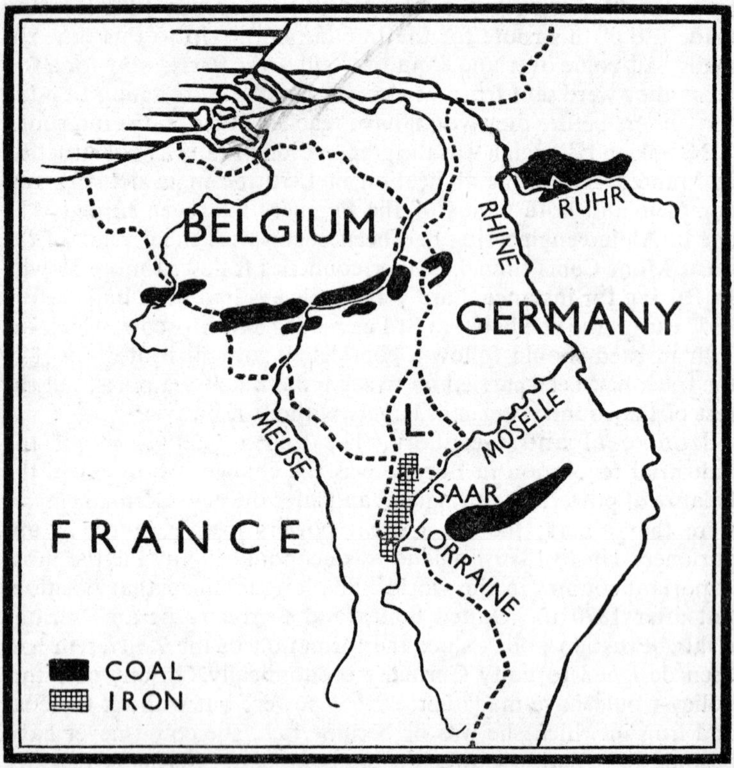

MAP 11. COAL AND IRON IN THE RHINE VALLEY

military strength. The Ruhr, the Saar, Luxemburg[1] and Lorraine were knit together into one gigantic industrial unit. The Ruhr and the Saar sent their coal by rail and river to Lorraine and received back pig iron and semi-finished steel which were worked up in the Ruhr. Bismarck once claimed that Germany was made by blood and iron; it was made strong by coal and iron.

[1] Luxemburg was part of the German Zollverein until 1918.

26. THE TWO NATIONS

There are two nations in England—the rich and the poor. DISRAELI

The background of socialism. It is possible, though not sensible, to write a history of the nineteenth century without mentioning socialism. It is possible because every attempt at socialism was a complete failure, a failure which never looked like success. It is not sensible because the growth of socialism was one of the most significant social consequences of the industrial revolution and because in our own time socialism has captured one country—Russia—and profoundly influenced others.

The background of socialism is the poverty and destitution of the working class compared with the enormous new wealth created by the industrial revolution. Of course this contrast struck plenty of people who were not socialists. It was far too glaring to escape any intelligent person's notice and consequently nineteenth-century politics and literature are filled with references to the social problem. The title of this chapter is taken from a novel called *Sybil*, written in 1845 by Benjamin Disraeli, who was then a rising young Conservative politician. He was never a socialist, though, as *Sybil* shows, he sympathised with the poor; the great Lord Shaftesbury, who was responsible for the Factory Laws which first limited the right of manufacturers to exploit their work people, was an equally sound conservative. Sympathy with the poor is not by itself socialism. The working men who established trade unions to bargain with their employers for more wages and shorter hours were no more socialists than Disraeli and Shaftesbury were. Political and industrial organisation and action by the working class is not necessarily socialism. Both factory laws and trade unions are consequences of the industrial revolution, attempts to make it tolerable for those who worked in the great factories which power-driven machinery made necessary. It was, therefore, possible to be shocked by the excesses of the industrial revolution and to seek a remedy for them without being a socialist. Socialism is merely one type of answer to the problem of nineteenth-century poverty. At first it was the least effective answer, but it is worth studying because of its future importance.

Karl Marx and the Communist Manifesto. Modern socialism may be said to date from a journey taken in November 1847 by two young Germans from Brussels to London. The elder of the two was Karl Marx, a young man of twenty-nine. The younger, who had recently set up a close friendship with him, was Friedrich Engels. For some years Marx had been feeling his way from ordinary liberalism through a vague idealistic socialism to what he described as scientific socialism. It was to preach this doctrine that he and Engels had travelled to London to attend an international congress of working men. Before long their views completely captivated the delegates and before the congress broke up it ordered a pamphlet incorporating their ideas to be translated into as many European languages as possible. This pamphlet was *The Communist Manifesto*, one of the most important documents of the nineteenth century.

In 1847 the liberals had not yet won their battle—like the socialists they were still hated revolutionaries—but the authors of *The Communist Manifesto* declared war not only on Metternich and his friends but also on Mazzini and his liberal followers. They foresaw the eventual triumph of liberalism, but they refused to welcome it except as a necessary stepping-stone to the further revolution of their dreams, the socialist revolution which would overthrow liberalism. According to their view it was inevitable that 'feudalism'[1] should disappear before capitalism and equally inevitable that in the future capitalism should disappear before socialism. The capitalists were liberals because they wanted freedom to do what they liked, which included exploiting the proletariat, the ordinary workers who were so poor that they had nothing to sell except their labour. But in order to secure freedom to carry on their business as they liked they had to overthrow the existing absolute governments, for which they needed the support of numbers. They therefore welcomed the help of the workers, who had their own quarrel with the old absolute governments. In this joint struggle the workers would learn political tactics. While they were learning how to fight, they would also, Marx believed, learn whom to fight.

[1] By 'feudalism' they inaccurately meant the old organisation of society which placed many restrictions on trade and gave many privileges to particular families or places.

In *The Communist Manifesto* Marx argued that there was a steadily increasing gulf between the few upper middle-class owners of the big accumulations of capital, on the one hand, and, on the other hand, the many workers who owned nothing but their labour. He saw the factories filled with ill-treated men, women and children. 'All are instruments of labour, more or less expensive according to their age and sex.' Their numbers were constantly recruited from the lower middle class, which was, he thought, being crushed out of existence by the big capitalists. In the necessity for making profits he saw the cause of the terrible crises which from time to time disorganised production, causing unemployment and bringing famine. In the eighteen-forties such crises were unusually common and Marx may be excused for thinking that they would necessarily become more and more severe and frequent. He drew his own conclusions from the increasing political education and economic deterioration of the workers. 'What the bourgeoisie produces above all', he wrote, 'is its own gravediggers. Its fall and the victory of the proletariat are equally inevitable.... The proletarians have nothing to lose but their chains. They have a world to win. Workers of the world, unite.'

The socialist failure in 1848. Marx and Engels carried the London conference with them, but they failed to impress Europe or even that section of it which in 1848 was engaged in making so many revolutions. Sometimes together, sometimes separately, they hurried up and down Europe trying without success to turn a liberal revolution into a socialist one. London, Paris, Cologne, Vienna, Brussels and a dozen more capitals were visited by one or other of the two friends in those exciting months. When all was over and the despots had returned, Marx and Engels came once more to England; this time, like so many political exiles, to make it their permanent home. At first they lived in great poverty—in 1852 Marx had to pawn his overcoat to buy paper on which to write a pamphlet and to borrow two pounds to pay for his daughter's coffin. His only regular income was from the occasional articles he contributed to the *New York Tribune*. Later, when Engels inherited some money, they were able to afford rather more comfort, but Marx lived and died a poor man. His immense literary work was only possible because he could use the British Museum reading-room.

The importance to history of these years of exile in London is double. First, Marx elaborated the theory behind *The Communist Manifesto* into his great work *Capital*, of which the first and most important volume was published in 1867. Secondly, he managed during these years to impose his views as thoroughly on the whole European socialist movement as he had previously done on the small London congress in 1847. In 1848 the only really important socialist attempt to seize power had taken place in Paris under Louis Blanc's leadership. This movement had owed little or nothing to Marx. It was the last socialist movement of which this could be said.

The First International. In 1864 the First International was founded at a meeting in London attended by British trade unionists, Polish refugees, and some Frenchmen, Italians and Germans. From the beginning Marx was its principal leader. For six years the International, to which many trade unions in various countries were affiliated, steadily gained in power, causing great anxiety among other sections of the community. Then in 1871 the disaster of the Commune[1] in Paris destroyed for the time being the international socialist movement in France and crippled it in other countries. In 1876 the First International was wound up at a congress in the United States.

Bakunin and Marx. During the lifetime of the International its members were engaged in a violent controversy between two groups—a socialist group led by Marx and an anarchist group led by a Russian, Bakunin. They quarrelled both over what they wanted and over how they proposed to get it. Marx wanted to socialise industry—to keep large-scale, factory production and merely to abolish private property in it. This meant keeping a well-organised society with plenty of power behind it. Bakunin wanted immediately to abolish the state and all compulsory powers and, presumably, to return to a simple voluntary communism of the sort primitive peoples are supposed to practise. Marx proposed to achieve his ends by mass organisation of the working class in trade unions and political parties. Bakunin would have nothing to do with politics in the ordinary sense and wanted to train small secret revolutionary bodies ready to work by murder and other

[1] The Commune, which is described in Chapter 31, was a short-lived socialist government in Paris after the Franco-Prussian War.

violent means. This controversy did as much as the disaster of the Commune to wreck the First International. It ended, however, with the triumph of Marx's views and left the way clear for socialism to become an open and organised political force. From this time on, socialist parties on Marxist lines were formed and prospered in most European countries. They were brought together in the Second International, which was founded in the 'eighties.

The class struggle and the Second International. The Marxian socialist considers that 'the class struggle' is the one outstanding fact of importance. In comparison with it national differences are unimportant. The gulf between Englishmen and Frenchmen is far less important than the difference between capitalists and proletarians, between employers and employed. The German worker has more interests in common with the English worker than he has with the German employer. Socialist leaders acted as if Disraeli's 'Two Nations' really existed and tried to build up through the International a working-class patriotism which would unite all the workers of the world no matter to what country they belonged. In socialist theory loyalty to the 'international working-class movement' outweighs all other loyalties. If this had been true in socialist practice the whole history of the world would have been different and the war of 1914 impossible. But when war came most German socialists threw in their lot with their own government and most English socialists supported theirs. The failure of socialism to become genuinely international was the end of the Second International as a real power.

Lenin and the Third International. One group of socialists, however, did stand out against the prevailing nationalism and opposed their own governments during the war. These were the men, of whom by far the ablest was Lenin, who began to call themselves communists instead of socialists and who after the war founded the Third or Communist International. For a long time before, they had been arguing with their fellow socialists about the way in which socialism could be achieved. As the socialist parties had grown more numerous and better organised, they had begun to play a great part in parliamentary life. It was natural for their leaders to look forward to the day when they would win a majority, and to believe that socialism could be brought about by parliamentary means. This tendency was particularly strong in

England, where the Labour movement was only half-socialist in any case, and in Germany. In countries where socialists were persecuted the older revolutionary view prevailed. It is not surprising, therefore, that the best known communist or revolutionary leaders were Russians.

As it happened, a socialist revolution became possible in Russia towards the end of the European war in 1917. The leaders of this revolution were genuine internationalists who would have been quite ready to merge Russia in a general socialist world state if that had been possible. The title they gave their country—the Union of Socialist Soviet Republics—is evidence of this. But in fact the socialist revolution was limited to Russia and, as time has gone on, Russian socialism has become more and more Russian and less and less international.

The comparative failure of socialism to fulfil Marx's hopes is due to two causes—the failure of capitalism to destroy itself and the failure of socialists in times of crisis to put their class before their country.

THE SOCIALIST MOVEMENT

The Communist Manifesto	1847
'June Days' in Paris (Louis Blanc)	1848
First International	1864–1876
Das Kapital (Marx)	1867
The Paris Commune	1871
German Social Democratic Party	1875
Second International	1889
The Labour Party	1900
Russian 'Revolution'	1905
First Russian Revolution (Kerensky)	March 1917
Second Russian Revolution (Lenin)	November 1917
Third International (Comintern)	1919

MAP 12. AMERICA: THE MELTING POT, 1 AND 2

1. These two maps illustrate the problems caused by the mixture of European races in the United States and by the large numbers of negroes. They show the position just before the war of 1914.

2. The first map shows the distribution of new settlers from European countries among the seven principal divisions of America east of the Rockies. Notice how they are virtually confined to the four northern divisions and how thoroughly mixed they are in each. But, while Irish and Italian settlers were numerous in the coastal divisions (I and II), they were few farther inland. Germans and Scandinavians, on the other hand, were few in New England (1), but many farther west. In the three southern divisions the negro problem took the place of the new immigrant problem in the north.

3. The second map shows the percentage of native born white Americans in each of the same divisions. The top line represents native born white Americans; the bottom line is made up, from left to right, of foreign born settlers, of people whose father and/or mother were foreign born, and lastly of negroes. In divisions I and II incompletely assimilated Americans were actually in a considerable majority; in every division they formed a substantial proportion of the population.

AMERICA

THE MELTING POT

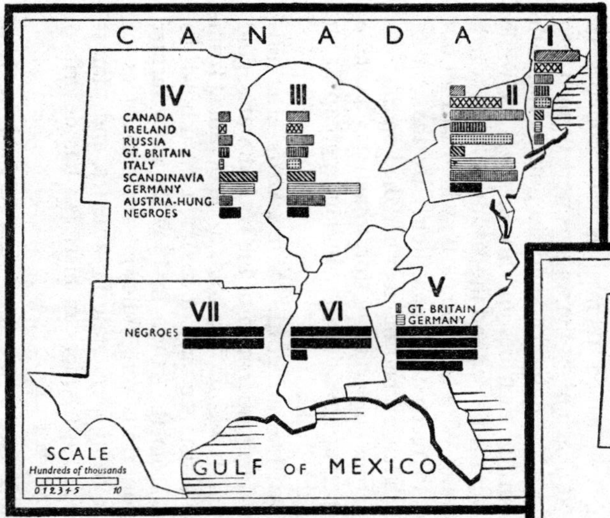

MAP 1

CANADA
GT. BRITAIN
GERMANY
IRELAND
ITALY
AUSTRIA-HUNG.
RUSSIA
SCANDINAVIA
NEGROES

I NEW ENGLAND
MAINE, etc.

II MIDDLE ATLANTIC
NEW YORK, PENNSYLVANIA, etc.

III EAST NORTH CENTRAL
OHIO, ILLINOIS, etc.

IV WEST NORTH CENTRAL
THE DAKOTAS, KANSAS, etc.

V SOUTH ATLANTIC
VIRGINIA, FLORIDA, etc.

VI EAST SOUTH ATLANTIC
TENNESSEE, ALABAMA, etc.

VII WEST SOUTH ATLANTIC
TEXAS, LOUISIANA, etc.

MAP 2

NATIVE-BORN AMERICANS, WHITE
ONE OR BOTH PARENTS FOREIGN BORN
FOREIGN BORN
NEGROES

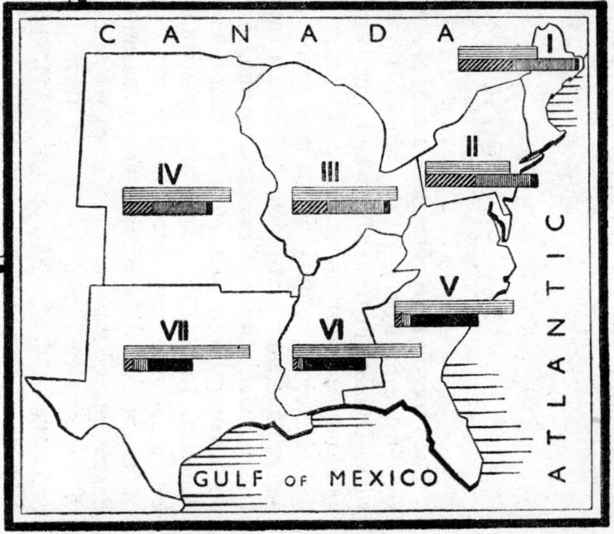

27. THE GOLD RUSH

The Oregon trail. The place-names of America, like its people, are European. Rome, Athens, Amsterdam, Calais, Toledo, Brunswick, Dublin, Belfast, Milan and Paris are not only European cities but American towns. Most of them grew up in the course of the nineteenth century as farmers gradually followed hunters across the great wheat-growing plains until at last they reached the Pacific coast, where at San Francisco and Los Angeles they met the northernmost outposts of the Spanish civilisation of Central America. In 1842 the people of Missouri and Illinois were seized with a desire to push on beyond the prairies to the rich new land of Oregon across the Rockies two thousand miles to the west. At Easter 'prairie schooners' (covered wagons) met at the township of Independence and marshalled themselves into convoys of perhaps a hundred each. Each night the wagons were formed into a hollow square in which the pioneers could sleep in the warmth of a great fire safe from Indian tribes and howling wolves, while sentries stood guard over the horses and thousands of cattle which accompanied the caravan. With luck the party, which had set out in April, might hope to reach their new home by the end of November, glad if only a few of their number had died on the trek. It was indeed a dangerous undertaking, for sometimes a whole convoy disappeared—perhaps murdered by Indians, perhaps from starvation after losing the trail.

Californian and Australian gold. A few years after this restlessness had seized the frontiersmen of the United States, gold was found in California. It was surface gold, to be obtained from the river beds with the assistance only of common wash bowls to separate the gold dust from the sand. In 1849 the early farmer pioneers of the Oregon trail were followed by multitudes of townsmen, adventurers from every state in America and almost every country in Europe. An American historian has described the wild gold rush of that year:

> Any and every route was taken by the 'forty-niners', round the Horn in the slowest and craziest vessels, across the Continent by the Oregon

or more southern trails, or, if pressed for time, by the deadly Isthmus of Panama. It seemed as if all the world was chanting, to the tune of 'Oh! Susanna'—

Oh! California
That's the land for me;
I'm off for Sacramento
With my washbowl on my knee.

By the end of 1849 thousands of Argonauts from every region of Europe, North America and the antipodes, were jumping each other's claims, drinking, gambling, and fighting in ramshackle mining villages such as Red Dog, Grub Gulch and Poker Flat. San Francisco arose in a few months from a squalid village to a city of twenty or twenty-five thousand, where eggs laid on the other side of Cape Horn sold for ten dollars a dozen, and a drink of whisky cost a pinch of gold dust; where Englishmen and Frenchmen, Yankees and Hoosiers, Georgia Crackers and Missouri Pikers rubbed shoulders with Indians, Mexicans, 'Sydney ducks' and 'Heathen Chinee'. Fortunes were made in the gold-diggings to be lost in a night at a 'Frisco faro palace; even more were made by speculation in goods and land.

(S. E. Morison, *History of the United States.*)

Two years later, in 1851, the gold rush was diverted from America to Australia. Melbourne was quickly empty and the governor of the colony reported 'cottages are deserted, houses to let, businesses at a standstill, even schools are closed. Fortunate the family, whatever its position, which retains its servants at any sacrifice, and can further secure the wanted supplies for their household from the few tradesmen that remain.' The hurry of Australians to the gold-fields was followed by a great inflow of adventurers from Europe. The average number of immigrants arriving in Australia for each of the ten years from 1841 to 1850 was 9000; between 1851 and 1860 the new-comers numbered 71,000 a year. Neither California nor Australia is to-day one of the world's principal gold producers, but each owe their original prosperity to this great burst of immigration caused by the discoveries of the middle of the nineteenth century. The men who came in the usually vain hope of getting and staying rich by picking up gold often stayed to make the country rich by developing its farming and its manufactures.

Poverty and migration. Of course gold was not the only reason for the enormous emigration from Europe in the second half of the nineteenth century. Poverty and politics also played their part. The mass migrations from Europe to the new worlds were caused even more, perhaps, by fear of starvation at home than by hope of abundance abroad. The great Irish migration in 1847 and the following years, a migration which halved the population of that country, was the direct result of the Irish potato famine which in 1846 had persuaded England to repeal the Corn Laws. The earlier emigration from Germany to the United States, which had reached the half-million mark before the middle of the century, was also caused by famine at home. It was the contrast between the miserable poverty of southern and eastern Europe and the rumours of prosperity in the New World which caused the big Latin and Slav emigration in the first quarter of the twentieth century. It is worth remembering that up to 1890 three-quarters of the immigrants into the United States came from northern and western Europe, and that after 1900 three-quarters of the new arrivals came from southern and eastern Europe. As late as 1870 nearly half the settlers entering the United States came from the British Isles; twenty years later the proportion was less than a fifth.

The political refugees. The political emigrants should not be forgotten. The failure of the revolutions of 1848 sent many thousands of liberals into voluntary exile. Their migration robbed Germany and Austria of many of their best men and helps to account for the failure of those countries to establish liberal governments when France and Italy did so half a generation later. If the German liberals of 1848 had been less adventurous and more home-keeping, like their French and Italian friends, the later history of central Europe might have been very different. On the other hand their migration enriched the cultural and political life of the United States. Those nineteenth-century exiles were the spiritual descendants of the Pilgrim Fathers. They strengthened the cause of the North in the American Civil War, fought in the eighteen-sixties to decide whether slavery should be abolished.

Refugees like Carl Schurz were a real enrichment of American public life. As a university student he had taken a prominent part in the liberal revolutions of 1848. When they failed he settled in America and there he became a close friend of Abraham Lincoln, a major-general in the Civil War, a United States senator and finally Secretary for the Interior in the cabinet. At the worst crisis in the Civil War President Lincoln was utterly worn out because he could not manage to sleep. One evening at the White House Schurz saw how exhausted the President was and began to play Beethoven to him until 'there fell upon him the first long and restful sleep he had known for days'. Less distinguished exiles turned from university life to farming without abandoning their educated tastes. In Texas an American traveller noted strange contrasts among these scholar-farmers: 'Madonnas upon log walls; coffee in tin cups with Dresden saucers; to hear a Beethoven symphony on the grand piano;...a bookcase half filled with the classics, half with sweet potatoes.'

These political refugees were important both to the New World and the Old because of their quality; but their numbers were relatively insignificant—a few thousands compared with the 60,000,000 odd who left Europe between Waterloo and the European War (1914), of whom more than half settled in the United States. The spur of poverty and the bait of riches were the forces which most moved these men. What happened to them in their new homes? Usually they secured a higher standard of living than they had been accustomed to before, but few of them achieved the riches of which they had all dreamed.

Wheat and cattle. At first most of the new settlers in the United States became farmers—the same is true of settlers in the other new lands. Once the first railway line from Atlantic to Pacific was completed in 1869 there was a great demand for settlers to people the completely empty spaces through which the track passed. The railway companies wanted settlers because they wanted farm produce to fill their waggons. They wanted settlers and they went to fetch them. Vast sums were spent in advertising for emigrants in all parts of Europe, in transporting them at cheap rates to the United States and in settling them on the land. Thus in 1874 the Santa Fé Railroad fetched 15,000 Germans to Kansas from southern Russia, where their ancestors had settled a hundred years earlier.

The new immigrants and the American-born settlers from the eastern states soon made their work felt. In the first twenty years after the original transcontinental railway was completed the wheat production of the United States was roughly quadrupled. After wheat came cattle-raising—the great age of the cowboys. Cattle-raising on a large scale was made economically possible by the invention of the refrigerator railway van in 1868, which enabled beef fattened on the Far Western ranches to be taken to the great cities of the East coast. A later but parallel development from the other side of the world was the arrival in England in 1882 of the first ship-load of frozen mutton from New Zealand.

Immigration and the industrial revolution. But the United States did not long remain an overwhelmingly agricultural country. In the generation after the American Civil War the industrial revolution changed America as completely as it had previously changed England. It is strange to think that as late as 1865 the United States was rather a backward nation economically; but in that year the American production of pig iron, for instance, was only one-fifth that of Great Britain. Her general industrial methods were also thoroughly old-fashioned. Before long, however, there was a great change. By 1887 the United States was the largest producer of pig iron in the world; by 1900 her production was roughly equal to that of England and Germany combined. European settlers played a large part in this transformation. One of the greatest centres for the production of pig iron was Pittsburg. It was here that thousands of the new Slav immigrants came to work the twelve-hour shifts which the iron-masters demanded and which American-born citizens refused to endure.

The families who immigrated to the United States in the second half of the nineteenth century came to help either with the development of agriculture or of industry. In the former case their new life was closely related to the economic life of Europe, for a great part of American wheat was grown for the European market. In the latter case their new employment was only indirectly connected with the needs of the Old World. Though the United States grew so rapidly into a great manufacturing country, it did not in the nineteenth century become more than a relatively small exporter of manufactured goods. The bulk of American industrial production was needed for her own needs, among which of course was the great railway development which made it possible for the prairies of the Middle West to become the cornfields of Europe.

The Rand. In all the later immigration to the United States the search for gold played an insignificant part. But before the end of the century there was another gold rush from Europe, though to a different part of the world. In the eighteen-eighties the great Rand gold field of South Africa was discovered. American wheat and Australian wool have a more obvious use than gold, which once again set men's minds on fire and drove them far afield—this time to Johannesburg in 1886. But the gold discoveries of the nineteenth century had a most important and valuable effect on the economic life of the whole world.

The importance of gold. In other chapters something is said of the enormous growth of industry and population in the nineteenth century. This growth in trade is a growth in the exchange of goods, which has to be accomplished by money. In the twentieth century we have grown used to money which is neither gold nor silver, but paper. Gold has not for us the importance which it had for our fathers and grandfathers. But in the nineteenth century it was impossible to think of money except in terms of the precious metals, since, though bank notes were in use, this was only for convenience and they could always be changed into gold or silver. An increase in population and in industry without an increase in gold with which to exchange goods meant an economic crisis and a severe trade depression.

Twice in the last century long periods of difficult trade conditions were ended by gold discoveries. The first period of difficulty lasted from about 1810 to 1852 and came to an end when the effect of the Californian and Australian gold production was felt. From 1873 to 1895 there was another difficult period, which was ended by the coming into full production of the South African gold-field. The following table shows approximately the effect of the gold discoveries on the economic life of the world:

	Discoveries	Gold money per head of world population £	s.	d.
1810			4	5
1847			4	2
1848–52	Californian and Australian fields			
1858–9	British Columbia field			
1868			9	1
1886	Rand mines, S. Africa			
1891			10	2
1918		1	1	6

The gold rush of the nineteenth century, therefore, had three great results. First, it helped to bring about a great emigration from Europe on a scale far larger than had ever taken place anywhere in the previous history of the world. Secondly, it helped to establish new countries overseas and to raise existing colonies into first-class great powers so that Europe ceased to be the one centre of white civilisation. Thirdly, it helped to raise the general standard of living by enabling men to exchange easily the enormous quantity of new goods, which could be produced as a result of the mechanical inventions of the industrial revolution.

PART FOUR

THE CONSOLIDATION OF NATION-STATES

28. BIRTH OF A NATION

Italy: 1849 to 1870

The price of foreign help. The disappointing experiences of 1848 and 1849 taught Italy an important lesson. Custozza and Novara drove home the humiliating truth that no Italian army could hope unaided to expel the Austrians. The man who first realised the necessity for foreign assistance, who set himself to obtain it and who put it to the best possible use, was Count Camillo Cavour. He understood, too, that Italy would have to pay a high price for this help, and he had the courage to pay it. The price proved to be the handing over to France of the coast town of Nice and of the French-speaking parts of Savoy which had been the original home of the Piedmontese royal family.

Cavour. The difference in character between Garibaldi, the soldier, and Cavour, the statesman of the Italian national revival, is clearly shown by their contrasting attitude to the loss of these two territories. Nice was the birthplace of Garibaldi; Savoy the family home of Cavour, who never learned to speak Italian without a marked French accent. Garibaldi was a romantic politician who never forgave the surrender of Nice though this small loss brought great gain to Italy. Cavour was a far-sighted realist—the loss of Savoy grieved him as much as the loss of Nice grieved Garibaldi, but he knew that it had to go. The difference in temperament extended even to their clothes. Garibaldi in his red shirt was obviously more than half a boy to the day of his death; Cavour with his round spectacles, which gave him the appearance of an amiable owl, looked the perfect bachelor civil servant.

Cavour, like Mazzini, had spent in England some of the tedious years of waiting for the false dawn of 1848. But while Mazzini had kept conspirators' company, Cavour had mixed with men of affairs. Mazzini had studied political theory in the British Museum reading-room, Cavour had studied politics in the House of Commons. And so when he was elected in 1848 to the new Piedmont parliament he understood, as few Italians did, the way to make parliamentary government work. It was not surprising to find him prime minister in 1852 at the early age of forty-two. The new prime minister was fortunate in the new king, Victor Emmanuel II, who had succeeded to the throne on his father's abdication after the second battle of Novara. Victor Emmanuel was not a great man; but he was a great constitutional king.

Cavour was able to play a decisive part in uniting Italy because he could rely on the support of his king and on the support of his country expressed in a parliamentary majority. The building up of this majority was Cavour's first task. His position was difficult, for the more radical politicians wanted a republic, while most of the conservatives disapproved of even the mildest constitution. In 1852, however, Cavour formed a coalition, called the Connubio ('marriage'), between various moderate parties, which kept him in office against the attacks of both Right and Left. From 1852 to 1858 he devoted himself to preparing Piedmont to become the nucleus of a constitutional kingdom of all Italy. He developed industry and agriculture by reducing tariffs, by negotiating commercial treaties, by building a railway from Turin to Genoa, and by setting up Co-operative Societies. The army was brought up to date and the great fortress of Alessandria re-equipped.

The quarrel with the church. Somehow or other these reforms had to be paid for. There was one obvious source of wealth which Cavour could tap without increasing the number of his enemies. The clergy of Piedmont were more numerous and more privileged than those of any other civilised European state. In Belgium and Austria there were roughly five hundred to six hundred laymen to every churchman; in Piedmont there were only just over two hundred. The Piedmontese clergy still had their own law courts and they still controlled all education. The smallest restriction of their privileges would bring excommunication, so Cavour decided that he might as well do something really worthy of that fate. Most of the monasteries and convents were dissolved and their property confiscated, the remainder were forced to pay a heavy capital levy and the special church courts were abolished.

These reforms made all Italian liberals sympathise with Piedmont. Cavour was doing what they would have liked to do. In this way he was able to convert many republicans to constitutional monarchy. They began to look upon Piedmont as a free Italian kingdom which might in time grow into a free kingdom of Italy. These domestic reforms therefore made Piedmont stronger in herself and more trusted and respected in Italy. They did nothing, however, to secure the foreign help which was essential if the dreams of Italy were to come true.

The Crimean War. Cavour's next step was intended both to strengthen still further Piedmont's standing in Italy and to make a start with the purchase of foreign help. That help could only come from either England or France: from England, which since Canning's time had been the unofficial patron of all liberal movements; from France because Napoleon III had inherited both the traditional French jealousy of Austria and his uncle's interest in Italy. As a young man he had even been a Carbonaro. In 1855 Cavour found an opportunity to please both France and England, who were engaged against Russia in the Crimean War. They welcomed Cavour's help all the more because they were exceedingly annoyed with Austria's refusal to join them. Piedmont had no quarrel with Russia and nothing directly to gain from the war. Indirectly she gained valuable experience for her army, greater prestige in Italy and, most important of all, the goodwill of France and England.

Orsini and the Pact of Plombières. The next thing Cavour had to do was to turn this goodwill to account. His chance came in a rather curious way in 1858 as the result of an Italian republican conspirator's attempt to assassinate Napoleon III. Orsini's attempt at murder might have turned Napoleon against Italy. It had the opposite effect. Napoleon was not frightened personally—he behaved with conspicuous courage—but he was afraid of a terrorist campaign which might destroy his dynasty. He thought it over and decided that the best way to end this danger was to solve the Italian problem which had created it. Besides, if he helped Piedmont against Austria he would be able to demand Savoy and Nice and thus make himself more popular in France. He was ready, therefore, to listen to Cavour's suggestions.

Cavour and Napoleon met unofficially, as if by accident, at Plombières, a spa in the Vosges, in July 1858. Napoleon promised Cavour a French army to drive out the Austrians. Cavour promised Napoleon Savoy and Nice. Napoleon promised to establish a kingdom of North Italy which should include Lombardy, Venetia and the minor north Italian states. Cavour promised not to ask for more and in particular not to disturb the Pope's possession of the States of the Church—for Napoleon dare not risk upsetting the French Roman Catholics whose support he had bought nine years before by his destruction of the Roman Republic. Both statesmen agreed to keep the agreement secret (Napoleon did not even tell his foreign minister about it until the end of the year); both statesmen agreed to provoke Austria into declaring war. Victor Emmanuel spoke to his parliament about the 'cry of grief' which he heard rising from unredeemed Italy; Napoleon told the Austrian ambassador that he regretted that his relations with Austria were so bad. The two conspirators got their way and Austria was rushed into a war for which she was not ready—the railway from Vienna to Italy was not quite finished.

Magenta, Solferino, Villafranca. Cavour was delighted when Napoleon joined him in April 1859. Everything promised well. The French and Piedmontese railways were more developed than the Austrian and by using them Napoleon was able to get his troops into the field in one-sixth the time that the old-fashioned marching would have taken. With such advantages the first Napoleon would have swept the Austrians from Italy before they had realised that the war had begun; but Louis Napoleon was a feeble general and, though he won two expensive victories at Magenta and Solferino, he was unable to dislodge the Austrians from the Quadrilateral.

At this point the arrangement between Cavour and Napoleon III broke down. It had only been based on self-interest and Napoleon III's interests had changed. Prussia had mobilised her army and Napoleon was afraid of an invasion of France. He dared not risk a continuation of the Austrian war and so he determined to make the best peace he could. In July he met the Emperor Francis Joseph of Austria at Villafranca and arranged an armistice, by which Lombardy was given to Piedmont but Venetia was kept by Austria. Piedmont had gained just half what she had been promised. Because the Franco-Italian alliance was purely selfish, Italy showed no gratitude for what she had gained, but only anger for what had been withheld. While Napoleon was in Turin a certain Mademoiselle Orsini was singing at the Opera. The crowd at once seized the opportunity for a vicious demonstration by loud shouts of 'Viva Orsini', the name of the man who had attempted to kill Napoleon.

Cavour himself was heartbroken by the failure of his plans and resigned.

Central Italy joins Piedmont. Napoleon had stopped the war, but he could not stop the national movement. The states of central Italy had already had revolutions of their own. The rulers of Tuscany, Parma and Modena were again in exile and the Papal authorities had been bundled out of the Romagna, the triangle bordered by the Apennines, the Po and the Adriatic. Nothing but union with Piedmont would content their inhabitants. Cavour became prime minister again at the beginning of 1860 in time to see this carried out and to pay Napoleon the necessary price by handing over Savoy and Nice. To make both transactions seem democratic, plebiscites were held. As is usually the case with plebiscites, the governments saw that the people did what they were told and Nice voted itself out of Italy by just as satisfactory a majority as Florence voted itself in.

The effect of these changes was greatly to strengthen Piedmont and to make it probable that her king would one day be King of Italy. This horrified Mazzini and the republicans. They decided that something must be done at once to counter-balance Cavour's success. The obvious place for their efforts was the kingdom of Naples and Sicily, whose people were always ready for a rebellion. This was hardly surprising, for few parts of Europe were so abominably governed. In 1852 Mr Gladstone had made some enquiries as to how 'Bomba' was treating the political prisoners he had made when he put down the revolution of 1848. He reported that the Neapolitan system of government was the 'negation of God'. There is no reason to quarrel with his opinion. In 1859 'Bomba' had died and been succeeded by his son, Francis II, nick-named 'Bombino', his father's equal in cruelty but his inferior in ability. Mazzini found it easy to start a republican revolution in Naples and in Sicily; but, as usual, he muddled the plans. The revolution broke out too soon in Naples and was suppressed. In Sicily he had a trifle more success. True, he achieved nothing directly, but he enlisted the support of Garibaldi.

Garibaldi and the Thousand. Garibaldi owed his success in 1860 and his failure in other years to the same qualities—boundless courage and boundless stupidity. Nothing could have been more foolish than the expedition which he led to Sicily in May 1860.

He had just over a thousand men; 'Bombino' had 100,000. He went to help a revolution which was almost extinct. He went on his own responsibility—officially his thousand was just a party of excursionists for whom two boats had been chartered. Cavour allowed him to go but took no responsibility. Legally the thousand were not soldiers but robbers; if 'Bombino' had caught them, he would not only have shot them, but been justified in doing so. Both Great Britain and Piedmont were 'accessories before the fact'; their fleets scared off the Neapolitan warships and saw Garibaldi safely across to Sicily, and then, when he had surprisingly conquered the island, they helped him in the same way to invade the mainland. Four months after he left Genoa Garibaldi entered Naples in triumph. By all the rules of war Garibaldi's expedition should have been a hopeless failure; but the rules of war did not apply because 'Bombino's' soldiers would not fight.

Cavour had gambled heavily in allowing Garibaldi to go to Sicily. There were two obvious dangers to face if the expedition was allowed to sail. First, it was impossible to say what the international complications would be. Secondly, Garibaldi was a republican who would never forgive Cavour for 'making him an alien in his native land'. Cavour saw two possible advantages to offset these difficulties. First, the adventure kept the Italian problem acute and, as long as there was trouble in Italy, Piedmont might hope to gain. Secondly, Garibaldi was not as obstinate a republican as Mazzini. If Italy could only be united under a king, he would be prepared to accept it. Garibaldi hated and despised Cavour, he admired Victor Emmanuel. Cavour was not proud.

Piedmont invades the Papal States. When Garibaldi got to Naples he announced that he would go on to Rome. The crisis had come. If Garibaldi attacked Rome, he would attack not only the Pope, but the French troops who were there to defend him. France might intervene again in Italy and this time Piedmont would stand to lose, not to gain, by her intervention. There was only one thing to do. Cavour must forestall Garibaldi. The invasion of the Papal States must be carried out by Piedmontese regulars, not by Garibaldi's Red Shirts, and it must be diverted eastwards from Rome itself. Napoleon was bound to disapprove of any invasion, since he was strongly influenced by his wife's Catholic sympathies.

Cavour had to gamble on his tolerating a carefully deflected Piedmontese invasion as the lesser of two evils. He acted quickly. On September 7 Garibaldi entered Naples; on September 18 the Piedmontese defeated the Papal troops at the battle of Castelfidardo. Fortunately for Italy the French contented themselves with withdrawing their minister at Turin. Then Garibaldi was checked at last by the Neapolitans at the line of the river Volturno. Soon afterwards Victor Emmanuel defeated the last Neapolitan army and rode on to meet Garibaldi, who saluted him as King of Italy.

Victor Emmanuel, King of Italy. Cavour had played a difficult hand with consummate skill. By heading Garibaldi off Rome he had both prevented Napoleon's interference and drawn Naples and the Papal States towards Piedmont and out of reach of Mazzini. In the autumn of 1860 these areas voted by overwhelming majorities for union under the House of Savoy. In 1861 the first parliament of a united Italy met at Turin and proclaimed Victor Emmanuel II king. Cavour's dream of a single constitutional monarchy had come true.

Italy was still without Rome, its natural capital, and without Venetia, one of its richest and most famous provinces. The latter problem had to be left until another favourable opportunity should occur to attack Austria. The former could be tackled straight away with fair hope of success. The Pope had been left in 1860 with the city of Rome and a narrow strip of country, the so-called Patrimony of St Peter. He could not defend it himself, but as long as French troops stayed there he was safe. Napoleon was quite prepared to withdraw his troops, whose presence annoyed French liberals, if the Pope would agree. Cavour offered the Pope highly favourable terms if he would abandon the States of the Church, which were now useless except as a source of wealth to the priests who misgoverned them. He not only promised 'a free church in a free state', but he was ready to undertake that the relatives of Pio Nono's favourite, Cardinal Antonelli, should keep the positions into which they had been put. By this time, however, Pio Nono was set in obstinate conservatism and foolishly refused the offer.

Rome and Venice. Only a miracle or the presence of the French could now preserve the Pope's temporal power. The latter was as unlikely to last as the former was to happen. Twice, in

1862 and in 1867, Garibaldi organised private expeditions in a vain attempt to solve the Roman problem in the way he had solved the Neapolitan. Then in 1870, nine years after Cavour's death, the Franco-Prussian War forced Napoleon to withdraw his troops. Victor Emmanuel's troops marched in. Italy had gained her capital, but only at the price of a bitter quarrel with the church which lasted until 1929, when the Pope accepted from Mussolini virtually the same terms as those offered by Cavour in 1860.

Rome was gained only as the result of an international war; so was Venetia. In 1866 there was a war between Prussia and Austria in which Italy joined Prussia and was given Venetia as a reward, although in the fighting she had experienced only defeats. A few Italian-speaking districts still remained outside the kingdom. These were added after the war of 1914, in which Italy deserted her allies and joined their enemies in return for the promise of these territories. The history of the making of Italy is, therefore, despite heroic episodes, predominantly the history of how a clever and unscrupulous diplomacy made use of other peoples' quarrels.

Cavour's legacy. This diplomacy, first practised by Cavour, robbed Italy of pride and self-respect. Cavour's legacy in internal affairs was equally doubtful. On the one hand he gave her an orderly, liberal constitution on English lines, for he was a true admirer of English ways. On the other hand he set her politicians an example of double-dealing, hypocrisy and lying. In 1860 an Italian statesman said that Cavour's word was no longer believed by anyone. Cavour himself once admitted that 'If we were to do for ourselves what we are doing for Italy, we should be great rogues'. His successors were just as unscrupulous, and without his excuse. Too often Italian politicians were in fact 'great rogues'. The best Italians found it impossible to respect their government. That is one reason why liberal Italy disappeared so easily before Fascism. The black-shirts seemed to revive the heroic tradition of Garibaldi's red-shirts in place of Cavour's sordid frock-coat parliamentarianism.

MAP 13. THE MAKING OF ITALY, 1859–1919

1. Look first at Piedmont, the nucleus of the future kingdom of Italy, in the north-west of the map. The arrow pointing eastwards shows the conquest of Lombardy (I) by the French and Piedmontese in 1859. The cession of Savoy and Nice was the price paid for French help.

2. Next came the revolutions in the central Italian duchies (II, IIA and IIB), which in 1860 voted for union with Piedmont.

3. Then came Garibaldi's conquest of Sicily and Naples (III), shown by the arrow pointing south from Genoa. This was an unofficial adventure, though it had the private sympathy of the Piedmontese government.

4. The first stage was completed by the Piedmontese invasion of the Papal States (IV) in 1860, shown by the third arrow. Note how it was deflected from Rome itself in order to avoid trouble with the French, who kept a garrison there to help the Pope.

5. The completion of the Kingdom of Italy was a by-product of three subsequent European wars:

(i) The Seven Weeks' War (1866) gave Italy Venetia (V).

(ii) The Franco-Prussian War (1870) brought in Rome (VI).

(iii) The European War (1914) gave Italy Istria (VII) and Tridentina (VIII).

MAP 13. THE MAKING OF ITALY, 1859–1919

29. BISMARCK AND WAR

Germany: 1851 to 1871

Prussia and Piedmont. Ten years after King Victor Emmanuel opened the first parliament of the kingdom of Italy, the German Empire was proclaimed in the Hall of Mirrors at Versailles. It took three wars to unite Italy and three to make the German Empire. The kingdom of Italy was an enlarged kingdom of Piedmont, the German Empire was an expanded version of the kingdom of Prussia. The Austrian Empire stood in the way of the formation of both the new states, and the Austrian difficulty was removed in both cases by the relentless and skilful diplomacy of really great statesmen in Bismarck and Cavour. The resemblance is close, but the differences are significant. Bismarck had at his disposal the finest army in Europe, while Cavour had to rely largely on foreign troops. Cavour, therefore, had humbly to ask for help, and the making of Italy had to wait on the convenience of other powers. Bismarck, on the other hand, was his own master, able to move when it suited him and not when it suited Napoleon III. Italy owed its unity to the patronage of France, Germany to the determination of Prussia. Because Piedmont was only a minor state its identity was easily lost in the new kingdom of Italy. Prussia, a great power, was not fully absorbed into Germany until the coming of Hitler. No Italian had a good word to say for Austria, but there were many Germans who bitterly resented her exclusion from the Empire. After all, the Austrians are a German people, and they had and have much more in common with the Roman Catholics of Bavaria than the Prussian Protestants have.

Prussia herself was a curiously assorted state. The Rhineland, Prussian since the Congress of Vienna, was a Catholic land of rich merchants and peasant proprietors. East of the Elbe lay old Prussia with its great, sandy plains, a countryside of large estates and few inhabitants, where the squires, or Junkers, still ruled in almost feudal power. It was this old Prussia which set the tone of the kingdom. The Junkers were harsh landlords and stern masters to any of their inferiors who had wills of their own; but they gave their king an obedience as unhesitating as that which they exacted from their labourers. Generations of service in the Prussian army had given them a high sense of personal uprightness, personal honour and an unflinching regard for duty, which were to be as great an asset to the new German Empire as their pride, their obstinacy and their stupidity were to be a liability. Side by side with the Catholic peasants and the Protestant Junkers were the new commercial middle class of Berlin and the great towns, whose numbers and influence had been steadily growing since the creation of the Zollverein.

Otto von Bismarck. Otto von Bismarck, who was born in the year of Waterloo, was a typical, burly Junker except in his far-seeing shrewdness. Like his neighbours he was a keen sportsman and a man of coarse but good-natured animal appetites—it was his ambition to smoke 100,000 cigars and to drink 5000 bottles of champagne before he died. He served Prussia with the intense devotion of his class, but he had no conception that he owed any duty to Europe. He believed that strict morals ought to regulate the behaviour of subjects to their king; but he believed equally that kings and their ministers were free from any moral obligations in their dealings with other states. All was fair in war and diplomacy. His freedom from any scruples was a great assistance in uniting Germany, but a great disservice to Europe. He and Cavour set a fashion which has never been discarded. Metternich with all his faults was a better European than they.

King William, Moltke, Roon. Bismarck was as lucky in the king he served as was Cavour. When Frederick William IV became insane in 1857, his brother William was made regent until in 1861 he succeeded to the throne. William was a man of firm, unbending character, a very different type from his romantic elder brother, who in 1848 had driven through the streets of Berlin wrapped up in a tricolour, while William was winning the nickname of 'Cartridge' by personally doing his share in firing on the mob. He was a plain, simple and devout soldier. Religion was the guiding principle of his life, but, without conscious hypocrisy, he was inclined to think of Germany as divine, and of the Divinity as German. His patriotic piety was parodied in England during the Franco-Prussian War in an imitation of a letter to his wife:

All praise, my dear and good Augusta,
We've had a most tremendous buster;
Ten thousand Frenchmen gone below,
Praise God from whom all blessings flow.

Cavour enjoyed the best of relations with his king; but he found Mazzini impossible and Garibaldi difficult. Bismarck was far more fortunate. The two other principal makers of German unity were his close associates. The Commander-in-Chief, Field-Marshal von Moltke, was a superb strategist and, though a Dane by birth, a great Prussian by conviction. His brilliant leadership was responsible for Prussia's lightning success in the three wars of German unity. The War Minister, Field-Marshal Count von Roon, had been Bismarck's friend since, as a young officer of thirty-one, he had met the future Chancellor in his student days at Göttingen. Roon had a Prussian tenacity, a fervent religious faith and the application of a great administrator.

The Prussian army was always first-class in fighting quality, but recently it had fallen sadly behind in organisation. Roon had been one of the first to realise this. In 1848 he met Prince William. The two men at once became intimate friends, and Roon soon persuaded the Prince that the army was inefficient. Neither was yet in a position to carry out the necessary reforms, but when William became Regent he asked Roon to draw up a complete scheme of army reform. For nine days Roon shut himself up and then produced a complete and masterly plan of re-organisation. It was soon clear that nobody but Roon had the capacity to carry it out and so in 1859 he was made War Minister. The plan provided for a bigger army by raising the period of compulsory military service from two to three years, a change which necessarily increased the cost of the army.

Bismarck becomes Chancellor. When Frederick William IV abolished the constitution he had granted in 1848, he left his subjects an almost powerless parliament as a slight consolation. In 1861 there was a general election which gave the liberals, who were opposed to the army reforms, a great majority. For a time it looked as if there might be a revolution. Neither William nor

Roon was prepared to allow parliament to have any say in the control of the army, which had always been directly under the king. The army itself would have resisted any change in its allegiance from king to parliament. But William knew that a country cannot be governed only by soldiers. If he was to keep his throne and secure the army reforms which he considered necessary, he must find a politician who would bring the parliament to heel. Few of the existing ministers were sympathetic and none of them were suitable. In despair William prepared to abdicate. At this stage Roon called in his friend Bismarck, who had served for some time as permanent Prussian representative on the German Confederation, where he had won a minor reputation for hostility to Austria. Since then he had been an ambassador, and in 1862 at the moment of crisis he was on leave at Biarritz. Roon sent him a telegram so carefully guarded that it seemed almost unintelligible, but Bismarck grasped its meaning and took the train to Berlin. The telegram ran: 'Periculum in mora. Dépêchez-vous. L'oncle de Maurice.' Maurice, Roon's nephew, had been Bismarck's friend at the university of Göttingen and it was he who had introduced Bismarck to Roon twenty-eight years before.

Bismarck found William in the depths of despair, with his abdication ready drawn up. The two men went for a walk in the royal park. The king came back, tore up the draft abdication, and made Bismarck prime minister. Neither man was confident of the future, but both were determined to do what they considered their duty. 'What will come of it?' said the king. 'Already I can see the place where your head and mine will fall.' 'Well', answered Bismarck, 'I cannot think of a finer death.' They compared their relation to that of Charles I and Strafford, but the sheer weight of Bismarck's personality decided the issue in favour of the king, so that Prussia remained a state where the king ruled, and parliament could only advise or criticise. Bismarck began by defying parliament and levying the army taxes without its consent. This established the right of the Prussian government to ignore the wishes of the majority of the assembly. This was the direct opposite of the British system for which the influential German liberals were working, and which had seemed within their reach before Roon sent his famous telegram to Bismarck.

From the beginning of his ministry Bismarck was determined to make the King of Prussia Emperor of Germany. Years before he came to power he had told Disraeli: 'If I am called to undertake the leadership of the Prussian government my first care will be to bring the army into such a state as to command respect, and then I will take any opportunity that offers itself to declare war with Austria, to bring the smaller states into subjection, and to give Germany a national union under the control of Prussia.' Disraeli was sufficiently impressed to warn his friends: 'Take care of this man; he means what he says.' He did. Within a week of becoming prime minister Bismarck told the members of the Prussian parliament: 'Prussia must collect her forces and hold them in reserve for a favourable moment.... Our frontiers have been ill-designed for a healthy body politic. The great questions of our time will be decided not by speeches and resolutions of majorities (that was the mistake of 1848 and 1849), but by blood and iron.'

The Schleswig-Holstein question. The 'favourable moment' soon came. In 1864 Bismarck temporarily dropped the policy of opposition to Austria; and, to manœuvre her into an unfavourable position, he even made her Prussia's ally in a war. The enemy was Denmark, and the dispute arose over Schleswig-Holstein, the German duchies which the kings of Denmark had ruled for many centuries in the same way as the kings of England had also been kings of Hanover. In 1863 a new King of Denmark proposed to merge the duchies in Denmark, taking them altogether out of Germany. Public opinion in Germany was furious, for Holstein was a purely German-speaking land, while Danish public opinion was enthusiastic, since North Schleswig was largely Danish-speaking. The German Confederation found a rival candidate for the position of duke in Frederick of Augustenburg, whose claim theoretically was nearly as good as that of Christian of Denmark. But, as Roon said, 'the Schleswig-Holstein problem is not a question of law or of pedigrees; it is a question of force, and we have it'.

In 1864 Prussia and Austria declared war and conquered the duchies with the greatest ease. Denmark had only been encouraged to fight out of a mistaken belief that she would be supported by England, whose prime minister, Lord Palmerston, encouraged her by a truculent speech. Palmerston had so often got his own way by threats—over the question of Belgium and the affair of Mehemet Ali, for instance—that he believed England had only to express a wish to see it granted. Now he discovered that England was not in a position to make good her word.

By his conquest of the duchies Bismarck gained four important advantages. First, he had secured a position of great strategic value, for, when the Kiel Canal was built, the Germany navy could move from the Baltic to the North Sea in a few hours. But Bismarck himself was never in favour of naval expansion, and this weighed with him less than the three other advantages. The second advantage Bismarck gained was the knowledge that he could safely ignore England as a danger to his plans. 'I have wasted four years of my life', he said later, 'in the belief that England was a Great Power.' Thirdly, he had staged a valuable dress-rehearsal for the reformed Prussian army, and lastly, he had drawn Austria into a highly embarrassing position. The real importance of the war was not that Prussia and Austria took two duchies from Denmark, but that they were bound to quarrel over the division of the spoil. The pretence that the war was fought to support Frederick of Augustenburg was quickly dropped. By the Convention of Gastein in 1865 Austria and Prussia came to a compromise which everyone knew was purely temporary. Austria took Holstein and Prussia Schleswig.

Seven Weeks' War. Bismarck's position was, however, precarious. The Prussian parliament continued to disapprove of the army reforms, and the necessary taxes had still to be raised without their approval. Even the king was doubtful about the wisdom of a war with Austria. The other German states were frightened of Prussia and growing more and more friendly with Austria. A single heavy defeat for the Prussian army would at this stage have meant the fall of Bismarck and a revolution in Berlin. But Bismarck was determined to have his war and determined to make it a success.

He took every precaution that he could. He had already in 1863 come to a friendly understanding with Russia, and he had discovered that he had no need to be frightened of England. Then in 1865 he went to see Napoleon III at Biarritz. Napoleon realised that France would be in danger if Bismarck's war was brilliantly successful, but he did not think it would be. He calculated that the war would make France relatively stronger by weakening both Prussia and Austria. Even if Bismarck did succeed altogether, Napoleon thought that he could safeguard France's interests by pushing her frontier nearer the Rhine, which Bismarck hinted that he might be willing to arrange. Napoleon, therefore, agreed to allow Prussia to go to war without interference from France. He did not trouble to get Bismarck's hints reduced to written promises, for he was confident that Prussia would be obliged to cede territory to France in 1866 as Piedmont had been in 1860. The next move was to enlist Italian support for Prussia. There was no difficulty in arranging for this in exchange for the promise of Venetia. Bismarck had failed to obtain German support for the war which was to establish German unity, but he had secured European toleration and even approval for it.

Bismarck was now ready to pick his final quarrel with Austria. At the beginning of 1866 he accused the Austrians of reviving the claims of Frederick of Augustenburg, and sent the Prussian army to take Holstein from Austria. Austria appealed to the other members of the German Confederation, who agreed to join in military sanctions against Prussia. Bismarck did not wait to be attacked. At once Prussian armies invaded Hanover and Bohemia, Austria's most northern province. The deciding battle was fought at Sadowa, or Königgrätz, less than a month after the beginning of the war. Prussia owed much of her success to a new type of rifle, the needle gun, which loaded at the breech instead of at the muzzle; but more to Roon's skilful use of the excellent railway system for mobilisation.

The North German Confederation. Austria was now ready for peace and Bismarck was wise enough to give it to her on terms which, while securing his aims, left the way open for friendship in the future. The Treaty of Prague was signed just seven weeks after the declaration of war, so that this second war of German unity is known as the Seven Weeks' War. Under the treaty Austria lost no territory except Venetia, but she lost her place in Germany, which was Bismarck's object. The old German Confederation was dissolved and its place taken by a new North German Confederation in which Prussia, its perpetual president, was the only great power. Prussia, indeed, was a far greater power than before, since she annexed Hanover, Schleswig-Holstein and some minor Rhineland and central German states. The three southern states—Bavaria, Würtemberg and Baden—were left isolated and embittered.

The dazzling success of the Seven Weeks' War at last reconciled Prussian public opinion to Bismarck and to the army reforms. His next task was to secure the co-operation of the disgruntled South German states, a hard task which was made easy by the unexpected stupidity of Napoleon III. He should have realised that it was now too late to ask for any territorial concessions from Prussia. Instead he instructed his ambassador in Berlin to ask Bismarck for Belgium and the left bank of the Rhine as his price for approving the Treaty of Prague. 'Do you require this under threat of war?' Bismarck asked. 'Yes', said the ambassador. 'Then war you shall have.' But Napoleon thought better of his threat and war did not come for another four years. Meanwhile Bismarck took care to see that Napoleon's demand for the Rhineland was published in the newspapers. Both Frenchmen and Germans were furious. The South German states made a military alliance with Prussia, and in 1867 they entered the Zollverein.

The Spanish throne and the Ems telegram. The final stage in the making of Germany came in 1870, when once more Napoleon gave Bismarck an opening. In that year Bismarck persuaded Prince Leopold of Hohenzollern to accept the offer which had been made to him of the throne of Spain.[1] Bismarck's object was to provoke a war with Napoleon, who was bound to be frightened by the prospect of Hohenzollerns to the east and to the south of France. Napoleon did what Bismarck wanted and threatened war. Prince Leopold did the last thing Bismarck wanted and withdrew his acceptance.

One night in July Bismarck, Roon and Moltke were dining together. They were gloomy company, for the prospect of war with France seemed to have gone. Then a telegram from the king was handed to Bismarck. They read it and Roon exclaimed 'Our God of old lives still'. Certainly their enemy had been delivered into their hands. Napoleon had sent his ambassador to Ems, where the king was, to insist that William should promise that no future invitations from Spain should be considered by members of his family. Bismarck selected the most insolent passages from the ambassador's demands, as reported in the telegram, and issued them to the press. Not only Prussia but all Germany was furious and insisted on war. The South German states were as enthusiastic as the North German Confederation. Mobilisation was ordered,

[1] For the history of Spain after the affair of Guizot and the Spanish marriages (Chapter 19) see the appendix at the end of this part.

and so perfect was the Prussian organisation that Roon was able to boast that he had never enjoyed so lazy a fortnight before. Bismarck wanted war because he knew that Germany was ready. Napoleon wanted war although he ought to have known that France was not ready.

The Franco-Prussian War and the German Empire. The story of the Franco-German War belongs to the next chapter. One of its consequences belongs to this. The separate German states which had fought side by side with Prussia in the French campaign realised that they must remain together in peace. Bismarck had little difficulty in persuading them to proclaim King William German Emperor. The new empire was not confined to the old German states. France was forced to surrender Alsace, a German-speaking province, and Lorraine, a French-speaking one, which gave Germany her natural strategic frontier along the crest of the Vosges. Both provinces had been German throughout their history until the seventeenth century, but since then they had become largely French in outlook. Germany never managed after 1870 to re-assimilate them. It is doubtful whether even the wisest government could have done so, but the harsh nature of the Prussian government made it quite impossible.

From the beginning Prussia was the leading spirit in the empire, and, although the separate kings kept their titles, they lost all but a purely local independence. An all-German parliament was set up, but the ministers continued to be responsible to the throne and not to the parliament. The new empire, like the old Prussian kingdom, was an aristocratic, military state. In it all Bismarck's dreams came true.

THE HOHENZOLLERNS

FREDERICK WILLIAM I
King of Prussia 1713–1740

FREDERICK II
(The Great)
King of Prussia
1740–1786

AUGUSTUS

FREDERICK WILLIAM II
King of Prussia 1786–1797

FREDERICK WILLIAM III
King of Prussia 1797–1840

FREDERICK WILLIAM IV
King of Prussia 1840
Insane 1857
Died 1861

WILLIAM I
Prince Regent 1857
King of Prussia 1861
German Emperor 1871–1888

FREDERICK III = Victoria, daughter of Queen Victoria of England
German Emperor 1888

WILLIAM II
German Emperor 1888
Abdicated 1918

MAP 14. THE MAKING OF GERMANY, 1864–71

1. The German national song 'Deutschland über Alles' describes what patriotic Germans think should be the boundaries of their country:

> 'From the Maas (Meuse) to the Memel,
> From the Etsch (Adige) to the Belt.'

These are marked A, B, C and D.

2. Bismarck had two ambitions: (*a*) to realise this greater Germany as far as possible, provided that (*b*) Prussia should be supreme in the new Germany instead of sharing the leadership with Austria.

3. The pivot of his schemes was the Seven Weeks' War (1866) against Austria and most of the other German states, nominally to decide what should be done with Schleswig and Holstein, which Austria and Prussia had conquered from the King of Denmark in 1864.

4. Prussia's victory in 1866 led to three things:

(i) The exclusion of Austria from Germany. The old German Confederation was wound up—that part of Austria which had formed part of the Confederation is marked with broken shading.

(ii) Prussia became supreme in North Germany. (*a*) She annexed a great deal of territory, shaded diagonally. (*b*) She formed the North German Confederation in which her influence was unchallenged—Saxony was the largest state after Prussia in it. It included all Germany north of the "South German" states.

(iii) The South German states were excluded from the new Confederation. This was necessary if Prussia, of which they were intensely jealous, was to establish its supremacy in the Confederation.

5. Bismarck's fear now was that the South German states would join Austria in a rival South German Confederation. To prevent this he got them into trouble with France (Franco-Prussian War, 1870), knowing that Prussia was bound to take the lead in any war between France and Germany. After the war the South German states joined Prussia, and the North German Confederation became the German Empire. Alsace-Lorraine, taken from France, was added to it.

6. In 1938 and 1939 Hitler recovered most of the Austrian parts of the old German Confederation.

MAP 14. THE MAKING OF GERMANY, 1864–71

30. THE SECOND EMPIRE

France: 1852 to 1871

Napoleon III. The Second Empire had every prospect of success. The professional republican and monarchist politicians were annoyed; everyone else was delighted. Victor Hugo might ask 'Because we have had a Napoleon the Great, are we bound to have a Napoleon the Little?' but in 1852 most Frenchmen did not expect that the new Emperor would be Napoleon the Little. They knew that he was courageous, intelligent and generous. They could not know that he lacked the wider qualities of statesmanship nor that he would increasingly suffer from an illness which made his policy at one moment rash, at another timid. The hopes with which France greeted him were themselves in part the cause of his undoing. She had been bored by Louis Philippe and shocked by 1848. She turned to the new Emperor simply because he was Napoleon, the heir of a majestic reputation. As Thiers said: 'The French were twice mistaken in him; once when they took him for a fool, and again when they took him for a man of genius.' The first mistake was unjust; the second fatal, for Napoleon III was never equal to the expectations which he raised.

He was perhaps himself to blame for attempting to repeat his uncle's victories; yet his people expected victories and he was bound to try and satisfy them. The trouble was that he neither won his soldiers' respect nor had he soldiers worthy themselves of respect. The army thought so little of him that he was once described by one of his generals as 'a man with a face like a bilious parrot'. His generals did win two battles in Italy in 1859, but their failure to drive the Austrians from the Quadrilateral left the campaign drawn. When in 1870 they were confronted with Prussians instead of Austrians, their bad generalship exposed France to a crushing defeat. Napoleon III could no more rival the victor of Marengo, Austerlitz, and Ulm than a nephew of Shakespeare could have written *Hamlet*, *Macbeth* and *Othello*.

'The Empire is Peace.' His only chance lay in trying to succeed where Napoleon I had failed, that is, in giving France peace and prosperity. This was his original intention. 'L'Empire, c'est la paix', he told a meeting at Bordeaux. Old Jacob Rothschild shrewdly turned the Emperor's epigram round to read 'Pas de paix, pas d'Empire'. Probably Napoleon in his heart knew that this was true, but he also loved peace for its own sake.

> I, like the great emperor, have great conquests to make.... I wish to conquer for religion, for morality, for prosperity that part of the population, still so numerous, which in the midst of a country of faith and belief scarcely knows the precepts of Christ; which in the heart of the most fertile country in the world can scarcely enjoy the first necessities. We have immense districts of virgin soil to clear, roads to open, harbours to dig, rivers to make navigable, canals to finish, our network of railways to complete. Such are the conquests I meditate, and all of you who surround me, who desire like myself the welfare of our country, you are my soldiers.

Hausmann and de Lesseps. In the twenty years of Napoleon's power France was able to gain many of these objectives. The main railway lines were finished, factories were established out of government loans, new banks were founded to encourage trade. In 1860 the Emperor made a commercial treaty with England, negotiated by Cobden, which greatly lowered tariff walls and increased trade between the countries. Nine years later the Suez Canal was finished and the first ship to pass through it carried the Empress Eugénie, a recognition of the fact that but for France the canal would not have been built. It was the work of a Frenchman, de Lesseps, who enjoyed the backing of his country in putting through his project against the bitter opposition of an England which could not forget that the first Napoleon had gone to Egypt with the intention of going on to India. Nor did Napoleon neglect the workmen, the soldiers in this war against industrial stagnation. Trade unions were made legal and large sums were spent on housing and other social reforms. Large public works were carried out to reduce unemployment. The most striking of these was the rebuilding of great sections of Paris under Hausmann, the vigorous Prefect of the Department of the Seine. The new Paris, the city as it is to-day, is a model of town planning, though unfortunately its architecture is fussy and tawdry. The plans were made to serve two ends—to glorify the Empire and to make revolutions more difficult. It was far easier to disperse a mob by 'a whiff of grape-shot' in the straight, broad streets which Hausmann planned than in the narrow, twisting lanes of the old city.

Constitution of the Second Empire. To remodel one's capital with a view to more efficiently shooting its population suggests a certain anxiety about the safety of the government. Napoleon III was never free from this fear. He had come into power by a *coup d'état*, more adroitly than honestly prepared, and he might be replaced by another, since Frenchmen had now plenty of precedents for revolution. The government which Napoleon established was a purely personal one. There was an elected assembly but it had little to do. It was not, for instance, allowed to question ministers, nor to criticise the policy of the government. The ministers were responsible only to Napoleon, who, however, on his side, did not always take them into his confidence. The Pact of Plombières, for instance, was kept secret for nearly six months from the foreign minister although during all that period it was the basis of French policy.

The newspapers, of course, were strictly censored and Napoleon's officials often showed themselves extremely touchy. Once, for example, a local paper got into trouble for reporting that a speech of the Emperor's was 'applauded several times', a roundabout way of saying that the crowd was not as enthusiastic as it might have been. At the same time Napoleon's rule was not brutal; it was not marred by political executions or party 'purges'. Compared with the Jacobin dictatorship during the French Revolution and with the Fascist and Communist dictatorships to-day, Napoleon's rule was mild. The opposition was allowed to exist, but that was all. Napoleon's great mistake was that he neither conciliated nor crushed his enemies.

In the later part of his reign Napoleon did, it is true, make concessions to the liberals. His first period of absolutism was followed by a long period of compromise with liberalism and then again by a short period of surrender to liberalism. The period of compromise started in 1860 when he allowed parliament the right to criticise the government. During the next nine years he made various other small liberal reforms. At last in 1869 he restored responsible government; that is to say, he gave parliament the right once more to control his ministers' policy by granting or refusing taxes. This great change was approved by a plebiscite in which Napoleon secured over 7,000,000 votes to 1,500,000. But by 1869 it was too late. Next year there came the Franco-Prussian War and the fall of the Empire.

The Emperor did not make liberal reforms because he wanted his subjects to enjoy representative government. He allowed them only because he could no longer prevent them. France knew perfectly well that Napoleon's ideal was what it always had been: an absolute, personal government resting on the approval of Frenchmen given from time to time in plebiscites. He wanted a state, free from political parties, but based on the will of the people. His ideal state was not far from that of a modern Fascist. But in the second ten years of his reign he was no longer strong enough to carry out his beliefs because he had suffered serious set-backs in his foreign policy. As early as 1854 Napoleon had abandoned the policy of peace which a Rothschild had said was necessary if the Empire was to continue. He was forced into an active foreign policy, which meant a series of wars, because Frenchmen, who were no longer allowed to debate home politics, had to be given some outlet for their political interests.

Liberals and Catholics. Unfortunately Frenchmen were sharply divided into two parties in their views of what foreign policy should be, and Napoleon in trying to please both succeeded only in annoying each in turn. Both groups of course wanted to see France strong, respected and, if possible, enlarged; but there agreement stopped. The liberals wanted to support liberal nationalist movements in other countries and notably in Italy. Italian nationalism had a double appeal to them, first because it was liberal and secondly because it was anti-clerical. They hated the church and they were glad to see the Pope, its head, humiliated. The conservatives on the other hand hated nationalism in general and Italian nationalism in particular for the very same reasons.

The conservative-minded Catholics had helped to make Napoleon Emperor and he dare not appear ungrateful; but the Second Empire, like the first, was the child of revolution. Its boast was that it alone could preserve the social reforms of the revolution by replacing anarchy with order. Napoleon could no more afford to quarrel with the liberals than with the conservatives. He was driven to follow two masters, which no man can do, not only by the political situation but also by his own inclinations. His 'Carbonaro' past and the spirit of the times, to which he was extremely sensitive, made him incline towards liberalism and nationalism; he was pulled the other way by his wife, the beautiful Spaniard Eugénie de Montijo, and by a natural desire to crown his social climbing with a Habsburg alliance. And so his foreign policy was ruined by hesitation and indecision; in trying to quarrel with no one he succeeded in offending all men, so that, when 1870 came, France was left friendless in Europe and Napoleon friendless in France.

Foreign policy: the period of success. In foreign policy as at home Napoleon's reign falls into three divisions—an early period of success, corresponding to the period of absolute government and lasting till 1860; a middle period of repeated failures, corresponding to the period of compromise and lasting till 1869; and a final period of catastrophe. The first period covers the expedition against the Roman Republic in 1849, the Crimean War of 1854–6 and the Italian War of 1859. The first of these wars left a thoroughly bad impression on Italian and French liberals, but won Napoleon the support of the Pope and rallied French Catholics to his side. The second again pleased French Catholics because it arose in part out of the French claim to act as the protector of Roman Catholics in the Turkish Empire, but it turned Russia into a dangerous enemy. The third war brought France Savoy and Nice, but otherwise was a disastrous failure. Italian liberals, and with them French liberals, were at first pleased by Napoleon's intervention, but they never forgave either his sudden withdrawal from the war or the French garrison which he left in Rome to preserve for the Pope the last few square miles of the States of the Church. This tardy repentance, however, was too late to placate the Pope for the loss of the greater part of his dominions or the Austrians for the loss of Lombardy. The Italian kingdom, which was born of the war, was far greater than Napoleon had planned, big enough to be dangerous. The war of 1859 gave Napoleon small gains but great losses.

Failures in Poland and Mexico. The second period covers the Polish rebellion of 1863, the Mexican adventure of 1864, the Seven Weeks' War of 1866 and the digging of the Suez Canal which was finished in 1869. In 1863 there was a rebellion in Poland which was put down with some difficulty. It brought Russia and

Prussia together into a close friendship, since both states contained large Polish areas; and it increased Russia's hatred of France, since Napoleon shared the traditional French sympathy for Poland and even considered intervention. He was, however, already too busily engaged in his Mexican adventure to undertake a Russian war. Napoleon's sympathy for the Poles increased the number of his foreign enemies without gaining him new friends at home, since Frenchmen despised the Emperor for his inability to help his friends.

In 1861 there had been a revolution in Mexico and the successful general repudiated his predecessor's debts. French and English financiers were hit and the two powers sent out an expedition to act as debt-collectors. That at least was all that England intended, but Napoleon had further plans. The United States were engaged in a long civil war and were unable to enforce the Monroe Doctrine. Napoleon hoped to found a French Catholic empire in Mexico with an Austrian Habsburg Archduke, Maximilian, as Emperor. This, he thought, would have four advantages. It would increase French prestige, give France commercial advantages, atone to Austria for the loss of Lombardy and please the French Catholics since the new party in power in Mexico was anti-clerical.

England withdrew from the expedition as soon as she learned Napoleon's plans. At first these were mildly successful. Maximilian went to Mexico and was recognised as Emperor—as long as there were French bayonets within reach. But the French did not stay long, for in 1865 the American Civil War came to an end and the re-United States at once ordered France out, an order which Napoleon was not strong enough to resist. Maximilian stayed behind, and was captured and shot in 1867. His widow, a Belgian princess, went mad. Europe was horrified at the tragedy brought about in the first place by Napoleon's ambitions and in the second place by his desertion of his friends. The Mexican adventure was not merely a defeat, it was a disgrace. France was humiliated and ashamed.

Isolation of France. Before the final tragedy in Mexico the Seven Weeks' War between Prussia and Austria still further weakened Napoleon's position. He failed to realise how powerful Prussia had become. He remained neutral, thinking that the war would weaken both states, instead of which it strengthened Prussia, who kept watch on the Rhine, at the expense of Austria, whose frontier nowhere touched France. Napoleon followed his mistaken neutrality by a worse mistake. When the war was over he demanded concessions from Prussia which he had no chance of obtaining. He asked for the left bank of the Rhine, which angered Germany, and Belgium, which infuriated England, where feeling was already sensitive at the French plan for a Suez Canal. Napoleon had made two more enemies. Thus by 1869 he had offended the French liberals, the French Catholics, the Pope, Austria, Russia and England. Prussia could reckon on their neutrality in the war which it was determined to make on France.

The surprising thing is that Napoleon should not have realised how isolated he was or that, realising it, he should not have made every effort to avoid war. Instead of trying to elude Bismarck, he went more than half-way to meet him. He courted disaster by the truculent attitude he took up over the Spanish marriage question after his original wishes had been met by the withdrawal of the German candidate. The Franco-Prussian War, which followed, was not a brutal assault by an unprovoked bully on an innocent victim. France welcomed a war which she was convinced she could win. The French prime minister announced the decision 'with a light heart'. Outside the crowds shouted 'À Berlin'. There they were crying 'Nach Paris'. The latter were right.

The Franco-Prussian War and the fall of the Empire. The French Minister of War declared that the French army was ready 'down to the last button of its gaiters'. It was true that the French had the new Chassepot rifle which they had tested on Garibaldi's volunteers in 1867, and they had the mitrailleuse, an early form of machine-gun. But their transport arrangements were so bad that a medical unit designed for the front finally arrived in Algiers. The French plans for concentration were so inefficient that not one of their Corps came into line either on time or up to strength. On the other hand the German mobilisation was perfect. The French had based all their strategy on the belief that they would invade Germany, but they quickly found themselves in retreat, an emergency for which they were unprepared.

Under Moltke the German armies swept through eastern France and bottled Marshal Bazaine up in Metz with 175,000 men. Marshal MacMahon, who had Napoleon III with him as a figurehead, attempted to relieve Bazaine but found himself surrounded at Sedan, a small town on the frontier. There was nothing to do but surrender, and on September 2 the Emperor of the French and 80,000 men were made prisoners. It is characteristic of Napoleon III, characteristic both of his courage and of his theatrical care for appearances, that he should have rouged his cheeks to hide the pallor of illness before riding out to give up his sword.

The disaster at Sedan had two immediate consequences. When the news reached Paris the Empress fled to England, escorted by her American dentist. Léon Gambetta proclaimed the Third Republic, which still lasts. For many years Eugénie had set the protection of the Pope above the interests of France. Now both causes were ruined together. On September 20, 1870, Italian troops marched into Rome, whence the French garrison had been recalled to the front. Italy had her capital.

The Third Republic and the loss of Alsace-Lorraine. The republic had come, but the Germans had not gone. They began to besiege Paris. In October Gambetta escaped from the city in a balloon and set up an energetic government, first at Tours, and later at Bordeaux. Thanks mainly to his inspiration France kept up a gallant struggle throughout the winter; but it was clearly impossible to relieve Paris after Marshal Bazaine had timidly surrendered Metz at the end of October and released further large German forces for the siege. At the end of January Paris capitulated. Next month France thought it time to come to terms. Gambetta gave way to Thiers, who thirty years before had written the proclamation which made Louis Philippe king. He was elected president of a provisional government, chosen by a hastily elected National Assembly, which met at Bordeaux and decided to open negotiations. When the French representative arrived at Versailles to ask for an armistice, Bismarck whistled a hunting tune and remarked, 'Gentlemen, the Kill'. It was. By the Treaty of Frankfurt France had to pay an indemnity of £200,000,000 and to give up Alsace, except Belfort, and most of Lorraine. France neither forgot nor forgave this loss. 'N'y parlons jamais, y pensons toujours'—France took Gambetta's advice.

France made peace not with the King of Prussia and his allies, but with the German Emperor. The Franco-Prussian War, which had killed Bonapartism, established the Third Republic and given

Italy her capital, also founded the German Empire. Ten days before the fall of Paris King William I of Prussia had been proclaimed German Emperor in the Hall of Mirrors at Versailles, where, nearly half a century later, the treaty was signed which gave back to France her lost provinces and condemned the third German Emperor as a war criminal.

31. THE THIRD REPUBLIC

France: 1871 to 1883

Political instability. In 1871 an old man might have been forgiven for prophesying a short life for the new French Republic. He could look back on six successive forms of government, each of which had claimed to be the final solution of France's difficulties. He had seen the failure of the First Republic, the First Empire, the Bourbon Restoration, the July Monarchy, the Second Republic and the Second Empire. Each had started with high hopes, none had lasted more than eighteen years. The Third Republic hardly pretended to be more than a makeshift, something to tide France over a peculiarly difficult period. It has already lasted nearly four times as long as any of its six predecessors. It started badly. When Léon Gambetta proclaimed the Third Republic he hoped to give France a fighting government which would carry on the war against Prussia, and save France from the invasion of 1870 as the First Republic had saved her from the invasion of 1793. Within a few months the Republic, which had been formed to fight, no matter what the cost, was forced to sue for peace at almost any price. The loss of Alsace and Lorraine was an inevitable but depressing birthday gift for the new form of government.

The Commune. When peace was made, the extreme republican party was furious and its representatives resigned from the assembly. Paris agreed with them and in March 1871 a new republic was proclaimed. The last of the revolutions of Paris had begun. The Commune, as the revolutionary organisation was known, was inspired by a strange medley of motives. It looked backwards to the first French Revolution and forwards to the Russian revolution. It revived the old Jacobin revolutionary calendar, but it adopted in place of the tricolour the red flag of socialism. It inherited the old Jacobin readiness for war to the knife—it regarded the members of the National Assembly as traitors because they had made peace with foreign invaders. It inherited also the old Girondin desire for the greatest possible measure of local self-government in place of the centralisation which the Jacobins had introduced and every succeeding French government had retained. The revolution was socialist as well as republican: that is to say, a large number of its leaders and supporters were socialists and would, if they had been successful, have tried to make Paris a socialist city. Karl Marx from London sent the leaders of the Commune innumerable letters of advice, and the Russian Soviet government has named one of its battleships *Paris Commune*. The story of the Commune belongs to the history of the international socialist movement as well as to the history of France.

The siege of Paris by the Germans in 1870 was followed in 1871 by a siege of Paris by Frenchmen. Two months after the Commune had been proclaimed the army of the provisional government at Versailles, recruited from prisoners of war released by the Germans, succeeded in fighting its way into Paris. Even then the 'Communards' did not give up the struggle. The events of 'Bloody Week', seven days of street fighting from May 21 to May 28, justified the boast of the revolutionaries that they would defend 'after our barricades, our houses; after our houses, our ruins'. There was no generosity to the defeated. The Commune had been unable to control its followers in Paris during the struggle and there had been a large number of atrocities: the Archbishop of Paris, for instance, had been shot. The National Assembly did not even try to control its followers in their victory. It is said that 20,000 rebels were massacred in the first few days after the fall of the Commune. The punishment which had begun in hot blood was continued in cold. For five years councils of war carried on the miserable work of persecution. Many were sentenced to death; 7500 were transported to convict settlements.

The failure of the Commune drove home the lesson taught by the failure of the Paris working class to defeat the middle class and the army in the 'June Days' of 1848. Under modern conditions of war an untrained mob has no hope of success against a well-armed, professional army. The building of railways had destroyed the old predominance of Paris over France, for the provinces could now quickly bring their weight to bear against the capital. A revolution can only hope to succeed in the modern world either when it has the army on its side or when there is no army to oppose it.

The monarchist danger. The Third Republic, then, started its career handicapped not only by the ignominy of an unsuccessful foreign war but by the disgrace of a peculiarly barbarous civil one. It now had to turn its attention to constitution-making. Here it met its greatest difficulty—the fact that it was a republic whose parliament had a majority of anti-republicans. The only consolation was that the monarchists could not agree among themselves. There were the Legitimists, supporters of the old line of the Bourbons, whose candidate was the elderly and childless Comte de Chambord, grandson of Charles X. His most dangerous rival was the young Comte de Paris, grandson of Louis Philippe. There were even a few Bonapartists left, whose candidate, after the death of Napoleon III in 1873, was his son, the Prince Imperial. As long as the monarchists quarrelled among themselves there was hope for the republic, since, as Thiers said, 'three people cannot sit on one throne'.

Thiers and Gambetta. The royalists' quarrels gave the republicans a chance to make the provisional republic efficient and popular. The first was the work of Thiers, the small man with large spectacles and tightly buttoned frock-coat who had been elected 'chief of the executive'. Thiers was at heart an Orleanist, a follower of the Comte de Paris; but he was a French patriot before he was a French monarchist, and he realised that the best hope of uniting Frenchmen lay in establishing a sound conservative republic. He set himself to this task. For two years he governed France with extraordinary ability and by 1873 he managed to pay off the large indemnity which France owed Germany under the Treaty of Frankfurt. The German army of occupation was then withdrawn.

The work of making the republic popular fell to Gambetta, the one-eyed, reckless leader who had rallied France when the disastrous news of Sedan became known. His father had been an Italian grocer in Genoa, but Gambetta was a thorough Frenchman of the South. He had one great advantage without which his passionate oratory would have been useless. Time was on his side. The monarchist majority in the assembly represented far more a desire for peace than a desire to have a king. In recent French history republics had been more warlike than kings, and Gambetta, the incarnation of the republican ideal, was also the great advocate for continuing the war. France had voted for monarchists because it felt no confidence that republicans would make peace. But war-weariness passes and Gambetta knew that, when this happened, his republican patriotism would become increasingly popular. He lived just long enough to see the republic fairly established, dying on the last day of 1882 as a result of an accident with a revolver. The next year his life work was consolidated by a law declaring that 'the republican form of government cannot be made the subject of a proposal for revision'.

MacMahon and the crisis of 1873. The most critical years for the republic were from 1873 to 1877. The chances of a restoration became exceedingly good early in 1873, when the royalist majority in parliament succeeded in forcing Thiers to resign. In his place they elected Marshal MacMahon, the descendant of an Irish Jacobite who had gone into exile with James II. He was a reasonably distinguished soldier, a man of fine character rather than of powerful intellect, and a convinced monarchist. With MacMahon at the head of the State, the royalists were half-way to victory. In August the Comte de Chambord was reconciled to the Comte de Paris, whom he acknowledged as his heir. The royalists now seemed certain of victory, for the Legitimists and the Orleanists ceased to exist as rival parties. The stage was set for MacMahon's resignation and the accession of the Comte de Chambord when a difficulty of the most trivial kind arose. The Comte de Chambord suddenly announced that he would not accept the tricolour in place of the old white flag of the pre-revolution days. Even MacMahon protested that this would make 'the Chassepots go off by themselves'. Under the tricolour France had won her greatest victories —to abolish it was unthinkable. Chambord had 'thrown away a kingdom for a napkin'. All that the royalists could hope was that he would die and leave the way clear for his more accommodating heir, while the faithful MacMahon still held the fort.

The republican constitution. But Chambord lived on and the republican cause was strengthened by his obstinacy. In 1875 France became officially what she was already in practice—a republic. In that year the National Assembly adopted, by a majority of one, the present republican constitution. It provided for a parliament of two houses, a Senate and a Chamber of Deputies. In theory the French constitution was very like the English, but in practice it has proved quite different. In England the existence of a strong two-party system, backed by the right of the prime minister to ask for a general election if he is defeated, has led to the rigid control of parliament by the cabinet. In France, where organised parties are weak and parliaments are never dissolved in the middle of their period of office, the deputies are able to control the government effectively and to change it frequently. In England we regard general elections as a means of choosing between two possible governments; in France this is impossible, for there are so many possible governments. The Second Republic had provided for the election of a president, who was to be the real head of the government, by a general vote of the whole country. This had made possible Louis Napoleon's dictatorship. The Third Republic guarded against this by giving the president no more power than a King of England and by allowing parliament to elect him.

The 'Seize Mai'. France had got her republican constitution, but she had still MacMahon as president. As long as he remained the royalists had hopes of success. The duty of the president of the republic to be a ceremonial figurehead, not a working politician, is not laid down in the constitution. It is a matter of custom, not of law. In MacMahon's time the custom was hardly established, and he took advantage of his position to dismiss the republican prime minister on May 16, 1877. He appointed in his place the royalist leader, the Duc de Broglie, the man who had engineered MacMahon's election as president four years before. It was necessary to hold a new election to give Broglie a majority. This was Gambetta's opportunity. He carried on a tireless campaign against the royalists and their allies, the leaders of the Roman Catholic church. 'Le clericalisme, voilà l'ennemi', he declared in a famous phrase. The cry was effective. In the elections MacMahon suffered his worst defeat since Sedan. Two years later he resigned in favour of a confirmed republican.

The crisis of the 'Seize Mai' settled many problems. It is the real turning-point in recent French history, for its outcome was the definite acceptance of the republic by an overwhelming majority of Frenchmen. Thanks to Gambetta's electioneering and the folly of its leaders the 'Seize Mai' marked the downfall of the church as well as of royalism. France decided in that momentous election that she was to be not only a republic, but a 'lay republic'—that is to say a republic which would reduce religion to a purely private matter. The victory of the republic over the church was a party victory, not a national one. It was due to the stupidity of the church in allying itself with one political party, and that the unpopular one. The church made it difficult for a man to be both a good Catholic and a good republican in the mistaken hope that Frenchmen would sacrifice their political loyalty to their religious. The anti-clerical party seized the opportunity to drive religion out of public life altogether. In doing this they weakened the republic, for, though most Frenchmen were now republican, barely half were anti-clerical. Those republicans who were also practising Catholics were left in a doubtful and difficult position. Two other results of the crisis are important. It decided that the president must not interfere in politics to over-ride the wishes of the parliament; and it decided that no French prime minister could appeal to the country over the head of parliament by holding a general election before one is legally due. Neither of these decisions was embodied in a law, but they are customs which no politician has felt strong enough to break.

32. BISMARCK AND PEACE

Germany: 1871 to 1890

Bismarck's restraint. The great statesmen are those who use force with restraint, who refuse to be stampeded by success into failure through feeling that their luck must hold. But in diplomacy, as in cards, it is often as fatal to under-bid as to over-call a hand. Above all, the statesman needs judgment. He must know certainly when he should act and when he should remain quiet; and he should never use more force than is strictly necessary to attain his end, for an excess of force will make him unnecessary enemies. Some he cannot avoid, but let them be as few as possible; and once he has gained his object, let him try to effect a reconciliation with his former enemies. This combination of qualities is excessively rare; it occurs hardly once in a century but it is apparent in all Bismarck's career.

His claim to be one of the greatest figures in European history rests equally on the wars he made before 1871 and on the wars he avoided after 1871. He had used the Prussian army brutally in his determination to absorb Germany into Prussia. Having achieved his end, he realised, as few politicians would have been cool enough to do, that he must never use it again. The 8,000,000 Germans inside Austria would have to stay outside Germany. The diplomatic situation which left Austria friendless in 1866 would never recur again. Europe was jealous of Prussian Germany, and another war of Germany against Austria would soon have developed into a war of Europe against Germany.

Relations with France and England. The Franco-Prussian War had, as Bismarck described it, made Germany a 'satiated state' which had reached the limit of safety in its expansion. That conviction shaped the whole of his subsequent foreign policy. Bismarck knew that France would probably find it impossible ever to forgive the loss of Alsace and Lorraine. He determined to make it impossible for her to regain them. His first thought was to crush her economically by the war indemnity he forced her to pay, but he had miscalculated her financial strength and found that

Thiers was able to pay the indemnity in full in less than two years. Having failed in his direct attempt he adopted a much more typical policy. He became France's friend and encouraged her to seek compensation for her losses in Europe by a forward policy in Africa. French colonial expansion had two merits in Bismarck's eyes. First, it gave the French something to distract their attention from Alsace and Lorraine, which might also conceivably land them in awkward native campaigns. Secondly, it could hardly fail to lead to difficulties between France and England and between France and Italy. It would thus rob France of possible allies who might conceivably have helped her to regain her lost provinces. 'I have sent the fiery steed of French ambition caracoling in the sands of Tunis', Bismarck cheerfully remarked, and added, 'They will find it heavy going.'

Bismarck attached the greatest importance to the friendship of England. He made up his mind never to trespass on either of our two preserves, the sea and colonisation. Here his perfect judgment is seen at its best. He set his face firmly against the creation of a big navy or the annexation of territory in Africa. His desire for English support even led him in the crisis of 1876 to 1878 to take an anti-Russian line and risk a Franco-Russian alliance which would have encircled Germany.

The Dreikaiserbund and the Triple Alliance. Russia's consistent friendship with Prussia had been Bismarck's most valuable asset in the years when he was building up German unity. Without her help he could never have dealt successfully with Austria; but, once the German Empire was established, it became apparent that the new state was much more powerful than either of its eastern neighbours. The undying rivalry of Austria and Russia made each determined that the other should not secure the support of the great new German Empire. Both states, therefore, wanted to be on good terms with Germany and were prepared, if necessary, to tolerate each other if that was the only condition on which Germany's friendship could be obtained. Bismarck took good care to see that this was the condition of his friendship, for he on his side had reason to fear equally either a Franco-Russian or a Franco-Austrian alliance.

An understanding between Germany, Austria and Russia was therefore a most important part of Bismarck's policy. During a great part of this period he was able to secure it. In 1876, however, a Russo-Turkish war shattered the Dreikaiserbund (League of the three Emperors) and forced Bismarck reluctantly to choose between Russia and Austria. He chose Austria, not only because she was a partly German country, but also because his policy of friendship with England suggested that he had better oppose Russian expansion. Bismarck did not like having to choose between his two eastern neighbours and did his best to restore the old grouping. In 1884 he was successful. From then until his fall from power in 1890 he was able to keep Russia free from French influence without sacrificing his friendship with Austria.

The crisis of 1876, however, had made it clear that, if the worst came to the worst, Germany would now choose Austria's friendship rather than Russia's, although Bismarck did all he possibly could to prevent that situation arising. The new German-Austrian friendship was expressed in the Dual Alliance made in 1879, which became the Triple Alliance when Italy joined it in 1882. Her adhesion deprived France of her last possible ally. It was brought about largely by Italy's disgust at the French conquest of Tunis in 1881, a place she had marked down for her own. Bismarck's policy of encouraging French colonisation thus justified itself. 'We have to prevent France finding an ally', he once wrote. He had succeeded.

The new Empire. Bismarck showed as much skill in his policy inside Germany as in his international diplomacy, though his success was less unbroken. His self-restraint here was nearly as great as and perhaps even more remarkable than in foreign affairs. He was forced to bargain with other countries, for they were independent powers; he could simply have given orders to Germany, for here he was dealing with subjects.

The new Empire was really the North German Confederation of 1866 enlarged to include the whole country. It was made up of four kingdoms—Prussia, Bavaria, Würtemberg and Saxony—eighteen Principalities and Grand Duchies, three free cities—Hamburg, Bremen and Lübeck—and the provinces of Alsace and Lorraine. Each of the states had its own separate government for local affairs and its own royal family. The King of Prussia was hereditary President of the Union with the title 'German Emperor'. The central government under the Emperor's control dealt with defence, foreign affairs, trade, railways, posts and telegraphs, coinage and customs. There was a parliament of two houses: the Reichstag, a democratically elected assembly, and the Bundesrath, to which each state government sent its delegation in proportion to its population. The Bundesrath met in secret. In theory every part of the Empire had a voice in its business, but in practice Prussia was supreme, not only because her king was the Emperor, but because her Minister-President was nearly always the Imperial Chancellor, the head of the federal government. Moreover, he was responsible to the Emperor and not to the Reichstag or the Bundesrath. Bismarck was Germany's absolute ruler as long as he and the Emperor remained friends—and they had known each other far too long for them to be likely to quarrel now.

In a country which for so many centuries had been no more than 'a geographical expression', centralisation under Berlin was bound to be unpopular. Bismarck realised this and did his best, by minor concessions, to make it as little burdensome as possible. He agreed, for instance, that Bavaria should control her own army in time of peace, that she should be consulted in foreign affairs, and that she and Würtemberg should manage their own post office. Bismarck's policy was not to use force unless it was necessary. He preferred people to agree with his policy. He was always ready to let Germany have Prussia's way—but it had to be Prussia's.

The Kulturkampf. It was not always plain sailing. In his home policy Bismarck had two difficult struggles—the first with the Roman Catholic church in the 1870's and the second with socialism in the 1880's. Bismarck as a typical Prussian Junker, brought up in Lutheran ways, had no love for the church of Rome. The new Empire by bringing in Bavaria greatly increased the Roman Catholic population with which Prussia had to deal. The problem of the relation of church and state became a serious one. Protestants were angered by the new doctrine of Papal infallibility which Pius IX had proclaimed in 1870. It was not only Protestants who were furious. Many of the most distinguished German Roman Catholics had opposed the new doctrine. They were now excommunicated and attempts were made to drive them out of the universities. They established a small independent body known as the Old Catholic church, which is in communion with the Church of England.

The clerical control of education in the Roman Catholic districts appeared to be a danger to the independence of the state. The Pope seemed to Bismarck to be attempting to dictate to Germany. He was determined to stop this. Remembering how in the middle ages a German emperor had been forced to go barefoot through the snow to Canossa to beg the Pope's pardon, Bismarck declared 'to Canossa we shall not go, either in the flesh or in the spirit'. At the same time the Roman Catholics were bitter at the thought that they were in a minority in Germany owing to the exclusion of Austria from the German Confederation in 1866. They belonged to the Great German party which wanted to see Austria included in Germany. They had established their own political party—the Centre party—and they were determined to oppose any changes which Bismarck might make in a Protestant direction. Both sides, then, were ready for the fight which came to be known as the Kulturkampf or fight for civilisation.

Between 1873 and 1875 Falk, the Prussian Minister of Education, secured the passing of the 'May Laws', which were designed to bring the Roman Catholic church under the control of the state as the Protestant churches had always been. The Jesuits were expelled, the other religious orders were forbidden to teach, education was put under lay supervision and all ecclesiastical appointments were to be approved by the government. But Bismarck had overestimated the strength of the Old Catholics, whose secession from Rome had encouraged him to undertake the struggle. The enthusiasm with which Protestants had rushed to support the Old Catholics only made most German Roman Catholics more loyal to the Pope than they had been before. A period of bloodless persecution followed, during which several hundred priests were sent to prison for disobedience, among them the Cardinal Archbishop of Posen. Then in 1878 Pius IX died. His successor, Leo XIII, was ready to meet Bismarck half-way. The resolute Falk resigned, and in 1879 most of the 'May Laws' were repealed; but the Jesuits were not readmitted, and the schools were kept under lay control. The Kulturkampf had ended in a draw, with the advantage on the whole in the Pope's favour. But Bismarck had extricated himself from a situation which might have led to a serious defeat for himself. Even if he had persevered and won he could only have weakened Germany. As it was, he had secured some of his points and preserved the unity of the German people.

Bismarck's economic policy. Bismarck needed the political support of the Roman Catholics in 1879 because in that year he changed his economic policy in a way which lost him the help of the liberals, on whom up till now he had relied. A Junker himself, he naturally sympathised with the great arable farmers of eastern Germany who were beginning to suffer from American as well as from Russian competition. He therefore decided to introduce a tariff on imported corn, thus abandoning Germany's relatively free trade policy and seriously annoying the liberal industrialists, who relied on cheap bread to make low wages tolerable.

Besides, Bismarck was glad to come to terms with the Roman Catholics because by 1879 the social problem was acute. In Germany the industrial revolution had been telescoped into a few decades. In the last few years the German people, with their outstanding genius for science and for organisation, had made immense strides in all branches of business, but especially in chemical and electrical industry. In Westphalia, in the Rhineland, and in conquered Lorraine rich mineral resources were exploited to the full by crowded factories and mines. But the speed of this economic expansion was dangerous. Boom periods were interrupted by disturbing slumps. There was one in 1874, caused partly by the inflation which followed the prompt payment of the French war indemnity. There was another in 1878.

The struggle with socialism. These depressions heightened the contrast between wealth and poverty, which was marked in the new manufacturing cities. Socialism had its chance. For some time its rise was hindered by internal disputes, but in 1875 a united Social Democratic party was formed. Its success frightened Bismarck, who in 1878 used an attempt to murder the Emperor as a pretext for a stringent law which severely limited freedom of speech and of the press, and gave the police the right to deport anyone they suspected. To these measures the majority of the German public, including the liberals, offered little resistance. The act remained in force for twelve years, during which time nine hundred people were deported and 1500 imprisoned. It failed, however, to kill the socialist movement, for in 1890 1,500,000 people voted socialist.

Social insurance. But Bismarck was too clear-sighted to be content with a merely negative social policy. All his sympathies were on the side of authority; but, unlike most conservatives of his day, he was logical enough to understand that authority must be exercised over capitalists as well as socialists. It was easier for a German than for an English conservative to realise this. The tradition of the Junkers meant nothing if not that the powers and duties of the state take precedence over any rights of any individual. It was natural then that they should assert the authority of the government over the wealthy new industrialists whose motives seemed to them to be purely selfish. It was natural that Bismarck should sternly suppress the impudent challenge of socialism to class-distinctions, and equally natural that he should himself adopt socialist principles by including the supervision of industry among the duties of the 'paternal' state. Accordingly he introduced a system of insurance for workers covering sickness (1883), accidents (1884) and old age (1889). The cost, as in English unemployment insurance, was shared between the employer, the worker and the state. His work was far from complete. Germany had as yet no equivalent to Lord Shaftesbury's Factory Acts and no provision was made for unemployment, but the fact remains that Bismarck introduced compulsory insurance against the worst evils of poverty more than twenty years before Mr Lloyd George persuaded Great Britain to do so.

33. THE EASTERN QUESTION. II

The Great Elchi: 1842 to 1856

Stratford Canning. In 1808, while Europe was at war with Napoleon, a Cambridge undergraduate was called away for the third time from his books to serve his country as a diplomat. For Stratford Canning, George Canning's nephew, this final interruption to his education was the beginning of his life work. He was sent out to be secretary to the English minister at Constantinople; two years later, at the age of twenty-three, he succeeded his chief. Altogether he spent twenty-two of the next forty-eight years as ambassador in Turkey. He was seventy-two before he resigned the Constantinople embassy in 1858, but to the last he was capable both of outworking and of terrifying his subordinates—one attaché never spoke to him without keeping his hand on the door knob, ready to make good his escape at the first sign of his chief's displeasure. Even at seventy-two Stratford Canning was not exhausted. He lived on in retirement for another twenty-two years, regularly speaking in the House of Lords—he had been made Viscount Stratford de Redcliffe in 1852—and translating 'Little Jack Horner' and other nursery rhymes into elegant Latin verse.

In these days when prime ministers can meet each other personally at a few hours' notice ambassadors rarely make history; in Stratford Canning's time an ambassador anywhere had to use his discretion almost daily; and an ambassador at Constantinople was virtually beyond control. In 1815 Foreign Office messengers often took a fortnight to reach Vienna. Farther east travel was slower still. In 1849 an English messenger with particularly urgent despatches rode the 820 miles from Belgrade to Constantinople in five days eleven hours. This was not only a record but easily one.

While Stratford Canning was ambassador, England's Turkish policy was at least as much his as his government's. His influence was due not only to his distance from London but to his personal distinction. He knew more about Turkey than any other Englishman; moreover he knew what he wanted and the Turks knew that he generally got his way. They called him the Sultan of Sultans and believed that 'if what he directed was inconsistent with the nature of things, then possibly the nature of things would be changed by the decree of Heaven'. To the English he was 'the Great Elchi', the ambassador of ambassadors.[1] At the start of his great embassy in 1842 he was allowed what no other ordinary ambassador has ever been granted, the privilege of writing his own 'instructions'. After his death Dean Stanley said in a sermon:

No one could enter into his presence as he sat on what may truly be called his throne at Constantinople...without feeling that they had seen a king of men. No one could hear the name of Stratford Canning named...without feeling that...the honour of England was safe in his incorruptible integrity, in his magnificent liberality, in his unshaken firmness. No one could hear his influence spoken of by Christian or Mussulman...without feeling that each man knew that he was a terror to evil-doers which no one could confront with impunity, a refuge for the desolate and oppressed which no one could seek in vain.

England's effort to reform and defend Turkey. Stratford Canning's policy had two fixed principles. He wanted to reform the Turkish Empire so that it might become a land fit for Christians; and he wanted to prevent Russia establishing a protectorate over it which would threaten English interests in the Mediterranean and in India. He had no illusions about the Sultan's government. When he first went to Turkey he wrote home to say 'it is rotten at the heart; the seat of corruption is the government itself'. Fifty years later he was of the same opinion, telling the Foreign Secretary of the 'continued weakness of the Ottoman authorities and their practical adherence to a system of corruption, injustice, oppression and religious domination'. The root trouble was the misgovernment of a large Christian majority in European Turkey by a vicious Mahometan minority. The governors of every province were Moslems, the army was entirely Mahometan, torture was freely used, no Christian could give evidence in a law court so that there was no redress against Mahometan oppression, all the taxes were paid by Christians, and the penalty for a Mahometan who turned Christian was death.

Between 1842 and 1852 Stratford Canning was able to get most of these abuses put right on paper. His first step was to force the Sultan to get rid of a reactionary minister in 1845. The new, more

[1] Elchi = 'ambassador' in Turkish.

liberal minister co-operated with Stratford Canning but, although the law was changed, the administration remained as bad as ever. Stratford Canning knew perfectly well that western influence would never reform Turkey while it was confined to Constantinople; it would have to be extended to every province before Turkey could become civilised. His efforts to do this, however, were hampered by his duel with Russia. He could not effectively bully the Sultan into carrying out the reforms, because the Sultan knew that in the end, however badly he governed, England would defend him against Russia.

Russia and the Orthodox church. Russia had a triple interest in Turkey. She coveted her territory and she sympathised with her subjects both as fellow Christians and as fellow Slavs. The Christians of Turkey in Europe, like the Christians of Russia, were Orthodox. There have been two great schisms in Christian history—the first in the eighth century divided Christendom into an eastern or Orthodox church and a western or Catholic church; the second in the sixteenth century subdivided the western church into a northern or Protestant half and a southern or Catholic half. England, as a Protestant country, had no objection to Russia's interest in the Orthodox Christians of Turkey, for Protestants in their hatred of Catholicism had come to tolerate the Orthodox church although they still considered it superstitious. What England objected to was the political expansion of Russia, a country which not only threatened English interests geographically but which practised the most bitter tyranny against liberal opinion.

The Tsar Nicholas of Russia was the man who had massacred Polish nationalists and destroyed the last relics of Polish independence. It was he who in 1849 had sent his army to help the Emperor of Austria re-conquer Hungary. The Christian subjects of Turkey might hope to see their religion respected under Russian protection, but they could not hope for any political freedom. Indeed when it was a question in 1848 of suppressing a liberal revolt which had taken place in a semi-independent Turkish principality under Russian protection, it had been the Sultan, under Stratford Canning's influence, who had stood for moderation, and the Tsar who had been in favour of brutality. Nicholas offered the subjects of Turkey only religious freedom; Stratford Canning offered them religious and political freedom.

Causes of the Crimean War. Nicholas himself was not anxious for a war with England—at any rate for the moment. He would have preferred a thieves' bargain by which he could take Constantinople while England annexed Crete and Egypt. In 1844 he visited London to try to arrange such a partition, but his proposals were rejected. When Russia alone remained unshaken by the revolutions of 1848, he felt strong enough to take the law into his own hands. In March 1853 the Tsar demanded that the Turkish government should definitely recognise his exclusive right to protect the Sultan's Christian subjects. Stratford Canning warned the Sultan that he would be a mere Russian vassal if he accepted, and he accordingly refused. Russia replied in June by invading the principalities of Moldavia and Wallachia, the northern outposts of the Turkish Empire. An attempt by Austria and others to mediate was wrecked, and by October 1853 Turkey and Russia were formally at war. Russia could count on winning this war if she were left alone, but neither England nor France would agree to this, and so six months later they joined Turkey and sent expeditionary forces to the Black Sea.

England went to war because she thought British interests in India and on the way to it were threatened. France joined her partly because Napoleon III saw an opportunity of securing glory for France, and partly because an alliance with England might safeguard his new dynasty. England had ruined Napoleon I; it should help Napoleon III. These were not the only reasons which moved the Emperor. Napoleon greatly disliked the Tsar, who despised the revolutionary origin of the Second Empire and whose predecessor had humiliated France in the Moscow campaign of 1812. A war against reactionary Russia would please French liberals; a war against Orthodox Russia would please French Catholics. France had long claimed to be the protector of the Catholic subjects of the Turkish Empire, a small minority in Palestine. The rival Catholic and Orthodox guardians of Christianity had begun to quarrel in 1852 over various sacred shrines in Palestine. Nicholas I held that these holy places were the property of the Orthodox church; Napoleon III held that they belonged to the Roman church. Here was a convenient subsidiary excuse for French intervention.

Austria's position. The future of Turkey concerned Austria at least as closely as any other European power—not only did she adjoin Turkey but many of her subjects were of the same races as the Sultan's Christian subjects. Nicholas hoped that Austria would repay him for his help against Hungary in 1849. On the other hand the Austrians could not afford to see Russia advance across the lower reaches of the Danube, because their economic life depended on keeping its navigation free. They did not want Russia, the militant protector of the Slavs, to liberate the Balkans from Turkey, because their own subject races would then be encouraged to rise. If gratitude should have made Austria an ally of Russia, policy would have made her an ally of England and France. In 1854, however, Austria was too weak to do anything but remain neutral, thus losing the good-will of both sides. She dared not fight lest, while her hands were full, Prussia should attack her position in Germany. She had to leave Great Britain and France to check Russia for her.

Austria's forced neutrality gave one of her most dangerous enemies a chance to gain the good-will she had lost. In January 1855 Cavour, the prime minister of Piedmont, sent a few thousand troops to join the Allies. His tiny country had no quarrel at all with Russia, but he realised that, when Italians and Austrians next came to blows, Great Britain and France would remember which nation had voluntarily helped in the Crimea, and which had stood apart.

'Generals Janvier and Février.' 'The Crimean War' is not quite an adequate title, for there was fighting to the west of the Black Sea and in the Baltic, but the principal attack by the allies was directed against the naval base of Sebastopol, in the south-west corner of the Crimean peninsula. The whole campaign was mismanaged from the start. The allies never had enough men to surround Sebastopol, which the Russians were able constantly to reinforce. In any case Sebastopol was an immensely strong defensive position thanks to its elaborate trench defences. The Tsar boasted that 'Generals Janvier and Février' fought for him, and certainly the bitterly cold winter imposed terrible hardships on invaders unused to the Russian climate. A generation of peace and economy had left the British army authorities utterly incompetent to provide adequate transport and supplies. Men froze and starved and died of cholera by thousands until the authorities were at last forced to remedy the situation after the damning exposures of William Russell of *The Times*, first of modern war correspondents. An efficient nursing service was created, to the scandal of the high command, largely by the ruthless vigour of one high-spirited, sharp-tongued woman, Florence Nightingale. When she first reached her hospital at Scutari the death rate was 42 per cent; she reduced it to 2·2 per cent.

The apparent stalemate of the war was broken when in 1855 'General Février turned traitor' and killed Nicholas I. His successor Alexander II was anxious to bring the war to an end. In September the Russians abandoned Sebastopol and the allies felt that honour was sufficiently satisfied. In February 1856 a peace conference met in Paris.

The Treaty of Paris. The nation which suffered most from the war was not defeated Russia but neutral Austria. Her loss of prestige encouraged Bismarck a few years later to plan her expulsion from the German Confederation. There was a more immediate misfortune. Not only was Piedmont represented at the peace conference; but Cavour was able to use this opportunity for a discussion of the Italian problem, which was strictly irrelevant to the business of the conference. Napoleon III gained in prestige from the war. The fact that the great powers of Europe came to his capital to make peace impressed his subjects considerably. England, too, secured her immediate object. Russian penetration into Turkey was checked for twenty years.

In the long run, however, the Treaty of Paris was as inconclusive as the war. It 'neutralised' the Black Sea by forbidding warships of any nationality to enter or leave it. It was to be open to merchantmen only. This was a severe blow to Russia, which was not even allowed to maintain a Black Sea fleet at all, and a corresponding gain for England, which was relieved from all anxiety about the eastern Mediterranean. On the other hand the rules of naval warfare were revised in a way which limited England's traditional policy of enforcing a strict blockade against neutral shipping. The navigation of the Danube was placed under an international commission, and the two principalities of Moldavia and Wallachia were made more independent of Turkey than before. Moldavia was enlarged by a slice of Bessarabia taken from Russia. Finally, the treaty guaranteed the integrity of Turkey, and deprived Russia of any right whatever to interfere in her internal affairs.

The way ought now to have been clear for Stratford Canning to take up again his task of forcing Turkey to carry out the reforms which he had extorted from the Sultan before the Crimean War. He implored the British government to do all it could to secure the united influence of the liberal European powers in favour of reform, but England and France felt that they had done enough by removing the Russian danger. They were content, when the Sultan promised to reform, to take his word for it and leave it at that. The Treaty of Paris not only provided no machinery for the continuous reforming pressure which Stratford Canning knew to be necessary, but even explicitly renounced all rights of intervention by European powers in Turkish internal affairs. The Sultan knew that he was once again free to rule as badly as he liked, and acted on his knowledge. Stratford Canning had succeeded in his less important aim, in checking Russian expansion; but the very efforts needed to achieve this limited success had so wearied Europe that she denied him the power to carry out the more important task of Turkish reform. The Treaty of Paris settled nothing permanently because it left Turkey as much 'the Sick Man of Europe' as she had been when Tsar Nicholas coined the phrase in 1853.

34. THE EASTERN QUESTION. III

The Congress of Berlin: 1856 to 1878

Russia's advance in Asia. The Crimean War and the Treaty of Paris, which closed it, had checked Russia's advance south-westwards into the Balkans towards Constantinople. It was natural that she should turn her attention to the south-east, and there during the reign of Alexander II she made rapid progress into the No Man's Land of central Asia, where fierce half-nomadic tribes, Mongol by breed and Moslem by creed, plundered and marauded. A great power living next door to barbarians is almost forced to embark on a 'civilising mission' to teach its troublesome neighbours better manners. The Russian advance into central Asia followed the three customary stages of penetration, conquest and settlement. Between 1859 and 1864 the Tsar's forces conquered the Caucasus. In 1868 they entered Samarkand. Five years later all Turkestan, Bokhara and Khiva were theirs, and they had reached the banks of the Oxus, the northern frontier of Afghanistan.

Englishmen were convinced that India was in danger and that all the gains of the Crimean War had been lost. Russia within reach of the Khyber Pass was more of a danger even than Russia at Constantinople. It was barely fifteen years since the Indian mutiny, and it was easy to reduce England to a state of panic about India. The use of large-scale maps would have shown that the danger really was small, but newspapers and atlases necessarily use small scales for their plans, and so amateur strategists shivered when they saw that Russia was only an inch or so away from India.

Worse was to come. By the middle of the 1870's it looked as if we should have to face Russia both on the Khyber and in Constantinople. The preface to this new chapter of dangers was written in 1870 when Russia took advantage of the Franco-Prussian War to get rid of the clauses of the Treaty of Paris forbidding her to keep a fleet in the Black Sea. The disarmament clauses of that treaty proved as unenforceable as the disarmament clauses of the later Treaty of Versailles. In both cases the method used to get rid of them was the one-sided denunciation of a treaty.

Germany, Russia, Austria. In defying international law Alexander was able to rely on the support of Bismarck, whose chief anxiety after the Franco-Prussian War was to prevent France gaining an ally who would help her to reconquer Alsace-Lorraine. This meant that he must keep on good terms with both Russia and Austria, and somehow prevent their quarrelling among themselves. In 1873, therefore, he took the lead in forming a triple alliance between Germany, Russia and Austria, called the 'Dreikaiserbund' or 'League of the three Emperors', which in one form or another continued for nearly a generation until it was eventually shattered by the incompatible ambitions of Russia and Austria in the Balkans.

Gladstone, Disraeli and the Bulgarian Atrocities. The 'Dreikaiserbund' had, however, hardly been formed before it was temporarily interrupted as a result of a Christian rebellion against the Turks. Between 1874 and 1876 the Turks suppressed this rebellion in Bulgaria by a policy of deliberate 'frightfulness', letting loose savage irregular troops on the whole Christian population. At the town of Batak several thousand men, women and children were burnt alive in the church, which was sprayed with tar and then set alight.

The news of these atrocities produced a first-rate political crisis in Europe and in England. For generations English policy had been to protect Turkey whenever possible while trying to persuade the Sultan to reform. Our weakness in not insisting on international control of Turkey after the Crimean War had resulted in the Bulgarian Atrocities. They in turn persuaded the leader of English liberalism, William Ewart Gladstone, to abandon our traditional pro-Turkish policy. When he heard the news in his retirement at Hawarden he forgot he was seventy, forgot the crippling pains of lumbago and embarked on a whirlwind campaign against the Turks. He wrote a pamphlet denouncing the Sultan and all his officers: 'Their Zaptiehs and their Mudirs, their Bimbashis and their Yuzbashis, their Kaimakams and their Pashas, one and all, bag and baggage, shall, I hope, clear out from the province they have desolated and profaned.' Gladstone was one of the few Christian statesmen who have tried to interpret politics in terms of their religion. He was doing so now. Our commercial and strategic interests in the Mediterranean seemed to him irrelevant. It was a simple case of right and wrong.

But Mr Gladstone was in opposition, and the prime minister, the Earl of Beaconsfield, took a different view. To him Mr Gladstone was just 'a sophisticated rhetorician, inebriated with the exuberance of his own verbosity'. The real problem, he maintained, was the effect of the Bulgarian disturbances on the balance of power in Europe. He saw at once that here was the old issue of the Crimean War over again. He suspected that Russia would use the general indignation against the Turks as an excuse for renewing her interference in the Balkans. He was determined that, persecution or no persecution, the Tsar should not establish himself in Constantinople disguised as a crusader. British opinion veered uncertainly between Gladstone and Beaconsfield as the course of events emphasised first the sufferings of the Christians, and then the undesirability of their rescue by Russia.

Beaconsfield's forbearance seemed to be justified by an important development in August 1876. The Turks themselves were now so exasperated by the corruption and incompetence of their government that there was a revolution in Constantinople, which made Abdul Hamid Sultan. The powers had sent representatives to Constantinople to devise common measures for policing the Turkish dominions and protecting the minorities. The new Sultan, however, evaded these proposed restrictions by appointing a liberal Grand Vizier, and by granting a model constitution—on paper. This was enough to satisfy most of the powers, and the international conference broke up without extracting any guarantees of reform. Abdul Hamid at once dismissed his liberal minister, who took refuge in Italy.

The Russo-Turkish War. This was altogether too much for Alexander. He saw that the conversations in Constantinople had done nothing to safeguard the victims of Turkish misrule in the Balkans. He determined to secure by arms what Europe had failed to get by diplomacy. In April 1877 he declared war on Turkey. The campaign consisted of a laborious advance on Constantinople by the Russian army in alliance with Montenegro and the semi-independent states of Serbia and Rumania. For five months the allies were checked by the obstinate gallantry of a Turkish general, who held out in the town of Plevna until he was wounded and taken prisoner in a final sortie. After the fall of Plevna the war was soon over. In January 1878 Adrianople was taken, and soon the Russians were at the gates of Constantinople.

San Stefano and the isolation of Russia. The Sultan now sued for peace, and Alexander II dictated the Treaty of San Stefano in March 1878. Montenegro, Serbia and Rumania were promoted from being Turkish vassal states to full independence and gained some territory. Russia took most of Turkish Armenia. A great Bulgarian state stretching from the Black Sea to the Aegean was set up under the protection of Russia. Russian influence was for the time being supreme in the Balkans.

She had been too greedy. It was bad enough that Russia should lurk behind the Himalayas, now she was virtually in the Mediterranean. The late Prince Consort had always said that this would be the danger, and his widow consequently felt quite sure of it. Queen Victoria bombarded Beaconsfield with notes pressing him to take the strongest possible line even if it meant war. '...And the language—the insulting language—used by the Russians against us! It makes the Queen's blood boil!' The public, too, had forgotten the generosity to which Mr Gladstone's Midlothian campaign had roused it. A music hall song—

> We don't want to fight, but by jingo if we do,
> We've got the ships, we've got the men, we've got the money too

—gave a new word, 'Jingoism', to our language.

Beaconsfield, however, had no intention of undertaking a second Crimean war if he could force the Tsar to surrender his gains without fighting. He sent the fleet into the Sea of Marmora, concentrated Indian troops at Malta, and called out the Army Reserve. He did not declare war. He collected sufficient allies to prove to Alexander that the only result of war would be defeat.

The first and most obvious ally was Francis Joseph of Austria, whose Slav subjects were dangerously excited by the triumphal progress of Russia. Then the creation of the immense new Bulgaria blocked any possibility of Austrian expansion towards the Aegean coast. Russian and Austrian ambitions in South-Eastern Europe cut across at right angles. The crisis shattered the 'Dreikaiserbund' and Germany had to decide which partner to support. Bismarck's own view was that the Eastern Question was 'not worth the bones of a single Pomeranian grenadier'. He only wanted to be the 'honest broker' between the various parties. He made up his mind, however, that if he had to choose, he would keep the friendship of Francis Joseph rather than that of Alexander II. It was likely to be more profitable. Pan-Slavism would leave no opening for German influence in the Balkans. As Serbia, Montenegro and Rumania were bitterly disappointed with their small gains at San Stefano, the Tsar found himself isolated, and so reluctantly he agreed to submit the whole matter to an international congress in Berlin.

The Congress of Berlin. He had the sense to give way thoroughly. Before the Congress began he privately arranged with England to give up his plan for a Bulgaria reaching to the Aegean. The Congress in fact had little more to do than recognise a settlement out of court. The actors were more impressive than the drama. Since Vienna in 1815 there had been no conference half so distinguished. From England came old Lord Beaconsfield and his new foreign secretary, Lord Salisbury. Russia sent Gortschakoff, the aged chancellor, lamenting that 'we have sacrificed a hundred thousand soldiers and a hundred millions of money for nothing'. Turkey was represented by a Greek doctor, the Sultan's physician; and France by M. Waddington, who had been at Cambridge and rowed in the Boat Race. Their host and chairman was Bismarck.

The Congress was Bismarck's triumph. 'What with deafness, ignorance of French and Bismarck's extraordinary mode of speech, Beaconsfield has the dimmest idea of what is going on, understands everything crossways and imagines a perpetual conspiracy', wrote Lord Salisbury. All the same Bismarck and Beaconsfield were in general agreement. Their understanding began with a long private conversation. 'After dinner', Lord Beaconsfield wrote, 'we retired to another room, where he smoked and I followed his example. I believe I gave the last blow to my shattered constitution, but I felt it absolutely necessary. In such circumstances, the man who does not smoke has the appearance of spying upon the other's words.' The sacrifice won Bismarck's gruff approval: 'The old Jew,' he said, 'that's the man.'

Gortschakoff never had a chance against Bismarck and Beaconsfield. When he insisted that the Turks should leave their Bulgarian frontier unfortified, Beaconsfield ordered a special train to take the British delegation home. The Russians gave way. There was an odd comedy towards the end. By some mistake the British and Russian envoys got their confidential maps mixed. Beaconsfield had the Russian copy, Gortschakoff the British. They argued fiercely but at cross purposes until Bismarck sorted out the maps. At last, in July 1878, the Treaty of Berlin emerged.

'Peace with Honour.' The Bulgaria of San Stefano was reduced to a much smaller area bounded by the Balkan Mountains, south of which the Turks were left in possession. The independence of Serbia, Rumania and Montenegro was fully recognised, but they gained nothing else. Austria was allowed to occupy, though not formally to annex, Bosnia and Herzegovina. Great Britain took over Cyprus on the same terms. 'It was necessary to bring back something, and that something was Cyprus.'

Lord Beaconsfield, indeed, claimed to have brought back more. 'We have brought you back, I think, Peace with Honour', he said. From his own point of view he was justified. He had set out to destroy the Treaty of San Stefano and he had succeeded. By making Bulgaria small he had checked the advance of Russian influence in the Balkans. She had won the war but nothing else.

'Backing the wrong horse'? But was Beaconsfield right in thinking that Russia was still the enemy? The ease with which she had been forced to give up her gains might have suggested that she was hardly any longer a power to be feared. Beaconsfield had not realised how important Germany had become, more important and more dangerous than Russia. It was she, indirectly, who gained most from the Congress of Berlin. From 1878 onwards German influence was more important at Constantinople than British—the Sultan had learned from Gladstone's Midlothian campaign that British support was no longer the sure defence that it had been in the days of Stratford Canning. A general election might put the enemies of Turkey in power in England. Bismarck's support of Austria against Russia had been invaluable to Francis Joseph, but he knew that he could only keep it if he did what he was told. The occupation of Bosnia and Herzegovina turned out in some ways an occupation by Austria for the benefit of Germany. There was some justification for Lord Salisbury's later view that 'we had backed the wrong horse'.

The Congress of Berlin entirely failed to solve the Eastern Problem. Turkey was not reformed and the small Christian states remained dissatisfied. Russia had surrendered her gains, not her ambitions. The net result was to introduce Germany, a new power, a new complication, into the already complicated Eastern Problem. The Congress of Berlin was followed by a generation of peace in the Balkans, but only because the powers were for a time too frightened of each other to risk another war.

Map 15. RUSSIA AND ENGLAND, 1850–1907

1. This map explains the suspicion with which Russia and England regarded each other in the nineteenth century. The broken arrow (1) in the north-west shows Russia's unfulfilled desire to take Constantinople and become a European power—a desire opposed by England in fear of her communications with India.

2. Arrows 2, 3 and 4 and the heavily shaded territory represent the gradual approach of the Russian and British Indian frontiers after 1850, when they had been separated by about 700 miles at the nearest point.

3. By 1870 Russia had absorbed the fertile country of Turkestan and reached the Afghan frontier due south of Samarkand—the distance between the Russian and British outposts had been halved.

4. This led between 1878 and 1880 to a second and successful attempt to establish British influence in Afghanistan—the first Afghan war (1838–40) had been a ghastly failure.

5. The Russian advance, however, spread westwards and eastwards until it had covered the whole of the northern frontier of Afghanistan. England pushed her territory northwards at the same time, until by 1900 the shortest distance between the British and Russian outposts was barely fifty miles.

6. The next move was rival British and Russian penetration of Persia, shown by arrows 5 and 6.

7. Soon, however, both countries became more frightened of Germany than of one another. The projected Berlin-Baghdad railway was a serious threat to British communications with India, while German influence in Turkey blocked Russia's hope of taking Constantinople. The result was a Russo-British agreement in 1907—a sequel to the Anglo-French Entente of 1904—which ended their rivalry and divided Persia into Russian and British spheres of influence (lightly shaded).

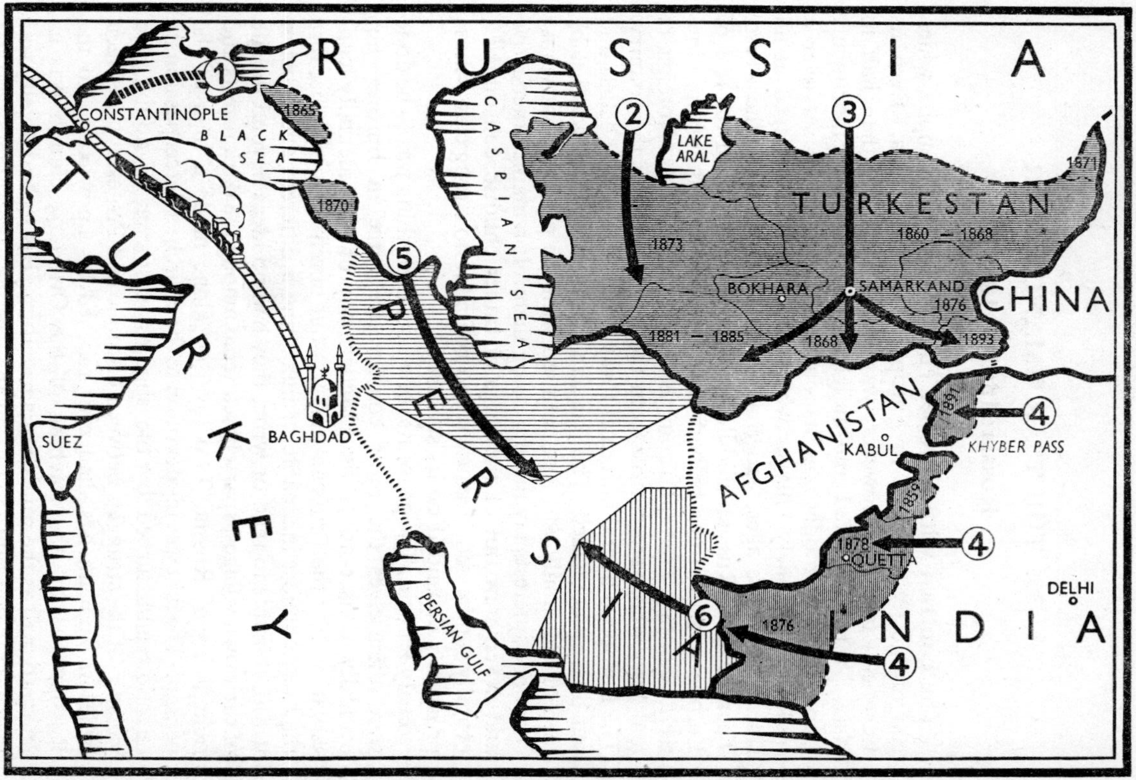

MAP 15. RUSSIA AND ENGLAND, 1850–1907

35. THE TSAR EMANCIPATOR

Russia: 1855 to 1881

The contrast with Europe. The Tsar ruled an empire which geographically was half in Europe and half in Asia. Even its European territory, however, was spiritually more than half in Asia. A tiny cultured and enlightened upper class, thoroughly European and modern in its outlook on life, was set over a multitude of lonely villages whose life, if it had changed at all since the middle ages, had changed in some ways for the worse. There was no Russian middle class like that which in the rest of Europe was the force behind the liberal movements of the nineteenth century. Trade and commerce were in the hands of foreigners and Jews who were cut off from the spiritual life of the country. The European liberal movement was led by men whose beliefs and self-interest coincided. The Russian liberal movement was led by men of old family who sympathised with the peasants but whose economic interests lay in oppressing them. It remained ineffective and never grew out of the stage of rather childish conspiracies, brutally suppressed, because it had behind it no strong class whose prosperity depended on its success.

Holy Russia. The peasants for whom everything was to be done were themselves the most conservative force in Europe—not cynically conservative, like Metternich; or sentimentally conservative, like the Prussian junkers; but unreflectingly so because they were almost incapable of imagining a different and a better world. Unable to read or write, they had no ideas beyond the life of their own villages, sparsely scattered among the dark forests and rich plains of Russia. Two loyalties alone made a man of the peasant. The first was his devotion to the church: a fierce, passionate devotion undisturbed by the slightest suspicion of doubt. No whisper of the struggles between Catholic and Protestant, between believers and atheists, had ever reached his knowledge. To the Russian peasant the teaching of the Orthodox church was as unquestioned as the routine of agriculture. His second loyalty was to the Tsar, the 'autocrat of all the Russias', the anointed of God who was the 'little father' of all his people. Church and throne stood together in the peasant's mind.

Condition of the Peasants. But, if the peasant had two loyalties, he had one grievance. The institution of serfdom was universally hated. Nearly fifty million Russians, about half the population, were serfs. They were bought and sold like slaves, beaten at the will of their owners, taxed and forced to work, not according to any fixed scale, but entirely at the whim of their lord or his steward. 'The negroes on the American plantations were happier than the Russian serfs' was the opinion of one Russian writer. Certainly the English peasantry in the middle ages and the French peasantry at the outbreak of the revolution were infinitely better situated than these unfortunate Russians. But it was because they were conservative, not because they were revolutionary, that the peasants hated serfdom. They regarded it as an innovation, although it was more than two centuries old, for peasants have long memories. At the end of the sixteenth century the migration of peasants from one district to another was hampering agriculture. To stop this restlessness, a decree had been issued attaching them to the same estates for life, a measure which had soon reduced them to slavery.

Pan-Slavs and westernisers. The peasant, then, hated serfdom but had no quarrel with autocracy—he did not want political liberty because he was not interested in politics, nor religious liberty because he did not know that different religious opinions were possible. The more educated section of the upper classes wanted to abolish serfdom for the peasants and to secure political and religious liberty for themselves. In the eighteenth century first Peter the Great and then Catherine II had contrived to cut Russian society loose from its half oriental ways and to attach it to the free-thinking civilisation of France and western Europe. The officers in the armies which had fought Napoleon in Germany and France came back to Russia more than ever convinced of the backwardness of their own country. These progressive officers found sympathetic allies among the higher civil servants, the doctors, the lawyers and university teachers and students. Westernised themselves, they wanted to westernise Russia.

The bishops and monks, the country squires and the less intelligent army officers, however, despised these westernisers as

Russians who were traitors to their country's traditions. Both sections were patriots, but they differed in their opinion as to what Russian patriotism should seek. The first wanted to modernise Russia; the second to expand the old Russia so that it should include or at least 'protect' all Slavs everywhere. Their aim was called Pan-Slavism. The greater number of Slavs outside Russia belonged to the Orthodox church so that the ambitions of statesmen and the prayers of the church went hand in hand. On the single issue of the abolition of serfdom the peasants and the westernisers were in agreement; on every other issue the peasants agreed with the Pan-Slavs.

The Autocrat of All the Russias. In the eighteenth century, when France was a despotism and French influence told in favour of the ruler's supreme power, there had been westernising Tsars. In the nineteenth century, when France was often a republic and its influence on Russia liberal, the Tsars were usually leaders of Pan-Slavism. They were in a position of enormous power, for the Orthodox church was ruled by a Holy Synod, which was little more than a government department, while the army had been successfully 'purged' of liberalism by Nicholas I. The most important instrument of the Tsar's power was, however, the secret police, officially called the 'Third Section of the Chancery'. Its head could arrest, imprison, deport or execute political offenders, without consulting any law court. Both under the Tsars and under the Soviets the secret police have been the real rulers of Russia.

The justification for such a rule in the nineteenth century could only be that it was efficient, that it enabled the army to defend and expand Russia. As long as the Tsar's army was considered invincible he could easily suppress the demands for reform. The Crimean War destroyed this legend of invincibility. Fighting on their own ground against troops who were tortured by the climate, the Russians nevertheless contrived to lose the war. The failure of autocracy to be efficient meant that it could no longer refuse reforms. Even if Nicholas I had lived, he would have had to give way. His son, Alexander II, actually wanted to.

Emancipation of the serfs. His first move was to recall from exile the 'Decembrists', the men who in December 1825 had vainly attempted to take advantage of the confusion which followed Alexander I's death to establish a constitutional monarchy.

Alexander II never intended to grant them a constitution, but he needed their help to bring the empire up to date. He realised that serfdom had to go. He made his plans carefully and in 1861 he was ready. That year he issued the Edict of Emancipation which turned nearly 50,000,000 serfs into peasant proprietors with their own land. Legally, of course, all the land belonged to the serfs' masters, but Alexander arranged that as much of it as was necessary to support the peasants should become their property. The former serf not only gained his land, he gained also freedom to move about freely throughout Russia, and freedom from the arbitrary punishments and taxes which his lord had had the right to impose. The landowners, who now had to pay for their labour instead of getting it for nothing, became keener and more scientific farmers. As one of them said 'formerly we kept no accounts and drank champagne, now we keep accounts and content ourselves with beer'. The serfs on the other hand had to pay extra taxes to the government to pay off the compensation which had been granted the landowners. Socially and personally they were much better off than before, economically their position was probably no better. The peasants still had their grievance.

Three years later Alexander carried his reforms a good deal farther. He thoroughly reformed the law courts, establishing trial by jury and making the judges independent of the government. In the country elected Justices of the Peace were appointed to try smaller offences, and everywhere a right of appeal was granted. It is quite unfair to think of the Russian criminal code as particularly brutal, except in the case of political offences. In some respects it was more humane than our own, which educated Russians thought extremely harsh. In the whole reign of Alexander II up to 1876 there was only one execution. Russian juries were lenient to the point of folly. It is said that one of them once returned a verdict of 'innocent, with extenuating circumstances'. A good deal of local self-government was established at the same time. In each Zemstvo or province a council was set up whose members were chosen by the directly elected smaller district councils. Their duties were much the same as those of an English county council and on the whole they discharged them intelligently and enthusiastically.

Alexander refuses a constitution. The westernisers wanted still further reforms. They had made considerable sacrifices when they agreed to the emancipation of the serfs; they had a right to expect in return their own political emancipation. In 1865, the year after the legal and local government reforms, the nobles of Moscow asked Alexander to set up a parliament. He refused. His refusal opened the second period of the reign, the period when he, like his uncle Alexander I before him, turned away from vague liberalism to the most stringent reaction.

Nihilism. The return of reaction made reformers give place to revolutionaries. Those were the years of the struggle with 'Nihilism', the creed of the rising generations whose more extreme members believed that there was no 'single institution in our present mode of life, in family or in social life, which does not call for complete and unqualified destruction'. That sounds a bleak and barren creed, but its followers did not want to destroy just for the love of destruction. They were individualists who thought that Russians were being smothered by the absolute authority of the state and the superstitious authority of the church. These must be cut away before men could be free to build anything worth having. Its leaders were socialists, but of the old pre-Marxian sentimental type. At first Nihilism was a university movement, drawing much of its inspiration from the scientific discoveries which seemed to be making nonsense not only of the divine right of the Tsar but of religion itself. Later its followers tried to make converts among the workmen of the towns and the peasants. They settled among them, sharing their poverty and teaching them Nihilism in secret.

Then deeds replaced words. There was a period of riots, which were intended to turn into revolutions, but which failed to do so because the towns of Russia were not yet big enough or important enough to upset the government. Russia would have to be more industrialised before it could have a socialist revolution. The secret police had all the cards in their hands; between 1863 and 1874 nearly 150,000 people were exiled to Siberia. The failure to produce a revolution by rioting led the Nihilists to try to bring one about by assassination.

Murder of Alexander II. The last three years of Alexander's reign were filled with murder plots. The long war between the Nihilists and the secret police ended in a stalemate in 1881. The police were powerless to protect the Tsar's life; the Nihilists were powerless to overthrow the Tsar's power. In March that year

Alexander II was driving through St Petersburg, then the capital, when a bomb was thrown at his carriage. It killed some of the escort, but Alexander II, who had survived several previous attempts, was again unhurt. He got out to attend to the wounded. A second bomb struck him. Horribly mutilated, he was brought back to the Winter Palace to die. On the scene of the crime they built a memorial, the Church of the Blood. It was a stern reminder to his successor that Russia was fated to be 'an absolutism tempered by assassination'.

The last sixteen years of Alexander's reign disappointed the hopes of its first nine years, and set the Tsars back on the course of unrestricted, unrestrained personal government which led to their final downfall in 1917. Why did Alexander change his course in 1865? The first reason is that Alexander only appeared to change. His whole reign is really one consistent whole. His subjects were mistaken in him—that is all. He was never a liberal, but only an autocrat in search of a contented people. He would agree to any reforms which did not curtail his own direct personal power. He freed the serfs and reformed local government because those changes strengthened Russia without weakening him. He drew the line at once when the reformers began to suggest that he should give up or share some of his own control over the central government. Even in the last years of political murders, when he was desperately looking for the support of public opinion, he refused to grant a constitution or liberty of speech. On the very day he was murdered he had, it is true, reluctantly agreed to call together a consultative council in which the Zemstvos and the towns would have been represented; but the new body would have had no power and would have satisfied nobody. Alexander II was never more than an enlightened despot—and the day of enlightened despots was past.

There are two other reasons, however, which help to explain the break in the reign. First, the Tsars lived a narrow court life surrounded by a small clique of extremely reactionary officials and grandees. Alexander II inherited Nicolas I's 'set'. He never changed them; but in the early days of his reign he imposed his will on them, while in later years he fell more and more under their influence. This was, partly at any rate, due to the unnerving shock of the great Polish rebellion of 1863. The tragic history of Poland, however, requires a chapter to itself.

POLAND 1772–1914

1. These maps show the various partitions of Poland. Poland proper is the land of the Vistula and its tributaries; but, as the large map shows, Poland in 1772 extended far beyond this to the east. The inhabitants of the territory annexed by Russia in 1772 and 1793 were nearly all Russians or Ukrainians and not Poles. This territory has remained Russian ever since without a break.

2. Russia's share of the third partition was also to a considerable extent non-Polish. The people were partly Russians and Ukrainians, and partly Lithuanians. This territory remained Russian until 1917–1920 when an independent Lithuania was established in the north, and Poland occupied a good deal of the southern strip.

3. The Prussian and Austrian shares of the old Poland of 1772 have had a more varied career. Napoleon put the greater part of them together to form the Grand Duchy of Warsaw (1806–1814). After the Congress of Vienna the Grand Duchy ceased to exist and a considerable proportion of its territory (once Austrian and Prussian) was given to Russia and known as Congress Poland. It enjoyed a limited Home Rule until 1831.

4. The Poland of 1918–1939 was made up of all Prussian and Austrian Poland, plus Congress Poland plus a part of Russia's share of the 1795 Partition. The 1939 partition leaves nearly all the Poles inside Germany and gives the Russians and Ukrainians to Soviet Russia.

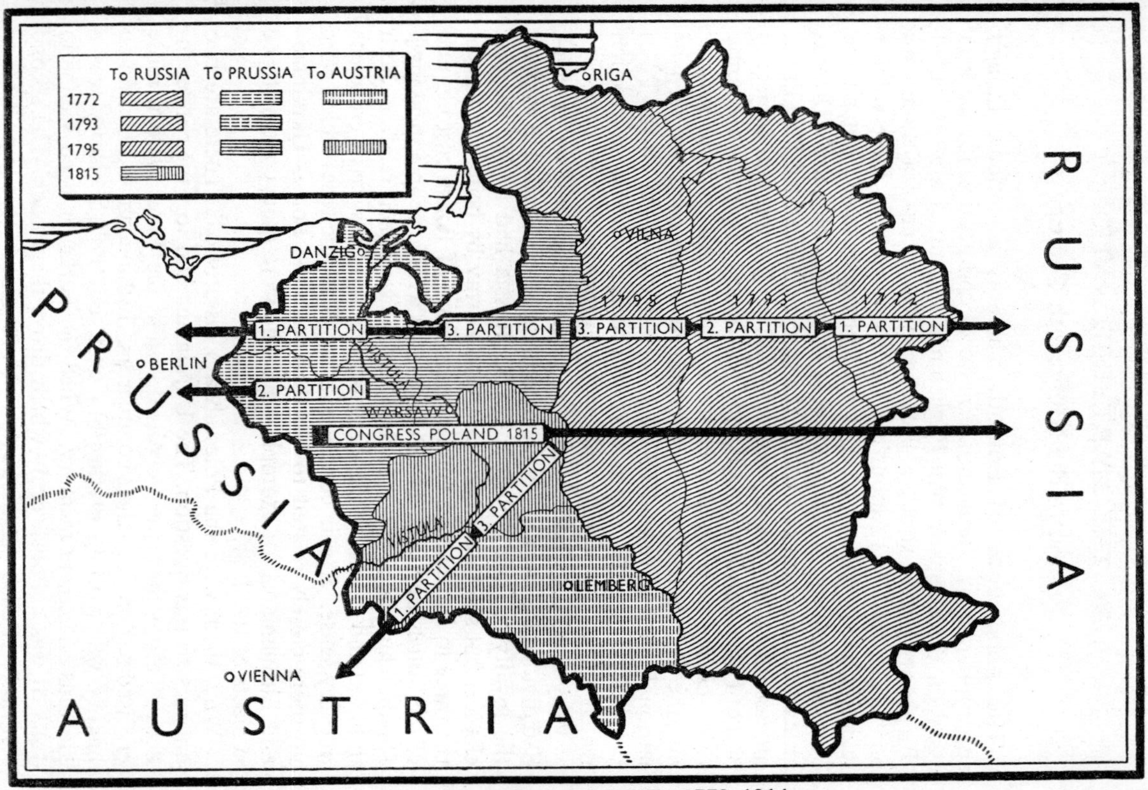

MAP 16. POLAND, 1772–1914

36. THE MARTYRED NATION

Poland in the Nineteenth Century

The persistence of Poland. There is a famous, though apocryphal, story about an international competition for the best essay on the elephant. An Englishman submitted a work called 'Elephants I have shot', a Frenchman contributed a discussion on 'L'Eléphant et l'Amour', a German sent in a three-volume scientific study called 'Über das Elephantenwesen', but all the Poles wrote about 'L'Eléphant et la Question Polonaise'. There is at least this amount of truth in the story: no educated Pole in the nineteenth century could long keep off the all-important subject to him of Polish nationalism. Poland had ceased to exist as a separate state, having been divided between Prussia, Austria and Russia; but the Polish nation was determined at all costs to keep its national spirit alive and to work for the day when all the Poles should be reunited in one independent state, a national resurrection which had to wait for the end of the European War in 1918.

Why the Poles hated Russia. None of the three states which had divided Poland among them permanently succeeded in winning the loyalty of their Polish subjects. Russia had by far the best chance, but she threw her opportunity away by being the first of the three to introduce a ruthless repression of Polish customs and of the Polish language. The Poles are a Slavonic people speaking a Slavonic language, and thus the Russians were their relations both in blood and speech, while the Austrians and Prussians were aliens in every sense. Russia contained many Slavonic races which were no more closely related to her than were the Poles—Ukrainians and Lithuanians, for example—but who were quite content to be subjects of the Tsar.

Why, then, were the Poles always dissatisfied? There are two answers to this question. The Ukrainians had lost their independence before they were civilised, while the Poles were a more civilised nation than their conquerors. Lithuanian and Ukrainian history was a thing of the dim past, but Polish history was a glorious and recent memory. When the Turks besieged Vienna in 1683, it was a Polish army which drove them back. The second reason for Polish dissatisfaction was religious. It is impossible to understand the nationalism of eastern Europe without understanding its religious nature. Pan-Slavism and the Orthodox church went hand in hand; but the Poles were not Orthodox. Once many of them had been Protestant, but the re-conquest of Poland had been the greatest achievement of the Counter-Reformation led by the Jesuits. In the nineteenth century there was no more ardent Roman Catholic nation in Europe than Poland, except possibly Ireland, whose history is rather like a simplified version of Polish history—simplified because the Irish were under the rule of one foreign country instead of three. The fact that the Russians were Orthodox and the Poles Catholic was sufficient to make an understanding between them so difficult as to be almost impossible.

Policy of Alexander I. In the years immediately after 1815, however, it seemed just possible that the Poles might settle down happily in Russia. The great majority of Poles were inside Russia, and the Tsar Alexander I seemed prepared to give the provinces he had gained at the Congress of Vienna, 'Congress Poland', a generous measure of home rule. The Polish army was kept distinct from the Russian army, the Polish language was recognised, the religion of the Poles respected and a Polish parliament was established, though the Tsar's ministers were not responsible to it. Poland enjoyed therefore much more freedom than Russia, and much more freedom than she had had under the partitions of the eighteenth century. The Poles in Russia were not only more free than the Russians, but also more free than the Poles in Prussia or in Austria: an important diplomatic point, for this strengthened Russia by securing Polish loyalty, and weakened Austria and Prussia by causing Polish disaffection.

Alexander, however, retained the government under his own personal control and so, when in 1819 he fell into line with Metternich and ceased to sympathise with liberals, he ceased also to approve of the concessions he had made to Poland. That year he imposed a censorship of the Polish press, the next year he suspended the Polish parliament or Diet for five years because of its criticism of his government. Alexander, as Byron said, had

> ...no objection to true liberty,
> Except that it would make the nations free.
> How nobly gave he back to Poles their Diet
> Then told pugnacious Poland to be quiet.

The Revolution of 1830. The Poles suspected that Alexander's suspension of their constitution would be followed by its abolition. The succession in 1825 of Nicholas I, who never had an instant's patience with liberalism, convinced them of it. They began to prepare for a war of independence. In 1830 they acted. The French revolution of that year, which set Louis Philippe on the throne, encouraged them to rise. The Tsar's proposal to send their army to restore Charles X of France forced them to do so. Not only did they sympathise with the French revolution, but they knew that once the Polish army was out of Poland the Tsar would be able to abolish the constitution without fear of trouble. The Poles chose as their leader an old and incompetent general who had once served Napoleon in a subordinate capacity. He negotiated with Nicholas, who was ready enough to talk, but only until he was ready to act. At the beginning of 1831 the Russian armies invaded Poland, in September they entered Warsaw and the rebellion was over.

The foreign intervention, on which the Poles had counted, had not taken place. England and France sympathised but did nothing. Austria for a moment thought that one of her Archdukes might become king of Poland, but soon thought better of it; and Prussia, was entirely hostile from the beginning. Nicholas capped his victory by abolishing the Polish constitution, the Polish army, and, as far as he could, the Polish language, which was forbidden for official use. Russian civil servants were sent to take over the whole work of government. The best Polish patriots took refuge in France. For twenty-five years the work of Russification went on without a break.

Alexander II and the Rebellion of 1863. When Alexander II became Tsar in 1855 he called back not only the Decembrists, the Russian liberal exiles of 1825, but also the Polish refugees of 1831. In Poland as in Russia Alexander was prepared to make some reforms, but not nearly enough to satisfy his more politically minded subjects. No Polish gentleman would be content with less than the constitution of 1815; many would acept nothing but complete independence. Since Alexander would only give them a few minor reforms, the Poles prepared for a revolution. The government knew that it was coming and attempted to prevent it by calling up the likely rebels as conscripts, but most of them escaped to the woods where they formed armed bands in preparation for the revolution which broke out in January 1863.

The prospects of success were infinitely worse than in 1830. Prussia again was hostile—Bismarck seized the occasion to conclude an alliance with Russia—Austria lukewarm, France and England sympathetic but ineffective. They offended Russia by protesting, but they did nothing else to help the Poles. In 1830 the Poles had had an army of their own; in 1863 they had nothing but an efficient body of conspirators. For over a year a fierce guerrilla war was carried on with the greatest courage but without hope of success. The country was organised under a secret committee of five whose names were unknown even to their immediate subordinates. Poland loyally obeyed their orders as long as it was physically possible to do so, but by March 1864 resistance to Russia was no longer possible. Another wave of refugees fled to France, while in Poland the policy of Russification was carried still further. This time the power of the Roman Catholic church was attacked by a wholesale suppression of monasteries, while the property and position of Polish nobles was also lessened. The Edict of Emancipation of the serfs, issued by Alexander in 1861, had been confined to Russia. Now it was extended to Poland in an effort which was partially successful to drive a wedge between the Polish nobles and the Polish peasants.

Poland turns to Austria. The effect of the two Polish rebellions of 1830 and 1863 and of the repression which followed them was to divert Polish sympathy away from Russia and towards Austria. The Poles loved none of their masters, but they seemed to get better treatment from the Austrians than from the other two. At least they were allowed the free use of their own language. Russia was plainly impossible and Prussia almost as bad. During both revolutions Prussia had supported the Tsar, while under Bismarck and his successors a policy of Prussianisation was carried on which was the counterpart of the Russification across the frontier. German was made the only official language in Prussian Poland, and great though unsuccessful efforts were made to induce Germans to settle in the Polish provinces and replace the Polish landowners and small farmers. Austria, therefore, was left to benefit by the support of the Poles. Galicia, the Polish province of Austria, was in 1867 given its own Diet or local parliament, while Poles remembered with gratitude that in 1830 and 1863 Austria had not been entirely unsympathetic.

In the great war of 1914 Austria reaped the benefit of her liberal attitude. The Tsar appealed to Polish patriotism: 'Poles! The hour has struck in which the sacred dream of your fathers and forefathers may find fulfilment....May the nation reunite into one body under the sceptre of the Russian Emperor. Under this sceptre Poland shall be reborn, free in faith, in language, in self-government.' The Poles, however, had had enough experience of Russian promises, and it was for Austria's service that the great patriot, Joseph Pilsudski, raised his Polish legion. Modern Poland was, it is true, established as an independent state by the allies after their victory, but that was only possible because the Tsar and his government had been swept away by the Communist revolution of 1917.

Diplomatic consequences of the Polish Question. The two great Polish revolts of the nineteenth century had an immensely important effect on the European balance of power. French sympathy with Poland, dating from Napoleon's Grand Duchy of Warsaw, was continuous throughout the century and led to a long quarrel between France and Russia, a quarrel which helped to leave France friendless in 1870, an easy victim for Bismarck's policy. This quarrel was not ended until the twentieth century. The unsympathetic attitude of Prussia and Russia towards the Poles was the basis of a firm friendship between the two countries which freed Bismarck from any fear about Prussia's eastern frontier. If he had not been able to count on Russian neutrality, he would not have been able to fight either the Seven Weeks' War or the Franco-Prussian War. Austria's tolerant attitude towards her own Poles and her tendency to encourage Russian Poles was one of many reasons for the bad relations between the two empires which was a constant factor in nineteenth-century politics. The ghost of Poland was never laid. It was one of the most persistent and troublesome features of the international situation between 1815 and 1914.

APPENDIX

Spain since 1846

Isabella's reign, after the settlement of the Spanish Marriages problem (see p. 146), was in theory constitutional, but was marked in fact by a series of military risings led by five different generals, who seized power in turn. Isabella was disloyal to each and finally the so-called liberals decided to get rid of her. She was expelled by the army and navy in 1868. The two leading generals looked for a suitable king and the throne was offered to a Hohenzollern prince (see p. 227). After his refusal it was accepted by Amadeus of Savoy, a good man but one who knew no Spanish. His principal backer was murdered on the day of Amadeus' arrival in Spain, and in 1872 Amadeus resigned.

There was then a short-lived republic, which had four presidents in one year, after which Isabella's son, Alfonso XII, was made king by the army. He might have made a good king, but unfortunately he died young in 1885 and his widow became regent for his infant son, Alfonso XIII. During her regency the last remains of Spain's empire in the New World (Cuba, Puerto Rico and the Philippines) were lost in the Spanish-American war, 1898. Alfonso XIII came of age in 1902. He reigned as a nominally constitutional king until 1923—Spain remained neutral in the war of 1914—when General Primo de Rivera established a dictatorship with the approval of the king. This was partly a result of the constant defeats which Spain was experiencing from the Moors in Morocco. The dictatorship lasted until 1930–1 when a revolution led to the overthrow of both dictator and king and to the substitution of a republic, which ruled all Spain until the outbreak of the civil war in 1936. As a result of this war the republic was overthrown in 1939 and Spain became a military dictatorship under General Franco. His victory was partly due to help from Germany and Italy.

PART FIVE

THE AGE OF IMPERIALISM

37. THE NEW IMPERIALISM

Patriotism and economic imperialism. For the greater part of the nineteenth century the history of Europe was largely self-contained. The nations were engaged in establishing their right to become nation-states, and as a result of the six wars[1] which filled the years between 1854 and 1878 two new states, Italy and Germany, were added to the number of great powers. In the 'eighties the European states carried their rivalries beyond their own continent, and the whole world became for a time an extension of Europe. The old rivalries on the Rhine and the Danube, the North Sea and the Mediterranean were not forgotten. They continued to poison European relations, but they were now reinforced by new disputes about the Congo and the Yang-Tse-Kiang, the great lakes of Africa and the rivers of Mesopotamia.

This colonial competition was the result of two principal factors. The first was the fact that the nation-states had nearly reached their natural limit of expansion in Europe and had quite reached a position of stalemate. Further expansion of Italy, Germany or France to bring the remaining Italians, Germans or Frenchmen within the national frontiers would have meant a general European war. No prudent statesman was prepared to take so great a risk for so comparatively small a benefit. It was impossible then for the nation-states to grow larger in Europe, but they were not satisfied with their present size. This was the first reason which made them turn aside from Europe to seek new colonies in lands which were still backward or uncivilised, where large gains in territory could be acquired at a relatively small risk of a European war.

More important perhaps than this sentimental reason for the new imperialism, which was supplementing the old nationalism, was the practical desire of business men for raw materials to supply the new factories, and for fresh customers to buy the goods which these factories turned out. A simple illustration will show how

[1] The Crimean War, 1854–6; the Italian War, 1859; the Schleswig-Holstein War, 1864; the Seven Weeks' War, 1866; the Franco-Prussian War 1870–1 and the Russo-Turkish War, 1877.

these various desires worked together. The invention of the internal combustion engine at the end of the century made the motor-car possible. The motor-car needed rubber for its tyres and oil for its petrol—both largely non-European products. Railways were needed to open up the countries where rubber and oil were found, and so the heavy steel and engineering industries, which had come near to exhausting the European market as far as fresh construction was concerned, found a new outlet for their goods. The natives of these colonial areas developed new wants for such things as bicycles and cotton cloth as a result of being brought into contact with Europeans, while the wages they got for their work on the oil wells or rubber plantations gave them the money to buy these luxuries. Thus in the case of England the Lancashire cotton industry and the Coventry bicycle manufacturers benefited from imperial expansion. It was not only raw materials which Europe needed from the colonies. A great many things, such as coffee, tea, cocoa, sugar and tobacco, which had once been luxuries were now necessary.

Joseph Chamberlain. It was this mixture of business and sentiment which led the enemies of Joseph Chamberlain, the apostle of England's new imperialism, to remark that 'the British Empire goes on expanding, and the Chamberlain family goes on contracting'. There certainly was a close connection between the two processes. Joseph Chamberlain, whose son Neville became prime minister in 1937, was a successful business man who made a considerable fortune from the manufacture of screws. He represented the new conditions of politics. Until his time English political leaders of both parties had been drawn from the small class of great landed proprietors, or from men like Disraeli who coveted their society. 'Joe' Chamberlain was an unrepentant business man, a dangerous radical who not only anticipated, but at one time would have welcomed, the fall of the monarchy and the establishment of an English republic. He was almost the first of the great political 'bosses'—men who organised the whole public life of their town, year in and year out and not only at election times, so that its political loyalty could be absolutely guaranteed. Under his control Birmingham did what Chamberlain ordered. For years it elected a solid body of radical members of parliament. Then, when Chamberlain became a Conservative, Birmingham obediently followed his lead.

It was the question of Home Rule for Ireland, to which Chamberlain was bitterly opposed, which made him change his party in 1886. For some years before this he had been exceedingly dissatisfied with the quiet, peaceful foreign policy of the liberal party. He regarded home rule for Ireland as a weakening of the British Empire, to whose extension and consolidation he devoted his life. His change from radicalism to conservatism is typical of the change which was coming over nationalist movements throughout Europe at this time. In the earlier part of the century, as we have seen, most nationalists were also liberals. The development of nationalism into imperialism naturally brought this alliance to an end. Liberals retained their belief in the right of each nation to manage its own affairs. The Irish, who wanted to be free of English rule, and the Boers, who wanted to avoid falling under it, seemed to liberals to have as good a case for freedom as the Italians or Germans had had when they fought for their liberty against the Habsburgs. To nationalists turned imperialist these and other similar peoples were simply 'lesser breeds' who ought for their own good and our glory to be brought under 'the law' as soon as possible. To them imperial expansion was the natural form of self-expression for a nation-state. They coveted colonies as some women covet jewels.

Colonies and tariffs. Seventeen years after his conversion from radicalism to conservatism Joseph Chamberlain underwent another conversion—this time from free trade to protection, or rather imperial preference. His object was to knit the whole British Empire together into one great customs union with a tariff wall against foreigners. He failed to persuade even the whole of his own party to support him, and the result of his campaign was to condemn the conservative party to a long period of hopeless opposition, for it was not until 1931 that the majority of English people could be persuaded to drop the free trade system under which they had lived since 1846. Nevertheless Joseph Chamberlain was only preaching what most other imperialist nations practised. His two conversions, then, were both typical of late nineteenth-century politics.

The new imperial patriotism had at first to fight for its existence against the apathy of governments which were accustomed to think of colonies as nuisances. As late as 1878 the British government, for instance, refused to allow an English company to acquire a thousand miles of the East African coast from the Sultan of Zanzibar. In Germany too, Bismarck was strongly opposed to the acquisition of colonies. The French were welcome to take what they could get in Africa to keep them from thinking too much about the loss of Alsace-Lorraine. But even Bismarck had to yield to the combined demands of business men and imperialists, and by the 'eighties Germany had started to look hard for land to colonise. The land which was still available was entirely unfit for white settlement. The non-European areas where white men could settle permanently were already either in the British Empire or else were independent countries like the United States or the South American republics. It was to these areas that Europe's surplus population had to go. The colonies which the various nations acquired in the last generation before the war of 1914 were of no use except for trade and sentiment. Individuals made fortunes out of the new colonies and large sections of the working class found employment as a result of their acquisition, but it may be doubted whether the economic benefits were worth the political jealousy which they occasioned. At any rate the smaller European countries, such as Norway and Sweden, which did not take part in the scramble, were able to enjoy as high a standard of living as Germany, France and Italy, which did. There is little doubt that the importance of these later colonies in political history is far greater than their importance in economic geography would warrant.

The colour bar. The economic benefits of the new imperialism may be doubtful; its political and social results were often undeniably bad. There were of course great administrators, like Marshal Lyautey of France and the Englishman Lord Lugard, who did much to improve the conditions of the natives under their care —it is possible in such cases to speak properly of 'the White Man's Burden'. But against this have to be set two great evils. First, the growth of colour feeling. The earlier colonists and traders in the East do not seem to have regarded themselves as being superior beings merely because their skins were white. The French have always avoided this stupidity, but the other nineteenth-century colonisers developed it very strongly, partly because of the increasing difference in power between Europeans on the one hand and Africans or Asiatics on the other, a difference due to the introduction in the West of power-driven machinery. One firm reason for the white man's feeling of superiority was the plain fact that, as the popular song said,

> Whatever happens, we have got
> The Gatling gun and they have not.

The feeling remains, though this material reason for it has disappeared completely, at least in the case of Japan. The bitterness caused among other races by the colour bar is one of the most dangerous features of the world to-day.

International rivalry. The second evil consequence of the new imperialism was the increased bitterness between the various European peoples themselves. Each came to think itself absolutely superior to its neighbours. Nationalism became more and more selfish and aggressive. In some cases ridiculous theories about racial origins were taught to justify the belief that a particular people were entitled to rule the world. Thus in Germany a namesake of Joseph Chamberlain, Houston Stewart Chamberlain, who was English by birth but German by adoption, taught that the greatest race in the world was the Aryan, that the Germans were the purest descendants of this people, and that in consequence Germany ought to be the leading power in the world. In England nationalism of the new imperial sort was just as blatant, though it did not justify itself by sham science. In the 'nineties a great writer, Rudyard Kipling, became exceedingly popular, partly because of his genius as a story-teller but partly also because his books calmly assumed the right of Englishmen to rule the world. This unconscious insolence reached its height in 1897 with the Diamond Jubilee celebrations. It coincided largely with the period of Joseph Chamberlain's greatest influence, the time from 1895 to 1903, when he was colonial secretary. It was perhaps the shock of the Boer War, a typical example of the working of the new imperialism, which first sobered England by forcing her to realise that as a result of her recent expansion she no longer had a single friend in Europe. In the next chapters we shall have to trace the growth of colonial jealousy among the powers until it culminated in the great war of 1914.

TEN BRITISH STATESMEN

Statesman	Office	Years	Party
William Pitt	Prime Minister	1783–1801 and 1804–1806	Tory
Viscount Castlereagh	War Secretary Foreign Secretary	1805–1809 1812–1822	Tory
George Canning	Foreign Secretary	1807–1809 and 1822–1827	Tory Reformer
Lord Aberdeen	Foreign Secretary Prime Minister	1828–1830 and 1841–1846 1852–1855	Peelite Tory
Viscount Palmerston	Foreign Secretary Prime Minister	1830–1841 and 1846–1851 1855–1858 and 1859–1865	Whig
B. Disraeli, Lord Beaconsfield	Prime Minister	1867–1868 and 1874–1880	Conservative
W. E. Gladstone	Prime Minister	1868–1874, 1880–1885, 1886 and 1892–1893	Liberal
Lord Salisbury	Foreign Secretary Prime Minister and Foreign Secretary (until 1900)	1878–1880 1885, 1886–1892 and 1895–1902	Conservative
Joseph Chamberlain	Colonial Secretary	1895–1903	Liberal Unionist
Sir Edward Grey	Foreign Secretary	1905–1916	Liberal

Map 17. NORTH AFRICAN RIVALS

1. First notice the north-western coastal strip, shaded diagonally, where the rainfall is good. Except for a small Spanish strip opposite Gibraltar virtually all this valuable land is French. Roughly speaking it is bordered on the south by the Atlas mountains. Algiers was French by 1848; Tunis was annexed in 1881, by which time it already had a considerable European population, nearly half of whom were Italians. Morocco was conquered between 1904 and 1912, when it became a French protectorate. Twice France and Germany nearly went to war over Morocco—Algeciras, 1906 and Agadir, 1911.

2. The rest of Africa is left white. Except for Abyssinia at the extreme east it has an average annual rainfall of less than ten inches. It is desert country of little use to any man, though most of it belongs to France. The one exception to the rule 'No rain, no good' is the irrigated valley of the Nile. The control of the sources of the Nile is therefore very important—the White Nile rises in British East Africa, the Blue Nile in Italian Abyssinia.

3. Egypt fell completely under English control in 1882 and the Sudan was added in 1898. That year English outposts moving south met French outposts moving east at Fashoda. War nearly followed, but in 1904 Franco-British differences were settled out of a common fear of Germany. France gave England a free hand in north-east Africa in return for one in the north-west.

4. In the north Italy had to be content with the desert of Tripoli, which she conquered from Turkey in 1912. Her anger at being forestalled by France in Tunis was one reason why she joined Germany and Austria in the Triple Alliance. Her disappointment was increased by the fact that she had a surplus population and sent out many emigrants, while the population of France was stationary.

5. Italy's other African colonies—Eritrea (1870) and Somaliland (1908)—were mainly useful as jumping-off grounds for the conquest of Abyssinia. Her first attempt at conquest was a disastrous failure —Adowa, 1896.

6. The inset map shows the African possessions in 1939 of France (black), Britain (vertical shading), and Italy (square shading).

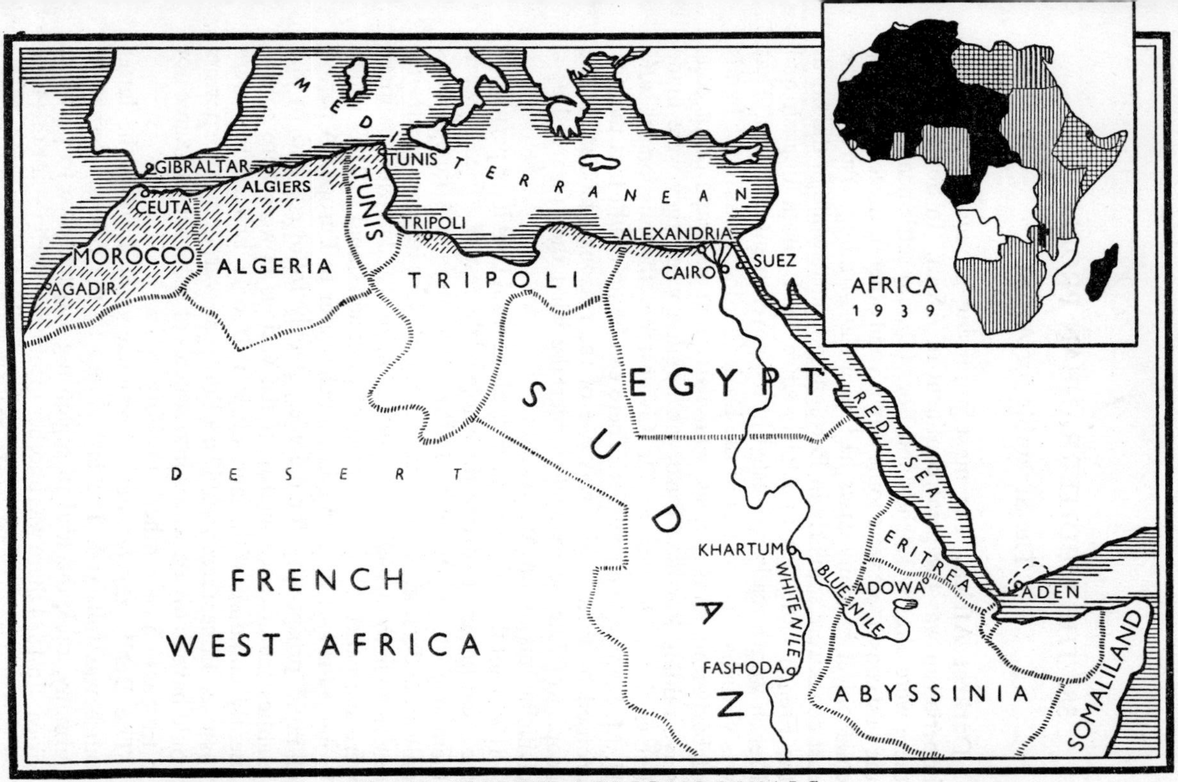

Map 17. NORTH AFRICAN RIVALS

38. THE MEDITERRANEAN PROBLEM. II

The Scramble for Africa

Darkest Africa. The most promising ground for the new imperialism was the great continent of Africa. In South Africa there were already English and Boer (Dutch) settlements. Along the west coast there were some English and other settlements, whose chief object in many cases had been slave-trading. Parts of the east coast were similarly in Portuguese hands; while in the north the French had been in Algiers since 1830. Both English and French were already interested in the government of Egypt. That was virtually all. The rest of the Dark Continent was in the hands of barbarous native tribes, whose frequent wars and outrageous slave raiding horrified European opinion when they were made known by such men as Dr Livingstone, the missionary, and H. M. Stanley, the traveller. There can be no doubt that the pity aroused by their stories was one of the reasons which made the peoples of Europe ready to undertake what they regarded as a civilising mission in Africa.

It was not the only reason. Pity alone would not have persuaded the great powers to divide the whole of the rest of Africa between them in less than fifty years. In the colonisation of Africa pity was only an understudy to the ordinary motives for imperial expansion described in the last chapter. The commercial and patriotic motives were sharpened by the opening of the Suez Canal in 1869 which brought the east African coast much nearer to Europe. A period of rapid prospecting for colonies quickly followed, and by the 'eighties the advance guards of the various powers were so frequently colliding with each other that only two courses were possible. Either there would be a European war, or else the powers would have to divide Africa into spheres of influence and agree not to trespass on each other's territory.

The Berlin conference. Fortunately for Europe the latter course was followed. The last comer to the colonial scramble, Germany, took the lead in organising a conference to partition Africa. In 1884 there was another meeting in Berlin—not this time a full dress affair of prime ministers and foreign secretaries, but a workaday gathering of experts. This conference, however, for all its lack of pomp, was quite as important as the famous Congress of Berlin. It settled the fate of western Africa south of the Sahara, while the conversations started there soon led to a similar agreement about east Africa. Britain got Kenya, Uganda and Rhodesia on the east, and Nigeria and the Gold Coast on the west. France was given most of the rest of West Africa as far as the Congo, as well as Madagascar on the east. Germany took Tanganyika on the east coast, and South-west Africa, Togoland and the Cameroons on the west. Portugal kept Angola and Mozambique, while the enterprising King Leopold II of the Belgians secured the immense Congo Free State.

Stanley and the Congo. The history of the Congo basin at this time is a good illustration of the various forces at work. Stanley's most famous journey, described in his book *Through Darkest Africa*, had been from the east coast to the west down the river Congo. Public opinion had been horrified by his account of the terrible condition of the people—it was reckoned that for every slave who reached the coast *en route* for Arabia or Mediterranean Africa six others had died. Leopold determined to use this pity to his own and his country's advantage. In 1876 he called a conference at Brussels which set up a charitable International African Association to civilise and save the people of the Congo. He sent out Stanley to found the Congo Free State nominally for the Association, but in reality as a Belgian colony. Moving up the river, Stanley met a French explorer on his way down prospecting for a French colony. It was this clash of rival imperialisms which was the immediate cause of the Berlin conference. The conference gave Leopold his way by recognising him as ruler of the Congo Free State. International humanitarian sentiment had been successfully enlisted to found a Belgian empire. That was the end of humanitarianism as far as the Belgian Congo was concerned. It soon gained a terrible reputation for its brutal treatment of the natives in the interests of the cheap production of rubber.

The Berlin conference not only partitioned central Africa; it also laid down the principle that all nations should have equal freedom of trade within this immense area. This second decision had much to do with the fact that the settlement was a lasting one. Another reason was that Africa south of the Sahara did not touch the vital interests of any European power. Its relative remoteness is illustrated again in the case of the Boer War, when England was engaged in conquering the independent Dutch republics of South Africa. Every European state sympathised with the Boers, partly for their own sake but still more out of jealousy, yet nobody thought it worth while to go to war with England. If the Boer republics had been in North Africa the story might have been very different.

The ease with which the affairs of central Africa were settled was in fact very far from being repeated in the case of Mediterranean Africa. Several times the great powers almost came to war, first over Egypt and later over Morocco. The Mediterranean problem in this period falls into two main stages—a time of Anglo-French rivalry lasting until 1904, and after that a time when England and France were brought together in opposition to Germany.

France, England and Egypt. It is easy to see why the English boast that 'the sun never sets on the British Empire' should have led a Frenchman to explain that this was because God couldn't trust the English in the dark. Geography seemed to have intended France to be the mistress of North Africa. Italy was hardly yet a great power, Spain was one no longer. In their default France was ideally placed to dominate the Mediterranean world from her great ports at Marseilles and Toulon. Yet France found that all her attempts to secure control of Egypt were frustrated by the English.

Mehemet Ali had managed to free himself from the control of the Sultan—though Egypt remained nominally a vassal state of Turkey until 1914—but he and his successors soon found that they had only exchanged an inefficient eastern master for an efficient western one. Mehemet Ali's policy of westernising his country was carried on by his successors, notably by Ismail, who reigned from 1863 to 1879. Westernisation costs money and the capital to develop Egypt was provided by French and English investors. Ismail hoped to play off one set of creditors against the other, and this he did for some time, selling his shares in the Suez Canal to England. There came a time, however, when the people who had lent him money persuaded their two governments to force Ismail to hand over the control of finance to their officials. A good deal of Ismail's expenditure had been on useful public works, but a large part had been wasted on personal extravagance which seemed to justify France and England in saying that he was totally unfitted to manage his own affairs, an argument which was strengthened by the appalling poverty and brutal treatment of the peasantry. At any rate in 1876 France and England combined to set up a system known as the Dual Control. An Englishman was to superintend the collection of taxes, a Frenchman to superintend their expenditure. Law suits between foreigners and Egyptians were to be tried by special Mixed Tribunals of Egyptians and Europeans instead of in the ordinary courts. In 1879 the two powers went further and persuaded the Sultan of Turkey to depose Ismail in favour of his son Tewfik.

Three years later a rebellion in Egypt broke this temporary friendship between England and France. The Egyptian army bitterly resented the European control of their country, and in 1882 it rose under Arabi Pasha, determined to throw the foreigners out. There were only two courses open to England and France. They could abandon Egypt as a total loss or they could go to war, crush Arabi and take over the whole administration of Egypt, since plainly the compromise of the Dual Control would never work again. The two partners chose different courses. France withdrew. She had just annexed Tunisia, losing thereby the friendship of Italy, who had indignantly joined Germany and Austria in the Triple Alliance. France was frightened, and reluctantly made up her mind to play for safety in Egypt. England took the other line and decided to fight. The Mediterranean fleet bombarded Alexandria, and an expeditionary force defeated the rebels at Tel-el-Kebir. From then onwards a permanent British army of occupation was kept in Egypt.

In 1884 Lord Cromer was appointed British Consul-General in Cairo. His position was odd. Egypt was not formally annexed to England and Cromer nominally was only a diplomat, but in fact he ruled the country like a dictator. Efficient, upright, domineering, he restored the country to financial solvency, and commercial prosperity. Naturally the benefits of Egypt's new prosperity went to England which had fought in 1882, and not to France, which had withdrawn. It was equally natural, if a little unreasonable, that the French should feel dissatisfied and cheated of their rights.

Arabi's rising therefore had the effect of restoring the old antagonism between France and England over Egypt, an antagonism as old as Napoleon's Egyptian expedition of 1798. The bitterness about Egypt was increased half a generation later by a new grievance over the Sudan.

The Sudan. Economically Egypt depends entirely on the Nile. Without its regular floods there could be no agriculture in Egypt, which would rapidly become as barren as the surrounding desert. But considerably less than half the Nile is in Egypt. Any Egyptian government is therefore bound to be interested in Abyssinia and the Sudan through which the upper Nile runs. At the time of Arabi's rising in Egypt the Sudan was aflame with a holy war preached by a fanatical Mahometan prophet, the Mahdi. Arabi was a Europeanised Egyptian who objected to the English and the French because they were foreign, not because they were modern. The Mahdi objected to any sort of modern civilisation and was as much opposed to Arabi as to England. Since Mehemet Ali's time the Sudan had been governed by Egypt, but the Mahdi's revolt was entirely successful in ending Egyptian control. The English government, therefore, was faced with the problem of the Mahdi as well as the problem of Arabi. It decided that Egypt was quite enough anxiety by itself and made up its mind to abandon the Sudan for the time being. General Gordon was sent up to arrange for the evacuation of the remaining garrisons. He was the least suitable man in the world for the task and, instead of withdrawing, he dug himself in at Khartum, where he was besieged by the Mahdi. The English relief expedition, sent to rescue him, arrived to find Khartum taken and Gordon murdered. English prestige suffered a heavy blow.

The trouble with the French over the Sudan arose from its reconquest fifteen years later by General Kitchener. After the decisive battle of Omdurman the Sudan passed into joint Anglo-Egyptian occupation, though as England already controlled Egypt, Egyptian influence hardly counted. Under its new masters the Sudan rapidly exchanged slave trading for cotton growing, and began to enjoy a peace and prosperity which it had never known before. This extension of English power southwards, however, made the French furiously jealous. In 1898, the year of Omdurman, they sent a small expedition under Captain Marchand to claim the territory south of Khartum for France. In this way they hoped to stop further English penetration and to keep the way clear for a French advance into Abyssinia. Kitchener, moving with a scouting force up the Nile, met Marchand at Fashoda, some six hundred miles beyond Khartum. After an awkward pause the Frenchman withdrew. In itself, Fashoda was only a chance collision between two advance guards, but it deserves to rank among the decisive battles that were never fought. After a dangerous moment, when war seemed possible and even probable, diplomacy solved the crisis. Really friendly relations, however, were not established until 1904.

The Entente with France. The agreement of 1904 divided north Africa into French and English spheres of influence. England was left free to enjoy the possession of Egypt and the Sudan; while France was left in undisputed control of Algeria, and given a free hand in the derelict native state of Morocco. But Morocco was not in 1904 in the possession of the French. There were other competitors, for Morocco was supposed to be rich in minerals. The most obvious claimant was Spain, which already held a narrow strip of the coast. A strong Spanish Morocco would obviously be a danger to English communications, largely neutralising the benefit of Gibraltar. There was, however, no question of this, since that ill-governed country was barely able to maintain a footing in her narrow coastal zone. A much more serious competitor was Germany, immensely proud of the powerful navy she was building and immensely anxious to acquire important new colonies. The experience of the Boer War, which proved that England had not a single friend in Europe, had made the British government decide that it had better try to make some. There was a possible choice between France and Germany. Past history indicated Germany as our natural ally; present fears for the supremacy of our navy and the future of Morocco indicated France. That is the background of the bargain with France in 1904, a bargain which completely changed British foreign policy.

The desperately lawless state of Morocco was a constant menace to the peace of Algeria, which the French had now thoroughly colonised. Algeria was treated as an integral part of France, sending its representatives to the National Assembly, and its troops to share in the normal garrison life of French provincial towns.

Its people, therefore, were encouraged to think of themselves not as subjects but citizens with an honourable share in the life of the mother country. The French are conquerors who, thanks to the absence of any colour feeling, have known how to reconcile the conquered to their fate. They were probably from the point of view of the Moroccans the best claimants to that land.

Tangier, Algeciras, Agadir. Naturally Germany did not share that view and the Anglo-French understanding of 1904 was soon put to the test. Germany's opportunity came in 1905 when a war between Russia and Japan deprived France of any possibility of help from her most obvious ally. The Kaiser made a state visit to Tangier, and assured the Sultan of Morocco of his support against French aggression, while his government bullied M. Delcassé, who was largely responsible for France's forward policy, into resigning. Germany wanted to find out whether Britain would stand by her new friend in a crisis. It was an expensive experiment, for at the conference of Algeciras next year Morocco was recognised, though with some reservations, as a French sphere of influence, while the French and British General Staffs began secret conversations about the best method of co-operation against a German attack. Germany's action in the Morocco crisis not only failed to weaken French influence in Morocco, but also changed the purely colonial understanding of 1904 into an informal alliance between France and England.

The result of the Algeciras conference provoked Germany to try her hand in Morocco once more, though with no better success. The occasion for this second venture came in 1911 when the French dispatched a punitive expedition to Fez, the capital of Morocco. Germany by way of protest and to extort concessions from France sent a gunboat, the *Panther*, to Agadir, a port on the Atlantic coast of Morocco, which England was determined should never pass into German hands. The threat of war implied by the visit of the *Panther* to Agadir was answered by an equally threatening speech from Mr Lloyd George, in which he warned Germany that she had nearly exhausted Britain's patience. Peace, however, was patched up and Germany was given part of the French Congo in return for leaving France a free hand in Morocco. After long negotiations Morocco was divided into three zones, a large French protectorate, a small Spanish one, and the international city of Tangier, administered jointly by France, Spain and Great Britain.

Neglected Italy. French Africa was not only growing westwards from Algeria into Morocco, but also eastwards into Tunisia, another of the decrepit Mahometan states of North Africa. It was Bismarck who encouraged France to annex Tunis in 1881, hoping in this way to bring about such jealousy between France and Italy that he need never fear a Franco-Italian alliance. He was entirely successful, for the Italians regarded Tunis as a natural outlet for their fast growing population. Events justified this view for, when settled French government made it possible for Europeans to settle in Tunisia, it was Italians, not Frenchmen, who first went out as colonists. To this day Tunis remains a French colony with as many Italians as Frenchmen among its European inhabitants—a perpetual grievance to Italian patriots.

When Italy was strong enough to take colonies for herself, she found that all the best of North Africa had already been annexed. From the first her attention was centred on Abyssinia. A beginning was made as early as 1870 by the purchase of Assab, at the bottom corner of the Red Sea on the African side. More land was gradually acquired until in 1890 the colony of Eritrea had grown to its present size. Six years later Italy advanced from this base in an attempt to conquer Abyssinia, an attempt which came to a terrible end when the Emperor Menelik cut up the Italian army at the battle of Adowa. Italy, however, still persevered. Between 1905 and 1908 she bought her way into Somaliland, another district on the frontier of Abyssinia. Later still, in 1912, she secured her first Mediterranean colonies after a war with Turkey, from whom she took Rhodes and the Dodecanese islands, a useful naval base, and the sandy desert of Libya with its capital at Tripoli.

Germany's failure to establish herself in Morocco was a setback to her prestige, but nothing more. No vital German interest was harmed, for Germany is not naturally a Mediterranean power. Italy was in a much more unfavourable position as a result of her failure to secure a single Mediterranean colony really worth having—Rhodes and the Dodecanese are valuable in case of war, not for themselves. Italy is the only one of the great powers which is a purely Mediterranean country. If she is to expand it must be along the coasts of that sea or not at all. Her relative failure before the war of 1914 did not change the balance of European alliances or power, for in those days Italy was still hardly a first-rate state. It left her, however, with a strong sense of grievance which has been the cause of great unrest in the world of to-day. The virtual division of the Mediterranean between France and England by the agreement of 1904 seemed at the time to be a complete solution to a century of conflict. It lasted thirty years, but events since 1934 have shown that it was only a temporary solution because it neglected Italy.

THE COBURGS

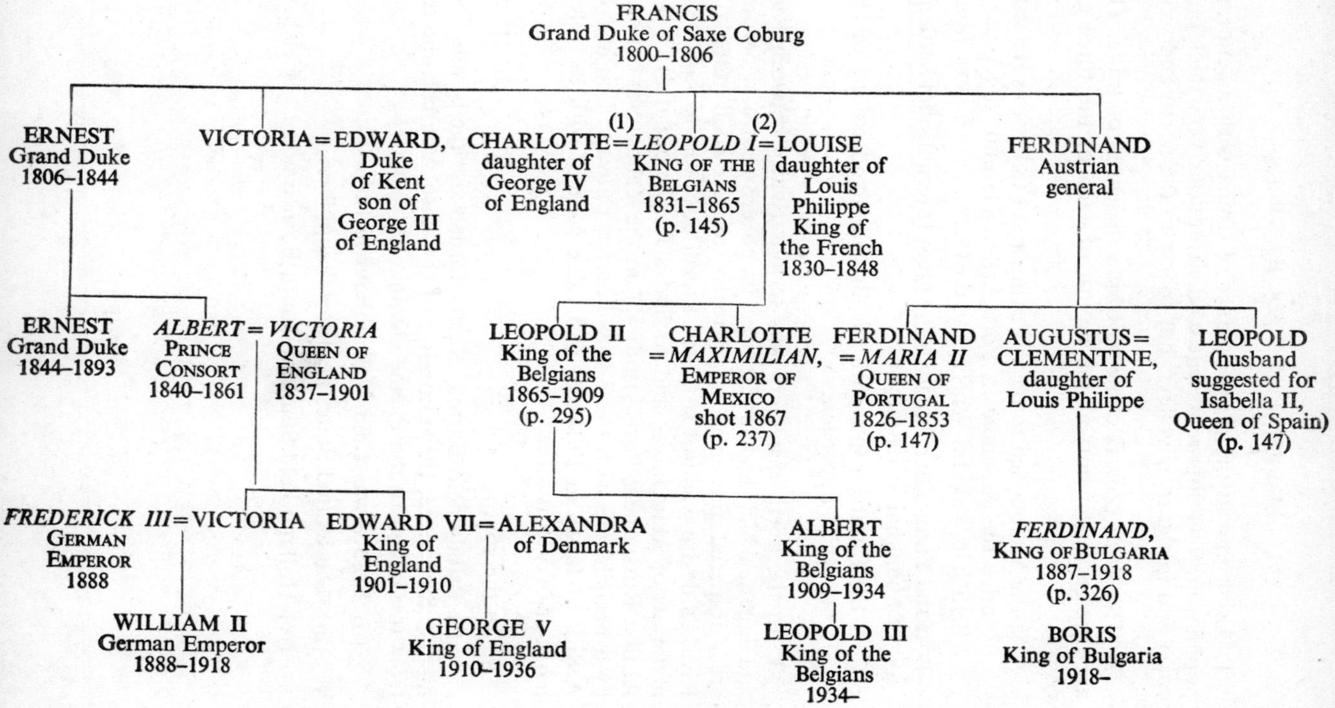
FRANCIS
Grand Duke of Saxe Coburg
1800–1806
ERNEST
Grand Duke
1806–1844
VICTORIA = EDWARD,
Duke
of Kent
son of
George III
of England
(1)
CHARLOTTE = *LEOPOLD I* = LOUISE
(2)
daughter of
George IV
of England
King of the
Belgians
1831–1865
(p. 145)
daughter of
Louis
Philippe
King of
the French
1830–1848
FERDINAND
Austrian
general
ERNEST
Grand Duke
1844–1893
ALBERT = *VICTORIA*
Prince
Consort
1840–1861
Queen of
England
1837–1901
LEOPOLD II
King of the
Belgians
1865–1909
(p. 295)
CHARLOTTE
= *MAXIMILIAN*,
Emperor of
Mexico
shot 1867
(p. 237)
FERDINAND
= *MARIA II*
Queen of
Portugal
1826–1853
(p. 147)
AUGUSTUS =
CLEMENTINE,
daughter of
Louis Philippe
LEOPOLD
(husband
suggested for
Isabella II,
Queen of Spain)
(p. 147)
FREDERICK III = VICTORIA
German
Emperor
1888
EDWARD VII = ALEXANDRA
King of
England
1901–1910
of Denmark
ALBERT
King of the
Belgians
1909–1934
FERDINAND,
King of Bulgaria
1887–1918
(p. 326)
WILLIAM II
German Emperor
1888–1918
GEORGE V
King of England
1910–1936
LEOPOLD III
King of the
Belgians
1934–
BORIS
King of Bulgaria
1918–

MAP 18. THE FAR EAST, 1914

1. Look first at the great expanse of China with its enormous population. The dots represent its distribution: very thick on the coast and in the river valleys, but sparse to the north and west.

2. The great powers came to China in search of trade. There are foreign settlements at the mouth of each of the three great rivers:

A is the Yellow Sea route to Peking and the Hwang Ho valley. Notice the Japanese in Port Arthur (1) and in Korea; the British in Wei-Hai-Wei (2) and the Germans in Kiao-Chou (3).

B is the Yang-Tze route with the great International Settlement of Shanghai at its mouth.

C is the Canton route with the British island colony of Hong Kong at its approach.

3. In the north-east there was serious rivalry between Russia and Japan, which had modernised its government and become a great imperial power with a greatly overcrowded population. Russia's desire was not so much to sell goods as to tap the raw materials of Manchuria, and to find a Pacific outlet for Siberia. At first Russia held Port Arthur (1) on the Yellow Sea connected by a branch line with the Trans-Siberian railway. She lost Port Arthur to Japan in 1904 and had to be content with the inferior harbour of Vladivostok.

4. In the south:

(i) Britain held Burma, North Borneo and the Straits Settlements. Singapore, the 'Gibraltar of the East', controls the route from the Indian to the Pacific Ocean.

(ii) France was China's neighbour in Indo-China.

(iii) The United States held the Philippines.

(iv) Holland held Sumatra, south Borneo and the Celebes.

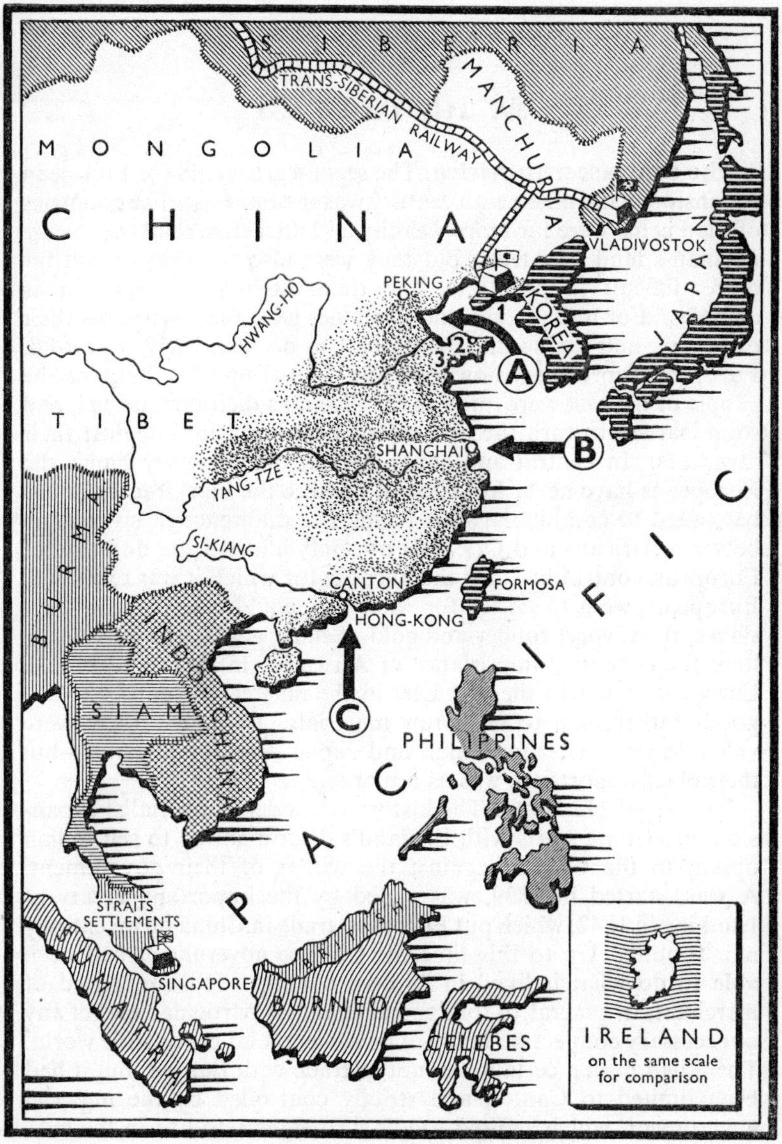

MAP 18. THE FAR EAST, 1914

39. THE FAR EAST

The contrast with Africa. The second great field for European expansion in the nineteenth century was the Far East. The countries of the Pacific were far more wealthy and attractive than the savage no man's land of Africa, but they were also for that reason far more difficult to penetrate, and quite impossible to retain as colonies. For a time guns and machines gave the Europeans their own way, and China and Japan had to do what they were told. European supremacy, however, was only temporary because the people of the east were quite as intelligent as their conquerors, and soon learned to turn the new mechanical inventions against their inventors. In central and south Africa, on the other hand, the Europeans have never lost control because the black races are too backward to combine against them. The difference in civilisation between Africans and Orientals not only affected the duration of European control but also the purpose for which it was exercised. Europeans went to Africa for what they could take from it—first slaves, then ivory, rubber and gold. Selling goods was only a side line; the economic importance of Africa lay in its raw materials. Europeans went to the Far East in the nineteenth century to sell goods rather than to fetch raw materials. Of course there were valuable imports from China and Japan—tea for instance—but their chief importance was as a market.

Treaty of Nanking. The history of modern imperialist expansion into China starts with England's determination to sell Indian opium to the Chinese against the wishes of their government. A war, started in 1839, was ended by the important Treaty of Nanking in 1842, which put European trade in China on an entirely new footing. Up to this time the Chinese government had been able to do what it liked in its own country. It disapproved of merchants in general, of foreign merchants in particular, and of any unnecessary connection between its subjects and the outside world. There had been a certain amount of trade with Europe, but it had been limited to Canton and strictly controlled by the imperial government, which refused to allow foreigners to live in the city. They had to go away as soon as their immediate business was done.

The war changed the position altogether. England gained the town of Hong-Kong as a colony of her own, and the richest part of China was also opened to European trade by the creation of 'Treaty Ports' where foreigners were allowed freely to settle and trade. The most important of these was Shanghai at the mouth of the Yang-tse-Kiang, China's greatest river and chief means of transport. In the next few years the English and French had twice to send expeditionary forces to fight the Chinese government, which still disliked the presence of foreign merchants and would have liked to get rid of them altogether if it could. It not only failed to do this, but had to recognise further treaty ports as a result of these expeditions, the last of which, in 1860, penetrated to Peking, where the Emperor's summer palace was looted and destroyed.

In the early stages of western influence in China the European powers were not, on the whole, looking for colonies or districts which they could regard as their own preserve against other European states. Those were the days when free trade doctrines were generally believed, and all that the European merchants wanted was the right to trade freely in China. English merchants wanted to sell their goods to the Chinese, they did not try to stop French and American merchants doing the same. The situation brought them into conflict with the Chinese government, but not with one another. Every extension of western trade and western influence was the result of a humiliating defeat for the Emperor of China, who began slowly to lose his grip on his own country. The progress of foreign trade in China was, therefore, accompanied by a decline in the strength of the Chinese government and a growth of lawlessness, because, until it was too late, the Chinese authorities set themselves not only against foreigners but against foreign methods.

The condition of Japan. The same gradual lapse into anarchy might have been expected in Japan, which, like China, was opened to European trade against the wish of its government. In 1854 a United States officer was sent to Japan to insist that she should enter into commercial relations with the rest of the world. When he landed, the American was really an explorer stepping straight back into the middle ages. Time had stood still in the lonely islands of Japan even more effectively than in China. It was seven centuries since the emperor, the Mikado, had been edged away into a corner, kept as a religious and social figure-head but deprived of political power. The country had been actually governed by dynasty after dynasty of hereditary viceroys, known as Shoguns. Sometimes these hereditary viceroys had themselves been compulsorily retired, like the Mikado, from politics, and Japan had fallen under the rule of hereditary deputies of the hereditary deputies. There were, so to speak, two layers of sleeping partners in the firm or, to change the metaphor, a sleeping queen as well as a sleeping princess. Since 1603 Japan had been more effectively asleep than ever under a dynasty of Shoguns who not only refused to allow foreigners to enter Japan but also forbade Japanese to leave the country. These Shoguns and their great barons had given the country generations of relative peace and of absolute stagnation. Under their rule Japan had not the slightest hope of resisting western expansion.

It seemed certain that Japan would go the way of China; but the unexpected happened and foreign interference led not to a weakening, but to a strengthening of government. In 1867, there was a political revolution in Japan which restored the power of the Mikado and abolished the feudal system in favour of a strong absolute monarchy, determined to adopt western methods and to beat the foreigner at his own game. It was rapidly successful. Japan became a powerful modern state before ordinary middle-class Europeans had begun to take her more seriously than Timbuctoo. By the time England was laughing at *The Mikado* (1885), which treated Japan as a rather picturesque South Sea island, she was already well on the way to becoming a world empire.

Japan succeeds; China fails. The reason for Japan's success in adapting herself to western civilisation are worth considering. The first reason was the peculiar position of the Mikado. His family had reigned since before the birth of Christ. The emperor is 'the Mikado', the son of Heaven, a god on earth. His service is religion. Shintoism, the religion of Japan, is one which worships, besides the emperor, the natural forces and features of the homeland so that, again, patriotism and religion are one. The natural inheritance of every Japanese is a complete and unthinking devotion to his country and his emperor. It only needed the reappearance of the emperor as a political power for all this loyalty to be available for the westernisation of Japan at his command. The second reason for Japan's success lies in the tradition of the Samurai, the knightly class of old Japan. In 1867 they sacrificed their feudal privileges because they saw that this was the only way to save Japan from becoming dependent on the European powers. They gave up their power, but they brought to the service of the new Japan their code of behaviour, called Bushido. According to Bushido the good soldier is the ideal man. Courage, honour, discipline, physical fitness, chivalry are the most prized virtues. Both Chinese and Japanese had the intelligence to adapt western civilisation; but only the Japanese had the will to do so and the disciplined tradition to create overnight a modern state and a modern army.

China failed to defend herself for the same reasons that Japan succeeded—the position of the emperor, the nature of its religion and the position of the soldier. By European standards the Manchus, the Chinese dynasty, had been reigning a long time; but to the Chinese they were still foreign invaders. The pig-tail, which seemed to Europeans the symbol of China, was to the Chinese the symbol of their servitude to the foreign Manchu conquerors. The first thing they did when the Manchu dynasty was deposed in this century was to cut off their pig-tails. Chinese religion is more concerned with the family than with the state. Its typical feature is ancestor worship, not emperor worship. Aggressive patriotism of the sort needed to resist western nations found no support from Chinese priests. In China too the soldier's was one of the less honourable professions, instead of being the noblest as in Japan. The typical Chinese gentleman was a cultured civil servant, not a knight.

The Sino-Japanese War. Nearly a generation passed between the restoration of the imperial power in Japan and the first trial of the country's new strength. During the interval she re-modelled her army on German lines, started a fleet on the English pattern and began to Americanise her industries. When Japan did go to war her first enemy was China, not one of the western powers. Japan, having decided to follow the industrial example of Europe, was beginning to experience the same need for colonial expansion. In her case the search for markets was reinforced by a search for colonies, where Japanese could settle, since she had already started on that break-neck increase of population which was to double her numbers within two generations and leave her with a density of population a quarter as great again as that of Great Britain. Her first chance for expansion came in Korea in 1894, a decrepit kingdom which was vaguely dependent on the Emperor of China. Japan decided to annex the country, and in the Chinese war which followed Japan won a decisive victory. She forced China to cede her some exceedingly valuable territory, including the site of Port Arthur, the most important strategic point in Manchuria. But Russia, France and Germany intervened and forced Japan to surrender all her gains except the island of Formosa.

Europe's search for colonies. It was not from love of China, but from jealousy of Japan that the European powers had intervened. The old period of opening China to general trade had by this time been followed by frantic efforts to obtain colonies or exclusive spheres of influence, ear-marked for the private use of a great power. The French had been particularly active in south China, where they were building up a great empire in former Chinese dependencies. England was more concerned with the centre of the country. Russia had reached the Pacific in 1860 and founded a naval base at Vladivostok in what had once been Chinese territory. Germany was beginning to look for colonies in China as well as in Africa. It looked as if the 'scramble for China' might have the same sort of result as the 'scramble for Africa'.

The situation was, however, now complicated by the intrusion of Japan. The Sino-Japanese War had given the European powers an unpleasant surprise, but their firm action had, they thought, put Japan back in a proper subordinate position. At the time her intervention seemed merely to have furthered European interests, since in the next three years they made China pay heavily for their help against Japan. Russia managed to secure Port Arthur as a second and better terminus for the Trans-Siberian railway, which had been finished in 1895. Vladivostok, its original goal, was ice bound in winter. Another Chinese port, Kiao-Chou, fell to Germany as compensation for the murder of two German missionaries. France obtained concessions in the south, and Britain declared the Yang-Tse basin an English sphere of influence and also seized Wei-Hai-Wei as an additional naval station. In 1899 the United States began to annex territory in the Far East by taking the Philippine islands from Spain.

The Boxer Movement. The Chinese people as well as their government hated the division of their country into European dependencies. In 1900 there was a violent rebellion against the foreigners. The complete failure of the Boxer[1] Movement, as the rising was called, to achieve more than a series of murders is a sufficient commentary on the helplessness of China. The rising began with the murder of the German minister in Peking, which was followed by a siege of the fortified British Legation. An international expeditionary force, which included a Japanese contingent, was sent out under the command of a German, Count von Waldersee. When the Kaiser said good-bye to his troops he told them of his hopes: 'Even as, a thousand years ago, the Huns under their king Attila made such a name for themselves as still resounds in terror through legend and fable, so may the name of Germany resound through Chinese history a thousand years from now.' It was an unfortunate comparison, for the name stuck and was used effectively as anti-German propaganda in the war of 1914. Certainly the Germans did behave like Huns in their treatment of Peking after the relief of the legations, but they were no worse than the other contingents. The end of the affair was the complete collapse of the Boxer Movement in 1901, the payment of a large indemnity by China and the keeping of European garrisons in Peking.

Japan's English alliance and Russian war. The next year an alliance was made between England and Japan, which at the time seemed extremely useful to this country. England was still seriously frightened of Russia, and regarded any extension of Russian power as a great danger. She had been bitterly opposed, for instance, to the Russian annexation of Port Arthur. But Russia was also the enemy of Japanese expansion in China—it was she who had taken the lead in forcing Japan to give up most of her conquests after the war with China. The Anglo-Japanese alliance, therefore, seemed to English statesmen an excellent way of removing the Russian danger. It had this further advantage. England was already terrified by the growth of German power in Europe and by German colonial ambitions. The Japanese alliance would leave us free from worry in the Far East if we became involved in a war with Germany. The alliance with Japan in 1902 was dictated by the same fear of Germany which led to the Anglo-French entente of 1904.

[1] 'Boxer' comes from a Chinese word meaning 'fist of righteous harmony'.

The immediate effect of the alliance was to give Japan a free hand in Manchuria. She determined to take back from Russia the territory she had unwillingly surrendered ten years before. The Russo-Japanese War of 1904–5 gave Japan all she wanted. Her armies easily defeated the Russians, and her navy under a great sailor, Admiral Togo, completely destroyed the enemy fleet in the Pacific and then sunk the Baltic fleet which had been sent round the world to the Pacific as a last hope. On the way it had nearly caused a war between Russia and England by shelling some English trawlers which it mistook for enterprising Japanese destroyers, lurking in the North Sea.

The Treaty of Portsmouth,[1] which ended the war, gave Japan Port Arthur and allowed her to take Korea. That was, perhaps, the least important result of the war. The total defeat of a great European power by an Asiatic state encouraged nationalist movements all over the East. What Japan had done, others might hope to do. The war destroyed Russia as an effective force in the Far East outside her own frontiers and made Japan the dominant power in the eastern Pacific. Ten years later she was able to use the European War of 1914, in which she only nominally took part, to add Germany's possessions in China to what she had already taken from Russia. Great Britain had made the Japanese alliance as an insurance against German and Russian dangers. Japan had used the freedom which that alliance gave her to become a greater danger to English interests in the Far East than either Germany or Russia had ever been. She was a bigger danger because she was the only great power on the spot. The United States and the European powers were left with no base from which they could defend their interests in China in the event of a war.

China's descent into anarchy. The whole position of Europe in China was reversed between 1902 and 1918. Before 1902 a weak China seemed to be to the advantage of the European states because they could extort trading privileges, colonies, concessions and spheres of influence from it. After 1918 a strong China became necessary for European traders because the survival of their trade depended on the Chinese government being able to resist Japanese aggression. The European powers were no longer able to check

[1] Portsmouth (U.S.A.) not Portsmouth (England).

Japan themselves, although they realised that their trade with China would disappear if the Japanese gained control.

A strong China had become a European interest, but it had ceased to be a possibility—largely because of the previous behaviour of the European powers. In 1911 a patriotic revolution broke out in China, the imperial dynasty was overthrown, and a republic was proclaimed in its place. But before any effective central government could be established, the country was plunged into an agonising civil war from which it has never recovered.

Map 19. RACIAL PROBLEMS IN AUSTRIA-HUNGARY IN 1914

1. Notice first how many races are found on both sides of the frontier. There were, reading clockwise from the north-west, Germans, Poles, Ukrainians, Rumanians, South Slavs, and Italians inside and outside the empire. Each part of these divided races was attracted to the other by the magnet of a common nationality. The problem was 'which way would the magnet pull?'—attracting those outside the empire to come in or vice versa.

2. Next notice that there are two nationalities entirely contained inside the empire. These are the Hungarians, and the Czechs and Slovaks, two closely related peoples of whom the Czechs lived in Austria and the Slovaks in Hungary.

3. Now follow the – – – line which after 1867 divided the empire into the watertight compartments of Austria and Hungary. Croatia, which had formerly been loyal to the emperor, became actively disloyal after it fell under the power of the Hungarians, who were traditional enemies of the Croats.

4. Lastly notice how the occupation of Bosnia-Herzegovina in 1878 and its annexation in 1909 made the nationalities problem more acute by bringing more South Slavs into the empire.

5. Compare this map with a physical map. Racial frontiers do not coincide with natural frontiers. There are Germans well down the Italian side of the Brenner Pass, almost half the Rumanians live on the Hungarian side of the Transylvanian Alps, and there are several million Germans on the Czech side of the Erzgebirge and the Riesengebirge.

MAP 19. RACIAL PROBLEMS IN AUSTRIA-HUNGARY IN 1914

40. FRANCIS JOSEPH

Austria-Hungary: 1848 to 1916

Francis Joseph. In 1916 a very old man with mutton chop whiskers died peacefully in the palace of Schönbrunn, near Vienna. Death had been waiting a long time for him. For sixty-eight years Francis Joseph had been emperor, the head of the oldest and proudest ruling family in Europe. His subjects also had been waiting a long time for his death—not because he was cruel or corrupt, but because he stood in the way of their ideas of government. All the earlier years of his reign had been filled with bitter struggles for home rule among the various nations over whom he ruled; but latterly there had been almost a lull in the struggle for national independence. Affectionate respect for his age, his devotion to duty and the sorrows of his private life made men wish that he should live out his time in peace. When he drove through Vienna the streets filled with crowds eager to cheer their old emperor, even though, when he had passed, they might shrug their shoulders and mutter: 'When Franz Josef goes, that will be the end of things.'

There had been a time when the Habsburgs had been so much the greatest power in the world that the five vowels had been twisted into a family motto to express their glory: 'Austriae Est Imperare Orbi Universo.' The saying could equally well be translated into German, where it runs: 'Alles Erdreich Ist Österreich Untertan' (the whole world is subject to Austria). Those days were long gone, but Francis Joseph still claimed to be, and for the most part was,

by God's grace Emperor of Austria; King of Hungary, of Bohemia, Dalmatia, Croatia, Slavonia, Galicia, Lodomeria and Illyria; King of Jerusalem; Archduke of Austria; Grand Duke of Tuscany and Cracow; Duke of Lorraine, of Salzburg, Styria, Carinthia, Carniola and Bukovina; Grand Duke of Transylvania; Margrave of Moravia; Duke of Upper and Lower Silesia, of Modena, Parma, Piacenza and Guastalla, of Ausschwitz and Sator, of Teschen, Friuli, Ragusa and Zara; Royal Count of Habsburg and Tyrol, of Kyburg, Görz and Gradisca; Duke of Trent and Brixen; Margrave of Upper and Lower Lausitz and in Istria;

Count of Hohenembs, Feldkirch, Bregenz, Sonnenberg etc.; Lord of Trieste, of Cattaro, and above the county of Windisch; Grand Voivode of the Voivodina Serbia etc.

Chance, marriage and Habsburg cunning had brought these miscellaneous lands together into one empire. They represented not one nation but twenty. Nationalism was therefore the chief enemy of the Habsburg family, and one which they fought at all times. Patriotism in its ordinary meaning of devotion to country was a treasonable sentiment for the subjects of Francis Joseph. When as a young boy of eighteen he was called to the throne his advisers gave him a proclamation to sign. It referred to the 'Fatherland'; the emperor crossed out the dangerous word and wrote in its place 'patriots for me'. That was the only patriotism the Habsburgs recognised; it was the only patriotism which would hold their empire together.

The four pillars of the Monarchy. It seems, perhaps, a little strange that they should have found any patriots of this curious brand, but in fact the Habsburgs were well served by four groups of men, the four pillars of the Dual Monarchy as they have been called. The first pillar was the army, whose regimental officers were cheerful, well-mannered and not too snobbish, though the high command was a close preserve for the great and incompetent aristocracy. The second pillar was the civil service, a leisurely, fairly tolerant body of men whose minds were inextricably tangled in red tape. Before the caretaker of a village school was allowed to spend a pound twenty different officials had to handle his request, and months would pass before the result of his application was known. But in spite of this fussiness the Habsburg civil service had a genius for devising compromises which somehow worked and quietened political passions. The third pillar was the Roman Catholic church, whose powerful influence was thrown unhesitatingly on the side of the emperor. Its great religious order, the Jesuits, had almost a monopoly of the education of the sons of noblemen, and, since the Jesuits are great schoolmasters, this guaranteed the Habsburgs the support of their fourth pillar, the aristocracy.

The pillars were strong, but there were not enough of them. Habsburg patriotism, 'patriotism for me', was confined to those who were either members of the court circle or in the service of the emperor. This left the great mass of Francis Joseph's subjects untouched, exposed to the influence of their various national patriotisms. In the long run these separate patriotisms were the empire's greatest danger—no statesmanship could have satisfied all their aspirations within the framework of one state—but for the time being their variety was an asset, because the emperor could play off one against another.

The nationalities. It is worth taking a look at the various races of the empire. There were 11,000,000 Germans, 8,500,000 Hungarians, 3,500,000 Italians and Rumanians, and 23,000,000 Slavs. But the Slavs were hopelessly divided among themselves. In the north-eastern corner of the empire, for instance, lived both Poles and Ukrainians—very similar races, but bitter rivals. Ukrainian passengers had been known to break a Polish booking-clerk's head because he refused to book them for Peremysl, offering them instead a ticket to Przemysl, whose existence they denied though it occupies precisely the same ground as Peremysl. The Habsburgs were able to rely on Polish loyalty by allowing them to ill-treat the Ukrainians. Similar conditions existed in almost every part of the empire. The map shows each race occupying its own corner, but this is a misleading simplification; everywhere there are pockets of one race embedded in a district predominantly belonging to another—Hungarians among Slovaks, Germans among Czechs, Italians among Slovenes. When the old Habsburg empire fell apart after the European War into separate self-contained nation-states, the net result was that there were several 'Austria-Hungarys' instead of one, for each of the new or enlarged states had its own problem of minorities of another race.

An economic unit. The Habsburg empire was, however, as well suited to be one economic unit as it was ill suited under modern conditions to be one political unit. The two halves of the empire fitted well together. Austria was rapidly becoming industrialised, while Hungary remained largely agricultural. Austria therefore could clothe and equip Hungary while Hungary could feed Austria. Nearly three-quarters of Hungary's exports went to Austria, from which came also nearly three-quarters of her imports. Taken together, then, Austria and Hungary were not far from self-supporting. Under Francis Joseph they formed a customs union, so that in the middle Danube basin before the war people lived under much more sensible economic conditions than they do to-day, when seven states, each with high tariff walls, divide this natural unity.

Francis Joseph himself had no doubt what was the proper way to rule his curiously assorted empire. He must be the absolute master of all his subjects: to try to govern them constitutionally, to ask them what they wanted, was merely looking for trouble. He could never forget that when he came to the throne in 1848 his empire had been deep in civil war, his capital the scene of street fighting. To the day of his death liberalism suggested anarchy to his mind. Schwarzenberg, the statesman, and Jellacic, the soldier, however, had soon crushed the liberal revolutions of 1848 with the help of the Tsar. For the next twelve years Francis Joseph was able to rule his empire in the way he thought all empires should be ruled—in army fashion, by strict discipline and obedience to orders.

Foreign disasters; concessions at home. Then in quick succession two disasters in foreign policy forced him reluctantly to make concessions to the hated forces of liberalism and nationalism. Austria's prestige had been shaken by her timid neutrality during the Crimean War. A few years later France had defeated her at Magenta and Solferino, and in consequence she had lost Lombardy and the control she had exercised over most of central and southern Italy. Francis Joseph had to choose between revolution at home and granting a constitution. In 1861 he gave way and allowed a parliament to be elected to represent the whole empire. He meant this to be only a temporary measure until he was strong enough to restore absolute government. It was only a temporary measure because within six years he was no longer strong enough to refuse further concessions.

The 'Ausgleich'. The Seven Weeks' War of 1866, in which Austria lost her place inside the German Confederation and almost her last Italian province, Venetia, once again nearly overthrew the empire. All the nationalities were restless and Hungary was dangerous. Francis Joseph bought off Hungary by making a compromise in 1867 with the most moderate of the Hungarian leaders, Francis Deak. By this 'Ausgleich', to give it its German title, all the Habsburg dominions were divided into two parts—the Empire of Austria, comprising the seventeen provinces in the west and north-east, with its capital at Vienna; and the Kingdom of

Hungary, covering the eastern half, with its capital at Budapest. The whole was known as the Dual Monarchy, because Franz Josef was now to reign by a separate title as Emperor of Austria and as King of Hungary. Each of the two states had its own government, its own parliament, and its own administration, but in foreign affairs, matters of defence, and as much of finance as was involved with these, they acted together through joint committees of the two parliaments. Even this strictly limited amount of co-operation was disliked by the Hungarians, who insisted that, though the joint committees might meet together to vote, they must on no account discuss together what they were going to decide. Their sessions therefore took place in silence. Still, the 'Ausgleich' was a reasonably satisfactory compromise as far as the relations of the Habsburgs and Hungary were concerned. As long as the empire lasted there was no more talk of a Hungarian rebellion or war of independence.

This was because the 'Ausgleich' left the Hungarians in undisputed control of the whole of the ancient Kingdom of Hungary at its greatest extent. They would have been satisfied with nothing less. But the Kingdom of Hungary contained many millions of peoples who were not Hungarian at all. Their position was made worse as a result of the 'Ausgleich' because, while the Habsburgs had treated all their subjects equally, the Hungarians, as a race, oppressed the Croatians as a race and tried to turn them into Hungarians. In consequence the Croatians became rebellious and ready to take the opportunity of a Habsburg defeat in a foreign war to start a civil war of independence. The 'Ausgleich' gave the Habsburgs the support of the Hungarians who had rebelled in 1848, but it lost them the support of the Croatians who had then defended them against the Hungarians. It solved one nationality problem, but created many more, since besides the Croatians there were many other races which were now handed over to the mercy of the Hungarians, whose outlook was perfectly expressed in the comment on the 'Ausgleich' which one of their leaders made to an Austrian friend: 'Now you will look after your barbarians, and we will look after ours.'

A Triple Monarchy? The only hope for the final survival of the Habsburg Empire lay in turning the Dual Monarchy into a Triple Monarchy of Germans (Austrians), Hungarians and Slavs, the most numerous of the subject races. It was a possible solution, given an emperor determined to force it through against vested interests. Francis Joseph himself was incapable of seeing that this change was either desirable or necessary, but his heir might have brought it about. Twice there seemed a hope that this would happen, but twice these hopes were killed by a personal tragedy which was also a political disaster. Francis Joseph's son, the Crown Prince Rudolf, was a liberal-minded man who might have saved the empire, but in 1889 he committed suicide under mysterious circumstances at Mayerling. His successor as heir to the throne was the Archduke Francis Ferdinand, a man in almost every way inferior to Rudolf—bigoted, surly, a rigid conservative—but one who sympathised with the Slavs and was ready, when he should come to the throne, to give them some form of home rule. His sympathy with the Slavs was probably due to the fact that he had married a Slav countess, a marriage which deeply offended the old emperor, since princes were not supposed to marry below their rank. Francis Joseph's natural dislike of political change and his unnatural dislike of his nephew combined to make reform impossible during his reign. It became impossible altogether when Francis Ferdinand and his wife were murdered in 1914 as they drove through the streets of Sarajevo, the capital of one of the Slav provinces. So intense was the old emperor's hatred of his nephew's plebeian marriage that he saw in this assassination the hand of providence: 'A Higher Power', he remarked, 'has restored that order which I myself was unable to maintain.'

Consequences of Sarajevo. His ministers at once saw in the event the hand of a foreign power. Across the frontier lay the small Slav kingdom of Serbia, a half-civilised state which hoped to grow larger by absorbing the people of kindred race who lived inside the vastly more civilised Austrian empire. It was difficult for Austria and Serbia to live peaceably side by side. The greater wealth, comfort and civilisation of their relations inside Austria was a constant temptation to the Serbians; the greater political freedom of Serbia was a constant temptation to the Austrian Slavs. Home rule for the Austrian Slavs was therefore the last thing the Serbians wanted, and they were prepared to go as far as murder to prevent it. If the Serbian government itself did not organise the killing of Francis Ferdinand, government officials knew about the plans and helped the criminals. To-day a statue in honour of

Princip, the murderer, stands in what is now the Jugo-Slav (Serbian) town of Sarajevo.

The murder of Francis Ferdinand was the death warrant of the Austrian empire. It killed all hope of turning the Dual into a Triple Monarchy, for concessions to the Slavs would have seemed like 'shaking hands with murder'. It encouraged those who wanted a war to crush for ever the Serbian danger to Austria. By the end of July they had got their way. If the war could have been limited to Austria and Serbia their policy might have been justified; but they should have known that this was impossible. When peace was made the Habsburgs were no longer reigning princes and the territories they had ruled were divided among seven different governments.

The war of 1914 became a general European war instead of a punitive expedition because Russia stood behind Serbia. The Habsburgs were obsessed by three great fears—fear of the Slavonic revival at home, fear of Serbia with her secret terrorist societies across the frontiers, and fear of Russia, the protector of all Slavs everywhere. If the first fear had been removed by a generous gift of home rule the other two fears might have disappeared; but home rule for the Slavs was only the policy of the Habsburg heirs, never of the Habsburg emperor. Since home rule was refused, the last two fears could only be removed by force. The problem, as many Austrian statesmen saw it, was to crush Serbia without bringing in Russia and causing a general European war.

Germany and South-Eastern Europe. The experience of the years between the Congress of Berlin in 1878 and the outbreak of the European war of 1914 gave Austrian statesmen some reason to think that this might be possible. They had during this period been able, with the support of Germany, to extend their influence to the south-east. Russia and Serbia had been furious, but powerless to stop it. In 1878 Austria was allowed to occupy the two Turkish provinces of Bosnia and Herzegovina and to rule them as temporary protectorates: an arrangement rather like that by which the English then held Egypt. The Austrians remained in control of the two provinces on these terms until 1908, when a revolution in Turkey gave Austria the opportunity to annex them permanently. Bosnia and Herzegovina were mainly Slav by race and they lay between Serbia and the Adriatic. The Serbians, and their friends the Russians, had therefore a double reason for objecting to their annexation to Austria. There was danger of war, but in the end nothing came of their opposition since Russia was too weak after her defeat by Japan and a revolution at home to risk a general war.

Germany's support for the south-eastern extension of Austria-Hungary started as passive approval but later changed to active encouragement. At the time of the Congress of Berlin Bismarck had been pleased to distract the Habsburgs' attention from Germany, from which he had recently ejected them, towards other parts of the world. His support for the occupation of Bosnia-Herzegovina was given for the same reason which led him to encourage French colonial ambitions. The French were to forget Sedan and the Austrians Sadowa by conquests in Africa and the Balkans respectively. But by the beginning of the twentieth century the growth of manufacturing industry had forced the great industrial powers to seek markets and raw materials beyond their own frontiers. Germany's natural line of expansion was towards the south-east. Francis Joseph's dominions, with their large German population and their military dependence on the German alliance, formed a convenient approach to the untapped resources of the Balkans and the near and middle east. German capitalists and politicians talked of a railway from Berlin to Baghdad in the same sort of way that their English contemporaries demanded a line from the Cape to Cairo. In the twentieth century, therefore, Austria's expansion towards the south-east was dictated by two motives—Austria's desire to free itself from the Slavonic danger, and Germany's desire to find an outlet for the forces which were driving every great power along the path of the new imperialism.

41. THE EASTERN QUESTION. IV

The Balkan Wars: 1878 to 1914

The Armenian massacres. Twenty years after the Bulgarian atrocities had roused Mr Gladstone's anger, Europe was again horrified by the news of Turkish massacres of their Christian subjects. This time the victims were the Armenians, whose home was in Asia Minor. The Sultan had promised the Congress of Berlin that he would protect this people, and the English government had tried for some years to secure the appointment of an international civil service to see that the reforms which he had promised were carried out. The attempt was soon given up, and in 1895 a rebellion gave the Turks an excuse for what the English Prime Minister described as 'not merely massacres, but horrors unutterable, unspeakable, unimaginable by the mind of man'. England had long been patient with Turkey, hoping to extort reforms from the Sultan in return for defending him against Russian ambitions. These hopes had proved vain time after time. Now England's patience was exhausted and Lord Salisbury confessed that she had 'put her money on the wrong horse'.

In 1895 and 1896 it was England who took the lead in demanding intervention in Turkey and a partition of her empire which would free all her Christian subjects from Mahometan rule. Germany took England's place as the Sultan's champion and it soon became apparent that if England insisted on interfering she might find herself involved in a European war. The English government was not prepared to take the risk and the Armenian massacres continued. From this time English influence almost disappeared in Constantinople, and Germany began with Turkish approval to carry out her plans for a Berlin-Baghdad railway which would have made Mesopotamia an economic, though not a political, German colony.

Serbia, Montenegro, Bulgaria. Germany, however, was as helpless as England had been to prevent the break-up of the Turkish Empire in Europe. Between the two decaying empires of Austria and Turkey there was a belt of Christian states, warlike, barbarous and intensely alert to seize any opportunity of enlarging their territory at the expense of Turkey. To the north-west lay Serbia, where a blood feud between rival dynasties led to frequent changes of ruler in the course of the century. Some idea of the insecurity of life in the Balkans may be obtained from the history of these changes. In the Serbian struggle for independence, which began in 1796, there were two famous leaders, Kara Georg and Milosh Obrenovitch. At first they were friends, but as the rebellion prospered they quarrelled as to which should rule the free Serbia. In 1817 Milosh Obrenovitch had Kara Georg murdered. The subsequent changes can be shown in a table:

Date	Ruler	Fate	Family
1817–1839	Milosh	Expelled	Obrenovitch
1839	Milan I	Died	,,
1839–1847	Michael I	Expelled	,,
1847–1858	Alexander I	Expelled	Karageorgevitch
1858–1860	Milosh (again)	Died	Obrenovitch
1860–1868	Michael I (again)	Murdered	,,
1868–1889	Milan II	Expelled	,,
1889–1903	Alexander II	Murdered	,,
1903–1921	Peter I	Resigned	Karageorgevitch
1921–1934	Alexander III (I of Yugo-Slavia)	Murdered	,,
1934	Peter II		,,

To the west of Serbia lay Montenegro, the smallest of the Balkan kingdoms and the only other one to be governed by a native dynasty. Its territory, which was entirely mountainous, touched the Adriatic but did not include any good harbours. It was separated from Serbia by the narrow strip of the Sanjak of Novi-bazar which the Austrians had occupied in 1878 at the same time that they marched into Bosnia and Herzegovina. Montenegro was the only Balkan state which had never fallen under Turkish rule.

The third Slav state was Bulgaria, which lay to the east of Serbia. Her history as an independent state, though it only began in 1878, was almost as savagely varied as Serbia's. Her first prince was Alexander of Battenberg, a nephew of the Tsar. Russia expected to treat Bulgaria as an unofficial protectorate, and at first she controlled the government and the army. The Bulgarians, however, had other views, and in Stephen Stambuloff they found a fierce and energetic leader. Before long Alexander quarrelled with his uncle, and Russia found that Bulgaria was now a barrier between her and Constantinople instead of an avenue leading to it. She, therefore, opposed the further growth of Bulgaria and was furious when in 1885 Alexander and Stambuloff seized the opportunity of a rebellion in southern or Turkish Bulgaria to double the size of the state. The next year Russia kidnapped Alexander in his own palace and deported him to Austria. A government friendly to Russia was established and all would have been well if Stambuloff had not organised a revolution, driven the Russians out again and recalled Alexander. The restored prince, however, thought he had better make friends with Russia, and so Stambuloff organised yet another revolution, expelled him and called in a fresh German prince, Ferdinand of Saxe-Coburg. For eight years Ferdinand and Stambuloff gave Bulgaria strong government and the country prospered. Then they quarrelled and in 1895 Stambuloff was murdered—apparently with Ferdinand's approval.

Rumania and Greece. There were two other Christian Balkan states—Rumania and Greece. Both were Orthodox in their religion. That is to say that, like the other Balkan states, they belonged to the same branch of Christianity as Russia and to a different branch from Rome. Neither, however, was Slav by race or language. The Rumanians speak a Latin language and claim to be the descendants of Roman legionaries. They are the only Balkan people to have a common frontier with Russia, and through the first half of the nineteenth century the two principalities into which their country was then divided were completely under Russian control. After the Crimean War, however, Russian influence came to an end and before long the two principalities were united. Since 1866 Rumania had been governed by a Hohenzollern prince who, on the eve of the Seven Weeks' War, had been smuggled into the country through Austria, where he disguised himself as a commercial traveller and narrowly escaped discovery and internment as a Prussian officer. Greece was a small mountainous kingdom under a Danish king—the German prince who had been its original ruler was forced to abdicate in 1862.

These Balkan states were founded after wars of independence against Turkey: in every case her Christian subjects had had to fight for their freedom. In every case they had enjoyed the sympathy and the support of Russia. In every case the other great powers had started by disapproving of these struggles and trying to find some solution which would have given the rebels less than the independence which they wanted. The great powers took this line largely because they were afraid that the partition of Turkey would inevitably mean the expansion of Russia, and that the small independent Christian states would be so completely under Russian influence that they would serve her as stepping stones to Constantinople. Russia herself shared this view, but it is clear that it was always wrong. At the beginning of the twentieth century Serbia was the only Balkan state that could reasonably be regarded as a firm ally of Russia, and this was due to her precarious relation to Austria described in the last chapter.

Greece was cut off from the other Christian Balkan states by a solid mass of Turkish territory—by Albania on the west along the Adriatic coast, by Macedonia in the centre and by Thrace on the east. Albania was a country where the Christian and Mahometan communities were fairly evenly balanced, and Thrace had overwhelming Christian majorities. Unfortunately the Christians, particularly of Macedonia, were anything but united. Greeks, Serbians, Bulgarians and even Rumanians lived side by side, inextricably intermingled, irreconcilably jealous of one another. All the Balkan states were agreed that Macedonia must be freed; none of them could agree as to who should rule it.

Such was the position in 1903 when two events ended the relatively peaceful period which had followed the Congress of Berlin. The first was the brutal murder of the King of Serbia and his wife, who were hunted through their palace before being shot and stabbed to death by supporters of the Karageorgevitch faction. The recent kings of the Obrenovitch family had been friendly to Austria—on which they depended for money—and the change of dynasty meant a Serbian 'forward' movement against both Austria and Turkey. The second event of 1903 was a renewal of Turkish massacres in Macedonia which led the great powers with the exception of Germany, Turkey's new friend, to insist on an international police force being sent out to protect the Macedonians.

The Young Turks and the Balkan League. This was another humiliation for the Sultan, whose army began to plot against him. A group of officers, led by Enver Pasha, formed a secret Committee of Union and Progress whose object was to remove the Sultan Abdul Hamid and to introduce western efficiency into the Turkish government. In 1908 these men, who soon became well known as the Young Turks, revolted, marched out of the capital and established a new government. In the following year they deposed Abdul Hamid. Their revolution was intended to strengthen Turkey, its immediate effect was to weaken her and give other powers the opportunity to assert themselves. In 1908, while Turkey was engaged in her revolution, Austria formally annexed Bosnia and Herzegovina, Ferdinand of Bulgaria declared himself a completely independent sovereign (Bulgaria had been nominally a part of Turkey until this time) and Crete proclaimed its union with Greece. The island had for some years been free of the Turks, but under a curious provisional government set up by the great powers.

At first the Young Turk revolution was welcomed by Europe as a liberal reformation; it was in fact a mutiny by exasperated patriots. Their intention was to copy western efficiency, not western civilisation. Hitherto the chief characteristics of Turkish administration had been laziness and irregularity. Local officials were left a free hand to make their fortunes out of taxes and the sale of justice. Often they were brutal tyrants; more often they were merely negligent and without any sense of duty. The Young Turks were determined to introduce modern methods of organisation. They planned a single standardised system which would control even the furthest provinces properly. The new efficiency, however, was designed to reinforce and not to replace the old brutality. The great powers had seized the opportunity of the Young Turk revolution to withdraw the international police force from Macedonia. As soon as it had gone the Turks started once more their policy of massacre. Greece, Serbia and Bulgaria were outraged by the sufferings of their kinsmen and fellow Christians; they were terrified of a Turkish revival. Immediately they began to form a league among themselves with the secret support of Russia. This Balkan League was largely the work of Venizelos, a Cretan who had made a reputation in his island's struggle for independence. In 1910 he became prime minister of Greece.

First and Second Balkan Wars. The new League had not long to wait for its opportunity. In 1911–12 Italy conquered Tripoli from Turkey. In 1912 the states of the Balkan League attacked Turkey against the wishes of most of the great powers. The allies were unexpectedly successful, for the new German-trained Turkish army was not yet ready for war. In a few months the Turks were almost driven out of Europe. The Balkan allies found making peace a harder thing than making war. Two points were in particular dispute. Monastir and central Macedonia were coveted by Bulgaria but had been conquered by Serbia; Greece had just beaten Bulgaria in the race to capture Salonica, the great Macedonian port on the Aegean. After great pressure from the powers, the Balkan League and Turkey signed the Treaty of London in May 1913, which in the main divided the spoil by letting each keep what it held at the end of the fighting. Bulgaria, which had undertaken most of the heavy fighting towards Constantinople itself, was profoundly dissatisfied and took the first opportunity of trying to obtain Monastir and Salonica, of which she felt she had been unfairly cheated. A month after the Treaty of London the second Balkan war broke out. Bulgaria suddenly attacked Serbia and Greece, but was defeated. Then Rumania, which had hitherto been neutral, invaded Bulgaria from the north, and the Turks took their chance to win back Adrianople and most of Thrace. The Treaty of Bucharest, which followed, merely recorded these changes. No one expected it to be permanent, and from the point of view of the Balkan powers the war of 1914 was merely the fourth in the series which had started in 1911.

The Serbians were almost as dissatisfied with the results of the Balkan wars as were the Bulgarians. They had made important gains in territory and had occupied, besides the greater part of Macedonia, the Sanjak of Novi-bazar, which Austria had evacuated in 1908. But they had found their way to the sea once more blocked by Austria's intervention. Serbia had hoped to annex much, if not all, of Albania; but Austria was obstinately hostile, and after a period of extreme diplomatic tension, and the intervention of the fleets of the great powers, it was agreed that Albania should become an independent principality.

The greatest loser by the Balkan wars was Austria, which took no direct part in them. On her southern border stood Serbia, twice as powerful as before, still as discontented, and more ready than ever to stir up trouble among the Slavs of Austria. Next to her a new danger was appearing in Rumania, whose success against Bulgaria had made her anxious to liberate her kinsmen who lived across the Carpathians in Hungary. The rising ambitions of Serbia and Rumania strengthened that party in Austria which wanted a 'preventive war' to check them before they grew too strong.

MAP 20. THE EASTERN PROBLEM, 1878–1914

1. Turkish territory is shaded vertically or marked black. It cut Greece off from the other Christian states. This Turkish belt was, however, narrowed when Greece took Thessaly in 1881 and when Rumelia was joined to Bulgaria in 1885.

2. Macedonia, marked black on the map, provided the most difficult and important problem during this period. Salonica, the best harbour on the Aegean, was coveted by Bulgaria, Greece, Serbia and Austria.

3. Macedonia was inhabited by an impossible mixture of Turks, Bulgarians, Serbians, Greeks, Albanians, gypsies and even Rumanians. Thus although Uskub had only 32,000 inhabitants, it was the seat of a Turkish army corps, a Greek archbishop, a Bulgarian bishop—both Orthodox—and an Albanian archbishop, who was a Roman Catholic.

4. North of Macedonia was the Sanjak of Novi-bazar, Turkish territory garrisoned by Austria from 1878 to 1908. This prevented the union of Montenegro and Serbia, which was cut off from the Adriatic by the Sanjak. The Sanjak gave Austria a corridor towards the Aegean, enabling her to reach the Vardar valley which enters the sea near Salonica.

5. The changes caused by the Balkan Wars (1912–13) are marked on the smaller map. Montenegro (1) and Serbia (2) divided the Sanjak, thus checkmating Austria's eastern policy. Serbia and Greece (6) divided Macedonia, leaving Bulgaria (4) to be content with relatively unimportant gains in Thrace. Bulgaria also lost some of her own territory to Rumania (3). Albania (5) was formed into a separate principality. Crete, which had been semi-independent of Turkey since 1898, was given to Greece.

Map 20. THE EASTERN PROBLEM, 1878–1914

42. REFORM AND REVOLUTION IN RUSSIA 1881 to 1914

Alexander III and Pan-Slavism. The Russia of Alexander III and of Nicholas II was quite as decrepit an empire as that of Austria, but its weakness was not quite so obvious because there was less freedom in Russia and therefore less opportunity for discontent to make itself known. Alexander II had reluctantly attempted to buy off revolutionary agitation with inadequate concessions and had been murdered by Nihilists. His son, Alexander III, was convinced that his father's mistake had been to make any concessions at all, and in consequence rigid reaction was the characteristic of his reign. Once more the rulers of Russia turned away from the west and back towards their own country's separate traditions. Pan-Slavism made the Tsars a disturbing influence in European affairs, a repressive influence in domestic politics. The one led to the war of 1914, the other to the revolution of 1917.

In this chapter we are concerned with the internal history of Russia, the preparation for the revolution, not the preparation for the war. If Pan-Slavism abroad meant the encouragement of all Slavonic nations, at home it meant the discouragement of all non-Russian nationalities, of which there were many in the empire. Some of these were Slavs, some of quite foreign stock. All were in some sense deficient from the point of view of Holy Russia, which set itself up to be the measuring rod of true Slavism. In Russia church and state were one—the Poles might be Slavs, but they were Catholics and not Orthodox and so they did not look up to the Tsar as their religious head, their 'Little Father', as true Slavs should do. The Finns were not Slavs at all; they spoke an outlandish language which was an offence to true Russian ears, and they belonged to the Protestant branch of the Christian church. Opposite the Finns, on the southern shore of the Baltic, lived Letts, Esthonians, Lithuanians and Germans. In all the large towns, particularly of the south-west, there was a considerable Jewish population—in some cities a third of the population were Jews. These non-Russian nationalities were among the most progressive and prosperous elements in the empire, which of course only made them the more offensive. Besides these and other subject nationalities of the western half of the empire there were in the east an enormous variety of Asiatic races.

It distressed Alexander III that so many of his subjects should be of different blood, speak a different language, and profess a different faith from his own. It spoiled that perfect uniformity which his mechanical mind expected from his vast dominions with their flat, monotonous plains. He accordingly started a systematic campaign of Russification, imposing on the lesser nations the Russian language, the Russian religion, the Russian culture and the Russian outlook. The ugliest side to this campaign was an outburst of anti-Semitism, or hatred of the Jews, who were treated with wanton brutality. The continuous persecution and occasional massacres to which they were exposed by the refusal of the authorities to protect them gave Russia a bad name in civilised countries. The word 'pogrom' for a massacre of Jews is a Russian word which passed into common use in English at this time. Soon, however, maps marked an independent Finland, Esthonia, Latvia, Lithuania and Poland. That is history's epitaph on Alexander III. He provided all these outlying provinces with grievances which they nursed until in the disasters of the great war they could safely turn on their masters and establish their independence. This partition of Russia is an aspect of the 1917 revolution which is often forgotten.

Socialism and Industry. It was possible to be a pure Russian by blood, but to fail to be a true Russian by conduct. Such men —liberals, Nihilists or socialists—suffered even more from the Tsar's repression than did the subject nationalities. They suffered personally and individually, but the cause for which they fought prospered. The period from 1881 to 1904, which covers the whole reign of Alexander III and the first half of that of Nicholas II, saw a great increase in the revolutionary movement. The peasants remained dissatisfied, the university students and professors and many of the civil servants continued to demand a constitution, but the Tsarist government had managed to suppress their activities in the past and could continue to do so in the future. A new revolutionary force, the socialist movement, was a much more intractable problem.

Russia's industrial development had started late and made relatively little progress. In 1914 four-fifths of Russia's population were engaged in agriculture, while in the United States five men were engaged in industry for every two employed in agriculture. Railways are a fair test of industrial progress, and they were an even better one before the development of the internal combustion engine. In 1914 Germany had 11·7 kilometres of railway to every hundred square kilometres of land, even backward Austria had 7, Russia had only 0·4. But though very little of Russia was industrialised, those parts of it which were industrialised had been developed on the most modern lines. The few factories there were in Russia were giant factories. In the great industrial centres of Moscow and St Petersburg half the factory workers served in plants employing over a thousand hands. For the whole country the figure was almost as great—41 per cent, compared with 18 per cent in the United States. Russia had almost jumped the early stages of industrial organisation. She had almost missed, too, the building up of a substantial middle class. Forty per cent of all the shares in Russian industry were owned by foreigners, in heavy industry the proportion was much higher still. The Tsarist government was confronted in the two capitals with a large, well organised proletariat which it had to face without the support of a numerous middle class such as existed in the western countries.

The sharpness of social contrasts in Russia was a powerful encouragement to socialism. Naturally the socialist movement in Russia could only work in secret, underground. To this extent it resembled the earlier Nihilist movement, but there the similarity ended. The Nihilists had worked by individual terrorism, the socialists realised that the system which they detested could not be changed by assassination. They did not renounce violence, but they realised that violence directed merely against individuals was useless, since there was always another man ready to take the dead man's place—and usually, as in the case of Alexander II and Alexander III, the change was not to the advantage of the murderers.

Nicholas II. As long as Alexander III lived the revolutionary movement had little hope of success. His successor, Nicholas II, however, was a much weaker man and one under whom the country might expect to get into serious trouble, which would give the revolutionaries the chance for which they were waiting. Had Nicholas been king of England he might have succeeded admirably, for he was dignified, virtuous and dutiful. But he was in no sense a leader of men. He kept a diary, but rarely had anything more important to record than remarks about the weather, and the names of the gentlemen-in-waiting. In middle age his favourite authors were Conan Doyle and Baroness Orczy. His hobbies were cycling and photography. Such placid homeliness would have suited a simple country gentleman; it would not do for the 'Tsar and Unlimited Autocrat of All the Russias'. By the nature of autocracy the existence of the empire depended upon the personality of the emperor. Nicholas had to exercise the power of a Napoleon with the powers of a mentally-retarded weakling.

A wiser man would have discovered the incompetence of his army and navy and the corruption of his civil service, and would have carefully avoided that forward policy in Asia which in 1904 caused the Russo-Japanese War and brought a succession of humiliating defeats on his country. As disaster followed disaster, indignation at home grew higher. In July 1904 the leader of the reactionaries was assassinated. In January 1905 a priest, who had started a movement for better conditions of labour in St Petersburg factories, led an orderly procession to the Winter Palace to present a petition to the Tsar protesting against industrial abuses. Without the slightest provocation the troops opened fire, and many hundreds of unarmed workers were shot down in the streets of the capital. The day became known as 'Bloody Sunday'. It taught the people to expect nothing from the government except treachery and massacre. The old feeling that the Tsar was the father of his people was virtually destroyed by this incident, out of which the revolutionaries made good propaganda. The slaughter was not the Tsar's wish, but it was ordered by his ministers in his name. From this time there was a gulf between him and his subjects which no statesman on either side was willing or able to bridge.

The 'Revolution' of 1905. As the year grew older disorders increased. In June a conference of representatives from the county councils petitioned the Tsar for a constitution, which he promised to give them; but when it was published in August it was found that the Tsar proposed to keep all power in his own hands. At this stage the socialists took a hand in the agitation, which passed out of the control of the moderate liberals. A general strike paralysed the country and cut it off from the outside world. There were mutinies in the army and in the Baltic and Black Sea fleets.

'Soviets', or councils of workmen's and soldiers' representatives, were set up in the big towns, and a successful socialist revolution seemed possible. The Tsar realised that he could only keep his throne if he gave way. The Japanese War was brought to an end by the humiliating Treaty of Portsmouth and a new constitution was hurriedly announced. The first Duma or Russian parliament ment in 1906. But by this time the socialist revolution had been suppressed. The short-lived Red Terror had been followed by a longer and fiercer White Terror, organised by private bands of reactionaries, known as the Black Hundreds, who set out to exterminate all the elements which were hostile to the traditional Russian outlook, and particularly the Jews.

The first three Dumas. Consequently when the Duma met, the Tsar and his advisers had driven a wedge between the liberal and the socialist revolutionaries. The liberals had a great majority among the deputies, but the White Terror had crippled the socialists, who alone could have given the liberals the physical support they needed for a struggle with the crown. The Tsar had, therefore, little difficulty in getting rid of the first Duma when it quarrelled with him. A second was equally unsuitable from his point of view, so he altered the right of voting before the third Duma was elected in November 1907. In this way he obtained a tame conservative parliament which put no obstacles in the way of his autocratic government. But this victory had been won by a trick and not by any change of opinion in the country. It was a trick which was only possible because of the temporary exhaustion of the revolutionary forces, an exhaustion from which, sooner or later, they would recover.

From 1906 until his murder in 1911 the Tsar's chief minister was Stolypin. He was too reactionary to please the liberals, for he severely limited the liberties of the universities, and too moderate to please the court, for he tried to satisfy the peasants by a new land settlement. In the end the disturbances of 1905 had discredited autocracy without establishing democracy. The reforms which the outside world, and particularly British public opinion, greeted with warm approval existed for the most part only on paper. The Duma, for example, was little more than a debating society. The Zemstvos, or county councils, which under Alexander II had been the most lively and progressive bodies in Russia, were strictly checked.

Freedom of speech, of worship, of the press, were granted in theory but denied in fact. The spirit of the reforms was expressed in the slight alteration of the imperial title from 'Tsar and Unlimited Autocrat of All the Russias' to 'Tsar and Autocrat of All the Russias'.

War or Revolution? The crisis of 1905 settled nothing. It left Russia floundering with half a constitution, which is worse than no constitution at all. The underground work of the socialists continued with renewed vigour. The events of 1905 were a severe defeat for them; but, once the first disappointment was over, they realised that it was a defeat which held the promise of a future victory. The situation developed into a breakneck race between revolution and war. It was war which won, but revolution came in a good second in 1917. It is wrong to regard the Russian revolution as simply the result of the European War. Both catastrophes had been long in preparation. To the Tsar's government a successful war seemed an alternative, the only alternative, to revolution. It was, therefore, ready to take the risk although it knew that another unsuccessful war would be fatal.

43. DIVIDED LOYALTIES

France: 1882 to 1914

From politics to art. Thiers had said 'a republic seems to be the form of government which divides us least'. He was probably right, but even a republic left Frenchmen hopelessly opposed to one another. The third republic could not help being an unexciting compromise between a restoration and a thorough revolution, either of which would have commanded a devotion which the compromise was unable to inspire. The trouble was that either would also have aroused fierce hatred among the followers of the other, a hatred which the insipid republic failed to provoke. After 1882 the majority of Frenchmen grew largely indifferent to public affairs, despised the Chamber of Deputies, and were sickened by its corruption; while a noisy minority became keen politicians competing with one another for power, often by fairly shady intrigues.

Men turned from politics, where the best that could be had was hardly a second best, to art and letters, where the French genius was the admiration of Europe. In literature Alphonse Daudet, de Maupassant, Zola and, a little later, Anatole France were writing novels which were read not only in France but wherever a well-told story was appreciated. But it was in painting that the most remarkable progress was made. Paris in the days of the Impressionists and Post-Impressionists was as much the artistic capital of Europe as Florence or Athens had ever been. Men like Pissarro, Cézanne, Gauguin, Manet, Picasso and Van Gogh taught the world to see differently, to value light as well as substance, to study pattern and shape for their own sake and to bring back imagination into art. The photographer had ousted the artist as far as a 'good likeness' went; but the French painters found new tasks for art at a moment when science seemed to have brought it to a dead end. Not all the painters were Frenchmen—Picasso was a Spaniard and Van Gogh a Dutchman—but their spiritual home was in Paris. The modern French schools rank with the great schools of Italy and the Netherlands as an unsurpassed summit in the art of painting.

It was not only artists who came to Paris. That city remained under the Third Republic the pleasure and luxury capital of the world. Foreigners were tempted to forget that Paris of the tourists was vastly different from France of the French. Paris had not escaped the decadence of cities which thrive on the extravagance of visitors; but foreigners who therefore assumed that France was decadent were profoundly mistaken. The spirit of Paris is in general mutinous against discipline, whether moral or legal; fickle in political loyalty; expert in all the ways of comfortable living; abreast of the latest fashions, not only in dress but in ideas; indifferent or hostile to organised religion. But the spirit of France as a whole is not Bohemian but bourgeois. The peasant proprietors of the villages and the middle-class tradesmen and professional people of the sleepy rural towns are solid men of property, firm upholders of an ordered society, often severely puritanical, anxious for commercial success, thrifty, industrious and highly domesticated.

Anti-clericalism. It is tempting, though misleading, to equate the fierce struggle over religion, which marked the turn of the century, with this difference between Paris and the provinces. But the provinces were by no means all homes of religion. Generally speaking the north and west of France was and is practising Catholic, but the south has for centuries been a centre of militant anti-clericalism. It returns to the French parliament an almost solid mass of deputies who regard the Roman Catholic church, the only church which exists in appreciable numbers in France, as their great enemy. In France the organisation of these enemies of organised religion is the Society of Freemasons, who are organised on different lines from the English society. Most French Freemasons are also 'free thinkers', people to whom organised religion is superstition. But not all anti-clericals are necessarily anti-religious. An anti-clerical is a man who objects to the interference of the church in politics. In France, because of what happened in the first French Revolution, the Roman Catholic church has usually allied itself with the Monarchists, and in consequence most, though by no means all, republicans have been strongly anti-clerical. To Englishmen this sounds as if republicans were not far removed from dangerous social revolutionaries, but in France the most aggressive atheists and agnostics have often been staunch defenders of private property and the existing social order.

After 1871 the church hoped to see a restoration, and its priests worked hard for Marshal MacMahon during the crisis of 'seize mai', 1877. It was then that Léon Gambetta told his countrymen: 'Clericalism, that is the enemy.' In 1892 Pope Leo XIII, seeing that the republic had come to stay, tried to undo the harm that had been done by ordering all French Catholics to accept the Republic and be loyal to it; but dislike of the church was too strong and widespread to be quietened in this way.

The education question. The main struggle took place over education. Naturally the church did all it could to see that children were brought up as devout Christians. Many of the religious orders were specially trained to maintain schools in which French boys and girls would learn to be good Roman Catholics. It was equally natural that the anti-clericals should object strongly to the control of the church over education; some because they thought that the church taught royalism as well as religion, others because it taught religion at all. For twenty-five years, from 1880 to 1905, the struggle went on. The church fought hard all the time, but consistently lost ground. In 1882 a law, for which Jules Ferry was responsible, made elementary education free and compulsory. No religious instruction of any kind was in future to be given in the state schools, which were paid for out of taxes; but if parents liked to send their children to private Roman Catholic schools, the government had no objection. At that time about a third of the children of France went to these private Roman Catholic schools. The next stage in the struggle came in 1904, when Émile Combes secured the passage of a law forbidding nuns or monks to teach in any schools in France. The part played by the private schools has gradually grown less, and by 1925 only a fifth of French children went to schools where any sort of religious instruction was given. The anti-clerical agitation was not confined to education. The movement led by Combes went further and secured in 1905 a law of separation abolishing the Concordat which Napoleon I had made with the Pope. The most important result for the church was that its clergy ceased to be paid by the state and have had since then to depend on voluntary contributions.

French socialism. This anti-clerical campaign, France's equivalent of the German Kulturkampf, kept pace with the steady swing to the moderate left which was a significant feature of French domestic politics after 1880. From this time the revolutionary tradition of the French workers was gradually diverted into constitutional channels. In 1884 Jules Ferry took the important step of legalising Trades Unions which had been forbidden by previous French governments. Thanks to him French workers were no longer driven in their plentiful disputes with their employers to fall back on the illegal use of force which might easily develop into revolutionary action. Gradually socialism, which had been the dreaded outlaw of the Second Empire, became recognised as a political nursery for the moderate conservative politicians of the Third Republic. Men such as Millerand, who rose to be President of France, and Briand, an outstanding diplomat in post-war Europe, started life as extreme socialists, but soon became socially rather more reactionary than an average English conservative member of parliament. Of all the well-known pre-war French socialist leaders only Jaurès, who was murdered in 1914, remained a member of the party to his death. It is roughly true that French public opinion moved towards the Left because socialism moved towards the Right by abandoning in practice its threatening revolutionary attitude in favour of a more sober programme of reform by stages.

Boulanger; and the Panama scandal. In spite, however, of this underlying stability, foreign observers and many Frenchmen persisted in believing that the republic was so ramshackle that it could not last. This was due partly to the frequent changes of government caused by the group system, partly to the fact that the leaders of French society remained royalist and boycotted the republic, but still more to three great and disturbing crises which punctuated French history in the generation before the war. The first crisis started in 1886 when General Boulanger, who had been commanding in Tunis, became Minister of War. He had been appointed because he was a republican leader, but he soon showed himself to be a candidate for dictatorship. The old Bonapartists, the Orleanists and the church hurried to his support, and a well-financed propaganda campaign gave him a large following. He was dismissed first from the Ministry of War and then from the army, but when Paris elected him to parliament he was within striking distance of a throne. His nerve failed him, however, and instead of organising a *coup d'état* he fled to Belgium when a rumour of his intended arrest reached him. His only outstanding quality was his vanity: clearly he was not of the calibre of which dictators are made, but the great support he received badly frightened republicans and taught them that a stronger man would have succeeded. In 1891, two years after his final failure and flight, Boulanger blew out his brains over the grave of his mistress in a cemetery in Brussels.

France had barely recovered from the Boulanger crisis when in 1892 the 'Panama scandal' was discovered. The company which Ferdinand de Lesseps, hero of the Suez enterprise, had formed to cut a canal through the isthmus of Panama was discovered to be not only bankrupt, but grossly corrupt. It had bribed ministers and members of parliament to support its plans. This was only the worst of several other unsavoury disclosures which destroyed public confidence in the honour of the statesmen of the Third Republic.

The Dreyfus case. Two years later began the astonishing Dreyfus case, which for a dozen years split France into two bitterly hostile sections and nearly caused a civil war. In this affair all the existing hatreds and divisions of French political life were swallowed up. Captain Alfred Dreyfus was a staff officer who was charged with selling military information to Germany because of a letter, said to be in his handwriting, which had been intercepted by the secret service. A court-martial found him 'Guilty' after a highly irregular trial in which some of the evidence was kept from the prisoner. Dreyfus was transported to Devil's Island off French Guiana. So far it was a sad but not an unusual story. A few years later a new man, Colonel Picquart, was appointed Chief of Intelligence in the War Office. New evidence of espionage and a further study of the documents convinced Picquart that the real traitor was not Dreyfus but a brother officer. Picquart was immediately sent off to a dangerous post in Tunis and replaced by a Colonel Henry, who had been concerned in the original prosecution of Dreyfus.

There is a double key to this overpowering desire to conceal the original miscarriage of justice: first, the professional reluctance of the army authorities to admit that they had made a mistake; secondly, the fact that Captain Dreyfus was a Jew. For some years a noisy campaign had been carried on against the Jews. This campaign, helped by the fact that some of those concerned in the Panama scandal were Jews, had converted the leaders of the army and church into keen anti-Semites. To them the case was simple: Dreyfus was a Jew, so Dreyfus must be guilty. The Dreyfusards maintained that the broken officer was a victim of this prejudice against his race and faith. The anti-Dreyfusards thought it monstrous that the honour of an unknown Jew should be set above the reputation of the high command. To begin with, the odds on their side were great. The whole machinery of the War Office was mobilised to prevent the case being re-opened; but at last Picquart the soldier, Clemenceau the politician, and Zola the novelist succeeded in obtaining a new trial. First of all they managed to bring the real culprit to trial. If he was guilty, Dreyfus must be innocent—but the War Office would never admit this, and so Colonel Henry forged papers to prove Dreyfus' guilt. Finally this forgery was brought home to Colonel Henry, who confessed and committed suicide.

It was no longer possible to refuse Dreyfus a new trial. He was brought back from Devil's Island, and in 1899 a second court-martial brought in the ridiculous verdict of 'Guilty, but with extenuating circumstances'. Everyone knew that the soldiers were merely trying to save their own face, and Dreyfus was immediately pardoned by the President. Finally, in 1906, a third trial by a proper law court acquitted Dreyfus of all guilt and he was restored to the army with the rank of major.

The republic as well as Dreyfus had been on trial. Behind the prosecution stood the high command of the army and the leaders of the church, both of which were hostile to the republic and hoped that it would soon be replaced by some form of government from above, such as a dictatorship or a monarchy. Liberals, socialists and radicals temporarily forgot their differences in defence of Captain Dreyfus. At the beginning of the case many socialists welcomed anti-Semitism as being a justifiable attack on the seamy side of capitalism, but when it was taken up by the church and the army they recognised it as a dangerous weapon against the liberties of the subject. Thus the fate of an obscure staff officer became the fate of France. Republicans and monarchists staked their whole future on the relatively trivial question of who wrote two sheets of note-paper. When Captain Dreyfus was at last set free, it was a victory for freedom of speech and thought over a bigoted church, and for the authority of parliament and of civilian ministers over the autocratic ambitions of the army.

Clemenceau and revenge. Dreyfus had many friends, but perhaps the man who did most to free him and save the republic was Georges Clemenceau, a statesman of genius whom France and the world feared as 'the Tiger'. It was a fitting nickname for a great patriot who was also a great radical. Clemenceau's passionate and bitter nationalism made it impossible for the reactionaries plausibly to claim that the only patriots were monarchists. His virile leadership inspired the jaded republic with a new faith. So long as the republic could find leaders of his quality it was ridiculous to argue that only a restoration could make France once more a great power in Europe. In fact it was the Third Republic which recovered in 1918 those provinces which the Second Empire had lost in 1870. Democrats had looked forward as eagerly as monarchists to the revenge on Germany. It is difficult for Englishmen to realise how much the redemption of Alsace-Lorraine meant to France in the years before 1914.

This longing for revenge was coupled with a fear of further punishment by Germany, so that Frenchmen had both strong positive and negative reasons for maintaining a great and efficient army. The foundations of a new French army had been laid in the years immediately before the Dreyfus affair. At this time a permanent general staff on the German model was established and conscription was made really universal. Military service had long been compulsory in France, but there had usually been a high proportion of young men who were excused service on one ground or another. From this time the French army became truly 'the nation in arms', a great levelling and democratic institution in which rich and poor had to serve side by side for three years. In 1905 the period of service was reduced to two years, but when the fear of Germany grew even more intense in 1913 it was once more raised to three.

Empty cradles. But France had one weakness which neither the strict new conscription laws nor the improved training of the regular army could remove. Her birth-rate was falling at a much faster rate than that of other western European countries. Her population was stationary, while that of Germany was still increasing, and foreign immigrants were being called in to fill up the vacant spaces which were beginning to appear in France. For this still unsolved problem she found a partial remedy in colonial expansion, which gave her a coloured army to reinforce her own. During this period Tunis, Madagascar and Indo-China were added to the French empire.

In 1914 France was still apparently a nation of deeply divided loyalties—Catholics against anti-clericals; republicans against monarchists. The parliamentary system was discredited, and many Frenchmen had simply given up all political interests in despair. By contrast Germany was a united and confident nation. The contrast was, however, not altogether fair to France. The people who thought France decadent because of the confusion of her politics were proved wrong when the whole country instantly united at the first coming of war. In 1914 France justified the old compliment that 'the French are always excited except in moments of excitement'.

44. IN SHINING ARMOUR

Germany: 1890 to 1914

German efficiency. Three military despotisms entered the Great War. None survived it. The much less imposing democracies, France and Great Britain, proved more lasting and emerged injured but intact. The empires of Russia, Austria and Germany disappeared in chaos before defeat in the field and revolution at home. It was not surprising that the first two failed to survive a war since even in peace they could hardly hold together. The eclipse of the Hohenzollerns was more surprising. As a family they were efficient, industrious, dutiful, patriotic, and, in their own way, devoted to the Protestant religion. Their record was relatively free from corruption, decadence and wanton oppression. Under them, and very largely because of their efforts, Prussia had grown from a primitive sandy waste into a great power, perhaps into the greatest of the powers.

Other nations admired Germany nearly as much as she admired herself. Her army was the most efficient in the world, her civil service the ablest, her universities the most learned, her people the most industrious and her business men the keenest and most persistent. English schoolboys might sneer at the initials D.R.G.M., which appeared on all goods made in Germany, and expand them into 'Dirty, rotten German make', but their familiarity to small boys showed how widespread German commerce had become. Nevertheless Germany had risen to unaccustomed wealth at a speed which had upset her moral balance.

The Kaiser and his court. The Kaiser's court and the Kaiser's Germany were abandoning the old Spartan simplicity of the Prussian squires and the comfortable, domestic life of the lesser German courts. They were turning instead to stupid and vulgar extravagance. At the court there was an expensive and largely unemployed crowd of minor servants. Six hundred lackeys were kept so that the great state banquets might be impressively served. None of them were on duty for more than a third of the year. The royal stables had to turn out two hundred carriages a day for the use of the Kaiser and his courtiers. The same passion for display marked the Kaiser's holidays. On his country estate he organised monster shooting parties for which all the game was carefully driven so that a big bag was inevitable. When he was forty-two the Kaiser erected a monument to record the slaughter of his 'fifty-thousandth animal, a white cock pheasant'. At sea on his yacht he fell back on cheap practical jokes, making elderly orderly officers practise physical exercises and pushing them over while they were doing a 'knees bend'. The bad taste of the court circle soon spread to the other parts of society. It was a warning that all was not well with Germany; but the impressive material achievements of the empire prevented men from learning the lesson.

In Germany everything depended on the emperor. His ministers were responsible to him and not to the Reichstag, and the character of William II was, therefore, of immense importance to his people. At first sight William II had much in common with our Henry VIII. He had the same impulsive interest in every side of life, displayed the same indiscriminate patronage of all the arts, was filled with the same restless energy and the same uncontrollable dislike of opposition. He had, too, the same faculty of being almost an expert in many techniques. His uncle, Edward VII of England, who heartily detested him, is reported to have described him as 'the most brilliant failure in history'. That is an exaggeration, but the contrast between his brilliance and his failure is remarkable. It is charitable to remember that he suffered from a withered arm. Perhaps his constant struggle with pain and ill-health helped to unbalance his character. It made him excitable, moody, ill-mannered and fatally reckless in speech. Unfortunately he was as sensitive as he was tactless and took offence as easily as he gave it. The best example of his fatal facility for saying the wrong thing is the celebrated interview which he gave the *Daily Telegraph* in 1908. The interview was meant to be reassuring to England but in it he said:

> You English are like mad bulls—you see red everywhere! What on earth has come over you, that you should heap on us such suspicion as is unworthy of a great nation?...I regard this misapprehension as a personal insult!...You make it uncommonly difficult for a man to remain friendly to England.

It was hardly surprising that his words had the opposite effect to that which he intended. The German ambassador in London pathetically told his staff: 'Now we may shut up shop.'

The repercussions of this interview were indeed so unfortunate that the Chancellor was able to persuade his master to keep silent on public affairs for nearly two years. In 1910, however, he broke out again, this time to preach the Divine Right of Kings, in which he believed with a kind of mystical rapture. This comfortable doctrine satisfied both his genuine religious instinct and his abounding conceit, for it meant that he could look on God almost as a Senior Partner in the Hohenzollern firm. His laboured militarism—in a public speech he described himself as standing 'in shining armour' by Austria's side—was a case of supply meeting demand. His people expected it as much as he enjoyed it. They had been taught to admire the hectoring manner which in an age of showmanship passed for majesty.

'Dropping the pilot.' It was plain that William's vanity and Bismarck's blunt mastery could never agree. Their differences came to a head two years after William's accession, in 1890. Bismarck wanted to strengthen the law against socialism. The emperor, who sympathised with the working class, prevented this. The emperor also demanded the right to deal directly with all his ministers; while Bismarck thought that the Chancellor should be the only means of communication between the sovereign and his government. Bismarck sent in his resignation. It was a habit of his, but always before he had been asked to withdraw it and given his own way. This time it was accepted and the maker of the German Empire retired to an embittered private life.

William II changed his policy as well as his minister. Bismarck's great object for the last twenty years had been to avoid trouble with France. He had managed to keep her isolated in Europe by making alliances with Italy, Russia and Austria. Towards the end of his ministry Russia was becoming increasingly restive and Bismarck, therefore, planned to replace her by England in his system of alliances. Nothing that William could do in the years that followed Bismarck's fall could prevent the bad relations between Austria and Russia growing worse. Russia therefore sided more and more with France and in spite of all Germany's efforts a Franco-Russian alliance was made in 1895. It was clearer than ever that Germany now needed England's friendship and should have done everything possible to obtain it.

William II was not naturally opposed to an English alliance—Queen Victoria was his grandmother and he had happy recollections of the country. The trouble was that he was unwilling to make any concessions to meet the English point of view. Bismarck knew perfectly well that England would never willingly allow the greatest military power in Europe to be also a great naval power. He had always, therefore, refrained from building a navy to compete with England's. The new Kaiser, however, determined to build an immense fleet at a time when Germany needed England's friendship far more than she had ever done during Bismarck's ministry. William was not building his navy for use against England and he thought that England should have been satisfied with his word. As it happened England herself was feeling the need of a friend on the continent at this time and twice, in 1898 and in 1901, she tried to negotiate an alliance with Germany. Both attempts failed because the Kaiser refused to accept any agreement to limit the size of his fleet in a way which would have satisfied England. It was pride and not policy which dictated his refusal; it was his stupidity, not his hostility, which drove England to make friends with France instead of Germany.

Tirpitz and the Navy question. The strongest character among the men who surrounded William was Grand Admiral von Tirpitz, one of the few straightforward and sincere men at the imperial court. He was an expert and had all the expert's passion for his own subject to the exclusion of all other considerations. A born sailor and a firm patriot, he was quite incapable of seeing that Germany might be safer and better without a navy than with one. His two great naval programmes—those of 1900 and of 1907—were the main obstacles to an understanding with Britain. The first was followed by the Anglo-French Entente; the second by the Anglo-Russian. When Germany suddenly set herself to create a navy virtually out of nothing, England inevitably thought that it was meant for use against her, an impression strengthened by the Kaiser's characteristic remarks that 'our future lies on the water', and that, 'the trident is within our grasp'. In England men retorted 'We want eight (battleships) and we won't wait'. The two governments, therefore, started an armaments race inspired by the lunatic logic of the old lady who said 'we must build our fleet up to the level that they said they would build theirs up to if we built ours up'.

Junkers and manufacturers. This naval expansion, which proved fatal to the Kaiser's Germany by losing it England's friendship, was a reflection of a change in her internal balance of power. Germany's home politics were determined by her recent industrial revolution, which had divided her people into three powerful groups engaged in a struggle for power. The first, the Junkers, were country squires and great landlords, soldiers and administrators, who had made Prussia the great nation that she was. They despised and feared the men of the second group, the 'new rich' of the big cities, the manufacturers and financiers, who were much their social inferiors but their equals in influence. These big business men had in 1892 and the following years secured an introduction of a new economic policy which seriously hurt the Junkers. Trade agreements were made with foreign countries by which Germany arranged to import more corn and export more finished articles than before. The manufacturers, of course, gained by this policy, but the landed gentry were threatened with ruin by the flood of cheap grain from Russia, Rumania and the United States.

The colonial empire. Another reflection of the increased power of the German manufacturing class was the demand for colonies from which raw materials could be procured and to which goods could be sold. Bismarck had been forced reluctantly to acquire colonies in the 1880's when most of the overseas German empire was secured. In 1884 South-West Africa, Togoland, and the Cameroons were taken under German protection, and the north of New Guinea was annexed. In 1885 a large territory in East Africa (now called Tanganyika) was added. Germany tried hard under William II to increase her empire and to make it pay, but on the whole she was not very successful. Compared with the wide and fertile estates which Britain and France held, Germany's overseas empire was hardly more than a meagre back-garden. For climatic reasons her colonies offered little scope for migration, and most of Germany's surplus population continued to go to the United States or Brazil. Neither as markets nor as sources of raw materials did they really pay their way. The total value of their trade was estimated at only about £3,000,000, and quite often they were run at an actual loss.

A more successful outlet for Germany's expanding industry was the diplomatic and economic domination of other independent countries which were too weak to resist her. Thus, the reign of William II saw Turkey brought more than ever under the influence of Berlin and Vienna, and away from that of London and Paris. The valuable privilege of building a railway to Baghdad was granted to Germany in 1902. Another aspect of this commercial penetration was the building of a great and efficient mercantile marine to which large government subsidies were given. Important lines, such as the Hamburg-Amerika and the Norddeutscher-Lloyd, were beginning to run great ships like the *Berengaria* and the *Majestic* (both confiscated and renamed by the Allies after the War).

Germany's interests outside her own frontiers and outside Europe were thus increasing rapidly in many ways during William II's reign. This expansion was the result of the energy of the new, rich manufacturing class. Its consequence was the demand for a big navy, a demand which appealed to patriotic sentiment as well as to business sense. The opposition to the big navy, and to the commercial policy which made it appear desirable, was weakened by the fact that it seemed unpatriotic to oppose Germany's growing power. The Junkers, who were the natural leaders of this opposition, were by training and inclination the most patriotic of Germans. Their objections to this dangerous new German development were, therefore, always half-hearted.

The socialist movement. Another circumstance threw Junkers and manufacturers into an uneasy alliance. This was their fear of the growing power of the socialists, the third powerful group in William II's Germany. The German trade unions were the best organised and most powerful, the German socialist party the most efficient in the world. Labour leaders everywhere held up Germany as an example to their followers. The same genius for organisation which had enabled Germany to make her army the envy of the world had given her workers the leadership of the international working-class movement. This leadership, however, was a thing which middle- and upper-class Germans hated just because it was international. The German miner was taught that he had more in common with the French miner than with the German mineowner.

The more efficient German socialism became, the greater the danger to the German state. It was the one form of national efficiency of which the country's leaders disapproved. As it happened their fear was ill-founded. When war came in 1914 the great majority of the German socialists supported the government. Their reason was characteristic of the state of Europe in 1914 and of its ill-assorted alliances. The German socialists cheerfully fought for Germany because it was a war against the Russian government, which had so brutally suppressed the attempted socialist revolution of 1905, just as the English socialists cheerfully fought for Britain because it was a war against the German government which had so brutally invaded Belgium.

THE APPROACH OF WAR

1882	Triple Alliance (Germany, Austria, Italy)
1895	Franco-Russian Alliance
1899	Boer War
1900	First Big German Navy Programme
1902	Baghdad Railway Concession
1904	'Entente Cordiale' (England, France)
1905	Rebellion in Russia Kaiser at Tangier
1906	Algeciras Conference
1907	Second Big German Navy Programme Triple Entente (England, France, Russia)
1908	Annexation of Bosnia-Herzegovina
1911	The 'Panther' at Agadir
1913	French Military Service raised to three years
1914	Murder of Archduke Francis Ferdinand

Map 21. CAUSES OF THE WAR OF 1914

1. This map shows pictorially the main problems which were disturbing Europe in the years just before 1914.

2. England, France and Russia—the Triple Entente—are shaded diagonally. Germany and Austria-Hungary, the two dependable members of the Triple Alliance, are left white. Italy, the unreliable member of the Alliance, is speckled. Germany's main fear was that she would be encircled by enemies.

3. England was worried by the navy which Germany was building (A) and by the railway (B–B) which she was making from Berlin to Baghdad. England feared for her trade, her Eastern empire and her command of the seas.

4. France still hoped some day to recover Alsace and Lorraine (C), which she had lost in 1870. She was worried by the contrast between her low birth-rate and stationary population (D) and Germany's high birth-rate and growing population (E).

5. France hoped to supplement her man-power by drawing on her North African colonies, to which she was busily adding Morocco (H). Germany, however, also coveted Morocco—the ship indicates the visit of the *Panther* to Agadir. England was interested in this dispute because of her Mediterranean communications.

6. Austria could count on German help because of that country's plans for south-eastern expansion (B). But she was worried by her minorities, who wanted to join their kinsmen in the Balkan States (F). Russia, Austria's old enemy, supported these plans. She was akin by race and religion to Serbia, and by religion to Rumania.

7. At the same time Russia herself was frightened by the possibility of a socialist revolution at home (G).

8. Italy was the dark horse of Europe in 1914. She is shown (K) walking a tight rope between Tunis (French but with a large Italian population) and Tyrol (Austrian but with a large Italian population). Her desire for colonies and particularly for Tunis made her join the Triple Alliance; her desire for gains at Austria's expense made her desert her allies when war came.

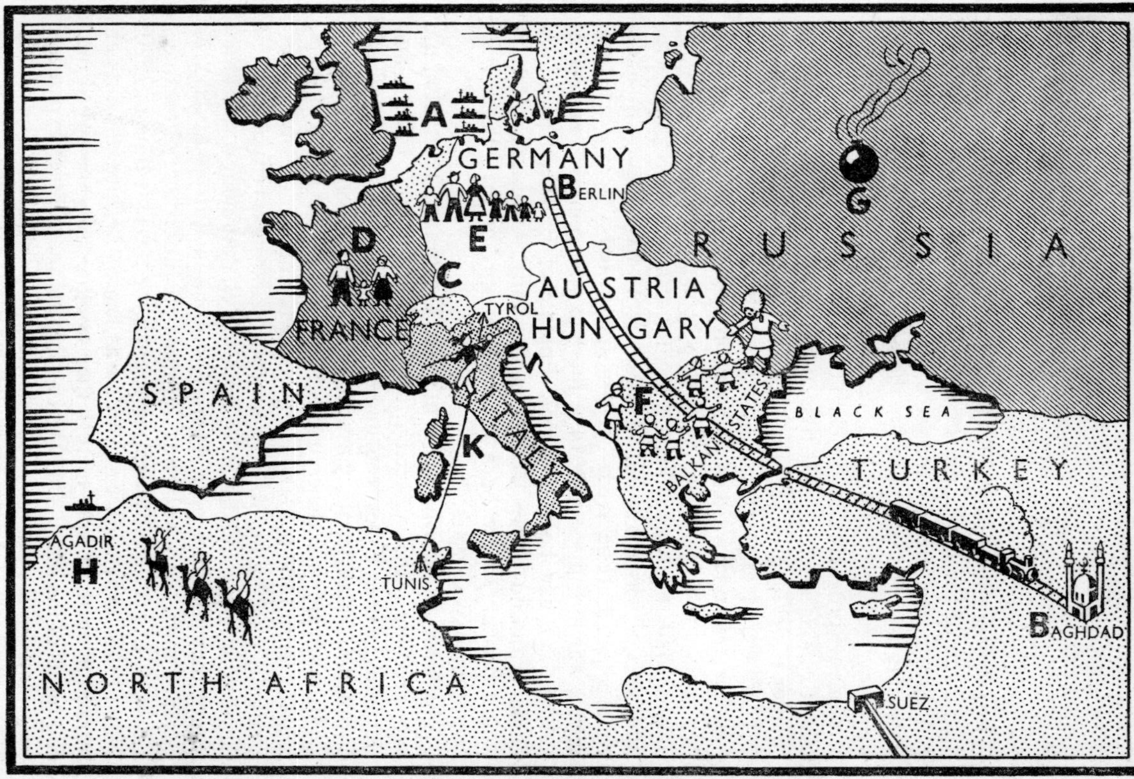

MAP 21. CAUSES OF THE WAR OF 1914

45. PARA BELLUM

For a generation before 1914 the great powers of Europe followed the advice of the old Latin saying 'si vis pacem, para bellum' (if you want peace, prepare for war). They saw war coming and each state determined to prepare for the war, not realising that in this way they were preparing war, making it inevitable by their rival armaments programmes. In each of the chapters in this part some aspect of this international rivalry has been described. This final chapter is devoted to knitting together these separate threads.

There were two great territorial problems in Europe before 1914—an eastern and a western problem. It was the linking up of these two problems, so that eventually they became one problem, which turned what might have been a local war into a world war. This 'linking up' was the result of the system of competing alliances under which Europe lived. There would probably in any case have been a war in Europe in 1914 as there had been in 1912 and 1913; but, if there had been no Triple Entente of France, Russia and England and no Triple Alliance of Germany, Austria and Italy, it would probably have been no more important than those half-forgotten Balkan wars.

The western problem: France and Germany. The first of the two problems, the western problem, was that of France and Germany. There is between undisputed French territory and undisputed German territory a wide belt of land which has changed hands many times in the course of history. For two hundred years, roughly from 1600 to 1800, France had been the aggressive power, gradually pushing her frontier eastwards, biting off little pieces of German territory and digesting them so well that in most cases they forgot their German past. There is no one in Europe to-day, for instance, more French and less German than the ordinary Lorrainer, although Lorraine only became French for the first time in 1670. This French expansion eastwards was lasting because the French were able on the whole to turn their conquests by the sword into conquests of the heart. This was possible because French expansion took place before the age of self-conscious nationalism. The people of the conquered regions were already French citizens at the time that men first began to take nationalism seriously as the only suitable recipe for making a good state.

The Germans were unwilling to recognise that they had permanently lost the land they had lost recently. They wanted to win back Alsace and Lorraine because they had not so long ago been German, in spite of the fact that they now wanted to be French. The possibility of further French expansion eastwards was removed when Bismarck in 1870 consolidated the separate German states into one strong empire, and even for the time being won back Alsace and Lorraine. From that time France and Germany stood suspiciously eyeing each other; Germany afraid that France would once again, as under Napoleon, try to make the Rhine her frontier; France fearful lest 1870 should prove to be only the first of a series of invasions.

The eastern problem: Austria, Russia, Turkey. The parties to the second great European problem, the eastern problem, were Austria, Russia and Turkey. In 1914 Turkey in Europe was confined to a small stretch of land near Constantinople. Her former vast European empire had been added either to Austria or to Russia or made into more or less independent and barbarous little kingdoms—the Balkan states. It was out of the relation of these little states to the two great empires of Austria and Russia that the war of 1914 immediately arose. The two empires were great in tradition and great in extent, but very small in power. They were only a little less sick than Turkey, the 'sick man of Europe'. It was out of their weakness—their fear lest any blow to their prestige might prove fatal by starting a revolution—that the war started. The Austrian government had one overmastering fear, the fear that her Slav subjects would revolt and throw in their lot with Serbia, the neighbouring Slav Balkan state. The Tsar's government had only one asset—the spirit of Pan-Slavism, the belief that Russia was the protector of the little Slav Balkan states against Austrian imperialism. When Serbia got into trouble with Austria in 1914, through the murder of the heir to the Austrian throne, war became inevitable. Russia dared not allow Serbia to be crushed by Austria since, if this happened, she could no longer claim to be the protector of the Slav states. Austria equally could not allow Serbia to persist in a campaign which before long would have led to the break-up of the Austrian empire.

Under Bismarck Germany's policy had kept the eastern and western problems separate. His policy towards France had been to encourage her to seek compensation in colonies outside Europe for her loss of Alsace and Lorraine. This policy reduced Franco-German friction and gave no cause for Anglo-German rivalry though it caused intense colonial jealousy between France and England. His eastern policy was to keep on friendly terms with both Russia and Austria by seeing that each kept on relatively friendly terms with the other. He therefore usually discouraged both Russian and Austrian expansion in the Balkans since this was likely to lead to war between them. He also gave to each a guarantee of help against the other.

The problems merged. Before his dismissal Bismarck had been forced by German public opinion partly to abandon both these policies. While Bismarck was still Chancellor Germany had started to acquire colonies and to make a closer friendship with Austria than with Russia. After his resignation the new Kaiser, William II, carried this change of policy much farther. He entered into a close alliance with Austria and Turkey, hoping to make them economically, though not politically, parts of the German Empire in the same sense that the Argentine is part of the British economic sphere of influence. This naturally frightened Russia, who turned towards France in search of an ally. The eastern and western problems had become joined in a way which was likely to lead to war.

This unification of the two problems proved fatal in 1914 because neither Germany nor France dared allow their weaker allies to be humiliated. Germany argued in this way. If Austria were humiliated by Serbia and Russia, then before long there would be a complete break-up of the Austrian Empire and Germany would lose her only faithful ally. Austria, therefore, knew that she could do what she liked and count on German support however much Germany might disapprove of her action. France argued in a similar way. If Serbia and Russia were humiliated there would be a Russian revolution and France would lose a valuable ally. In both rival alliances, therefore, the weakest partner could set the pace. Where Russia and Austria led, France and Germany had to follow. And unfortunately the politicians of Russia and Austria were as irresponsible as gamblers always are—both their empires were tottering and only a lucky chance could save them.

Italy's part. It is customary to speak of the Triple Entente of England, France and Russia facing the Triple Alliance of Germany, Austria and Italy. The Triple Alliance was, however, hardly more than a Dual Alliance of Germany and Austria since Italy was only a nominal member of it. She had joined in 1882 because France seemed to block her prospects of building up an African empire, not because she loved Germany or Austria. Austria in fact was her hereditary enemy and there were still in 1914 many Italians living under Austrian rule. If France blocked her expansion in Africa, Austria blocked it in Europe. It was thought probable that Italy would remain neutral at the beginning of a European war and eventually join the side which seemed most likely to win, if it was prepared to pay for her support. The prophets were right for that is what she did.

Position of England. The Triple Alliance did exist at least on paper from 1882 to 1914, but the Triple Entente was only formed in 1907. After 1895, when the Franco-Russian alliance was concluded, Europe was divided into two rival camps. Germany and Austria faced France and Russia. So far England was not concerned. She had difficulties with Russia about the effects of Russian expansion in Asia on British interests in India and China. She had difficulties with France over rival colonial claims in Africa. It did not seem probable that within ten years she would be allied with France and within thirteen with Russia. But the Boer War taught Britain how isolated she was, for all Europe sympathised with the Boers. It became necessary to abandon her policy of isolation and to find friends. Three things made her turn to France—the insistence of Germany on a big navy, the obvious intention of the Kaiser to build up a large colonial empire, and English anxiety lest the ports on the other side of the Channel should fall into the hands of the greatest military power on the continent.

After 1870 it was clear that Germany and not France was the greatest military power on the continent and British foreign policy was bound therefore to oppose any possible German control of Belgium, Holland and the extreme north-west coast of France. A war between France and Germany was, however, likely to have just that result. Most of France's land frontier north of the Alps is well protected by hilly country, which is extremely difficult for armies. There are only three routes possible for an invading army and, of those, two were sufficiently defended by fortifications. An international guarantee treaty, the treaty of London of 1839, was supposed to have closed the third by making Belgium neutral. It was, however, clear that if there was to be any war between France and Germany it would probably for strategic reasons be fought in Belgium. This was common ground to the general staffs of all the great powers. It was an important factor in establishing an informal alliance between England and France.

Before this happened, however, several attempts had been made to establish an Anglo-German understanding or even alliance. They all failed because Germany was unwilling to make any concessions about the size of her navy. These failures led to a settlement of Anglo-French difficulties in 1904. The two countries settled their colonial differences and agreed to give each other moral support in resisting the claims which Germany was beginning to make on North Africa. Two years later this understanding was made more far-reaching by secret military conversations between the French and English general staffs about plans in case of war. In 1907 England reached a settlement of her difficulties with Russia and a rather looser Anglo-Russian entente supplemented the existing Anglo-French one. Although there was no formal alliance with either country England was from this time on reckoned their friend and Germany's potential enemy.

The approach of war. A last attempt to stop the race in naval armaments was made in 1912. When this failed an agreement was made by which the French fleet was concentrated in the Mediterranean and the English in the North Sea. There was still no formal alliance with France, and England was still officially free to support France or not as she chose in case of war. In honour, of course, she could hardly refuse to help France, and the French were able to count on this half-promised assistance when they made their plans in 1914. Half the English government and all the English people had failed to understand the inevitable result of these arrangements, should war actually break out. Many people think that this secret half-alliance was one cause of the war. A full alliance, they argue, would have been better because Germany would have known that England would join in a European war; as things were, she might hope for England's neutrality and gamble on it. This in fact was what she did in 1914. Ordinary Englishmen, however, would never have allowed the government to make a military alliance with France against Germany.

The two main European territorial problems were, therefore, reinforced after 1900 by a third problem in which England and Germany were rivals. The existence of this problem, which was one of trade and naval competition, was the deciding factor in causing the war of 1914. It led England to throw in her lot with France and this made the Triple Entente strong enough to risk a war. If France and Russia had stood alone in 1914, as they had in 1895, they would not have dared to take a strong line in the crisis caused by the murder of Francis Ferdinand. Austria would have had her way, Serbia would have been crushed and there would have been no European war that year.

EPILOGUE

The End of the Story

I. The War of 1914

Since the time of Napoleon there had been no war so general as the one which started in 1914. For the first time for a century virtually all Europe was in arms. The actual duration of the struggle was relatively short—a few months over four years—but the pressure of fighting was so great and so continuous that the European war must rank as one of the most exhausting and destructive wars of history. In this epilogue no attempt can be made to tell the history of the years after 1914; all that can be done is to indicate what happened to some of the movements whose course we have traced through the century.

At the outbreak of war the balance of force was on the side of the central powers. The German army was more efficient than the French; and Germany was soon in possession of the industrial parts of France and Belgium so that the original economic superiority of the central powers was still further increased. On the eastern front Russia soon showed herself to be at least as unreliable as Austria. The two incompetents were evenly matched and roughly cancelled one another out. The war was not going to be won or lost by battles on the eastern front. Its importance lay in the fact that Germany was unable to concentrate all her armies on the western front. Had she been able to do so there can be no doubt that she would have won a quick and decisive victory. The only field in which the allies were unmistakably superior was at sea, but sea power did not become decisive until much later in the war.

The western front. Somehow or other the first German attacks in 1914 were checked: the thrust towards Paris was turned back at the battle of the Marne, and the subsequent dash to secure the Channel ports was stopped at the first battle of Ypres. The campaign in the west then settled down to trench warfare on a line reaching from the English Channel to the Swiss frontier. Each year attempts were made to break the stalemate, but each year until 1918 the defence proved too strong for the attack. The allies failed at Neuve Chapelle (1915), the Somme (1916) and at Arras, Passchendaele and the Aisne (1917); the Germans failed at Verdun (1916). At last in 1918 there was a change to open warfare from what had really been a gigantic siege. First, the Germans broke through in March and for some anxious weeks victory again seemed to be within their grasp. Then in July they were turned back from a point near the extreme limit of their advance in 1914. The counter-attack of the allies pierced the wonderful defences of the Hindenburg Line, and 1919 might even have seen an invasion of Germany but for the armistice.

The eastern fronts. Meanwhile on the eastern front the war had always been more open. At first only Germany, Austria, Russia and Serbia were concerned, but the number of countries engaged gradually grew until finally every state in eastern Europe was at war. Turkey and Bulgaria joined Germany in 1914; two years later Rumania threw in her lot with the allies. Greece was bitterly divided into sympathisers with the allies and sympathisers with the central powers. Finally the former prevailed and Salonica became the base for a long and difficult campaign against Bulgaria, fought in the barren mountains of Macedonia.

The most important eastern campaigns, however, were those which concerned Russia. The German conquest of Russian Poland in 1915 showed how weak Russia was. In the same year her ultimate fate was sealed by the failure of the allies to force the Dardanelles and open a route by which Russia could have been sent munitions and reinforcements. The end of the dynasty and the end of the war came in 1917 when one revolution turned the Tsar off the throne and a second led to peace with the central powers. Germany and Austria were thus able to send reinforcements to other fronts. Russia's collapse made the German offensive of 1918 in France possible. On the other hand 1917 also saw the beginning of the end of the Turkish Empire, for in that year English armies took Baghdad and Jerusalem. These victories were followed up in 1918, and before the surrender of Germany both Turkey and Bulgaria had been forced to ask for peace.

Italy. Besides the eastern and western fronts a southern front came into existence in 1915 when Italy, which had put herself up to the highest bidder, finally joined the allies. For some time this was pure gain, for the Italians kept Austrian armies occupied which otherwise would have been used elsewhere. In 1917, however, the great Austrian victory of Caporetto nearly brought Italy to revolution, and forced the allies to send her precious reinforcements. The tide turned again in 1918 and Italian and English victories in the Alps led to the collapse of the Habsburg empire.

The war at sea. Without the British navy and the United States the war, however, might well have ended in a stalemate. At the battle of Jutland in 1916 the German high seas fleet was fought off the sea and spent the rest of the war in port. But the period of England's greatest naval danger came after Jutland. Germany started a blockade of Britain by submarines. Merchantmen were sunk without warning by ships they never saw. The loss of life and tonnage was heavy and for some anxious months it almost seemed as if England might be starved into submission. At last, however, the danger was overcome and in the end it was Germany, not England, which suffered from starvation both of food and of industrial raw materials. The strength of both countries lay in their highly developed industry, but this was also their weakness because it meant that they had to import food. The British navy maintained throughout the war a highly effective blockade and by the winter of 1917–18 food was extremely short in Germany, so short that both the health and the courage of the population suffered. After their defeats in France in 1918 the rulers of Germany felt unable to face another winter of semi-starvation and this was one of the main reasons for their request for an armistice.

The United States. The submarine campaign not only failed to defeat England; it finally destroyed Germany by its repercussions on America. The loss of neutral lives and shipping so angered the United States that in 1917 they threw in their lot with the allies. They brought the promise of boundless resources both of men and of material. American troops reached France in time to take part in the fighting, but in too small numbers to contribute greatly to the victory. It was not what the United States did, but what they would have done in 1919 which was decisive. The exhausted central powers could not face another year's campaigning in which they would have had to face fresh American armies as well as English and French.

II. The Peace

In November 1918 the Kaiser abdicated and Germany asked the allies for an armistice. This was granted on very harsh terms, but on the understanding that the final treaty would be drawn up on generous lines. By this time the Russian, Turkish and Austrian empires had all fallen to pieces and Bulgaria had asked for terms. The statesmen who met at Versailles in 1919 had, therefore, at least as arduous a task as their predecessors at Vienna a hundred years before. They did not do their work as well.

The clue to a proper understanding of the peace treaties lies in the history of the war and particularly in the complete and sudden collapse of the central powers. Many people now think that a stalemate and a compromise peace would have been more satisfactory than the complete victory which was won with America's help in 1918. At the time, however, the absolute collapse of the central powers was deliriously welcomed as a chance to build a better world.

Germany. Germany's collapse meant that she was left without a government strong enough to negotiate; she had to surrender. A new German constitution, drawn up at Weimar, introduced a very elaborate form of democratic government. But this new liberal Germany began life at a serious disadvantage because the terms of peace, dictated by the allies, were so severe in three ways that they left Germans with a bitter grievance not only against their conquerors but also against the leaders of the new liberal government who had been forced to accept the peace treaty. The first and least important way was the taking away of all Germany's colonies—a blow to her prestige which was bound to be resented and to cause trouble if she again became a great power. The second was the compulsory one-sided disarmament of Germany and the occupation of the Rhineland by English, French and at first by American troops. The Rhineland is not only the centre of Germany's romantic medieval history, but it is also the goal of French ambitions. As long as the French were on the Rhine and she herself disarmed, Germany was bound to feel both humiliated and frightened for the safety of her frontiers. The third and far the most important point was the decision to make Germany pay as reparations the total cost of the damage caused by the war. This proved to be impracticable, but not before it had produced in Germany an economic crisis in which money lost all its value, and the entire savings of middle and working class families disappeared. This experience made so bitter an impression on Germany that its memory was one of the main conditions which made possible Hitler's rise to power. Besides these grievances the surrender of territory to France (Alsace-Lorraine), Denmark (Northern Schleswig) and Poland (the Polish corridor and Upper Silesia) were relatively unimportant.

The Austrian Succession States. The effect of Austria's collapse was to divide up the Habsburg territory between seven different governments—Rumania, Jugo-Slavia (Serbia), Italy, and the new states of Czecho-Slovakia and Poland, besides Austria and Hungary themselves. This partition was probably inevitable, but it certainly made the people concerned poorer by setting up many new customs frontiers. The new Austria, which was entirely German-speaking, was forbidden to join Germany, since the principal object of the peace was to weaken the latter country.

Russia and the Bolsheviks. The effect of Russia's collapse was to divide her western provinces between six different governments—all but one of them the governments of new states. From north to south Finland, Latvia, Estonia and Lithuania were carved wholly, and Poland largely, out of Russia, while Rumania gained the province of Bessarabia. More important than this partition, however, was the establishment in Russia of a Communist government in 1917. For the first time in European history a socialist revolution had been successful—and, since socialism professes to be international, this was a danger to the stability of all other European governments. It has to be remembered that the fate of Russia was not dictated by the allies from Paris, but arrived at by civil war, compromise and bargaining which western Europe was unable to affect.

The Russian revolution of 1917 took place in two stages. In March the Tsar's government collapsed under the weight of its own incompetence. It had failed to supply its soldiers with proper food or munitions and it had even failed to take precautions against the rising tide of indignation. No resistance was offered to the March revolution and soon a liberal government was in power. Its objects were to give Russia a constitution on western lines and to carry on the war. But Russia was thoroughly sick of fighting, and unless its new government dropped its second aim there was bound to be another revolution—to make peace. It came in November and, unlike the first, this was a planned revolution. Its leaders were the small Bolshevik party who, under Lenin's guidance, were determined not only to make peace—in which they had the support of the majority of the people—but also to turn Russia, by force if necessary, into a socialist State. Force was necessary because the desire for socialism was soon found to be confined to the small industrial working class.

Turkey in transition. The effect of Turkey's collapse was to end the Eastern problem in its old form. Before the war Turkey in Europe had already been confined to a small belt round Constantinople. Now Turkey lost as well most of her Asiatic provinces, where the Arabs claimed their independence. They were given this at first only in a modified form under 'protection' from the western powers, but before long they were able in most cases to turn this into virtually complete independence. Turkey even lost some of the coastline of Asia Minor to the Greeks, but before long this was reconquered. She ceased to be a great Empire, but she soon became a small but efficient nation-state—at least the equal in civilisation of any of those neighbours which had once been her subjects. An unsatisfactory part of the Turkish settlement was the arrangement made for Palestine. Palestine had been for many centuries a Mahometan, Arab country without ceasing to be a Holy Land for Christians and Jews. Before the war many Jews in Europe had become members of the Zionist movement which looked forward with seemingly ridiculous optimism to the recovery of Palestine as the Jewish national home. During the war the British government had secured the support of many Jews throughout the world by announcing in the Balfour Declaration of 1917 that it favoured the setting up of a Jewish national home in Palestine. At the same time Britain and France had secured much Arab support by promises which, to say the least, were difficult to reconcile with the Jewish view of what the Balfour Declaration meant. The peace treaty made Britain responsible for governing Palestine in such a way that the rights of the Arab majority were safeguarded while Jews were allowed to settle in the country and found a National Home there. Eventually Arabs and

Jews were expected to become so friendly that they could form together a self-governing state. The arrangement caused much dissatisfaction to both nations, each of whom thought that they had been promised more than they got.

Italian disappointments. The most disappointed of the allies was Italy which had joined the side she thought most likely to win and which promised her most. From first to last her participation in the war was a business venture. It was successful, but not as brilliantly so as she had hoped. She managed to free the last Italians remaining under Austrian rule and to annex a large number of Germans as well, but she failed to extend her colonial empire or to acquire the coveted eastern shore of the Adriatic, 'Mare Nostrum' both to ancient Romans and modern Italians. She was bitterly disappointed by the fact that Serbia had to be rewarded as well as Italy—Serbia gained Austria's Balkan provinces—and that England and France shared all the German and Turkish Empires between them. Italy was left in 1919 with small gains and with humiliating memories of the rather ineffective and selfish part she had played—her rout at the battle of Caporetto was a disturbing memory.

The League of Nations. On top of these territorial and financial changes there was erected in 1919 a superstructure of organisations which it was hoped might be the beginning of an international government. A League of Nations was established to which it was thought that in time all civilised states would belong. This could not, however, take place at once for the victors refused to allow Germany to join, while the United States refused to take up her membership.

This decision of the United States was so important that it is necessary to try to understand the circumstances which caused it. All through the nineteenth century the United States had been developing a civilisation similar in many ways to that of the western European democracies, but strangely remote from them. The New World and the Old were in politics two really distinct worlds. While economic circumstances were uniting the two worlds ever more closely, politics kept them apart. During the first half of the war of 1914 America had remained neutral, but eventually in 1917 she intervened against Germany. She alone of the principal combatants had taken no part in the two problems which had led to the original outbreak and she was, therefore, more free than any of the other states to help in a disinterested way to build a better world. It was for this purpose that her President, Mr Woodrow Wilson, had taken the principal part in establishing the League of Nations. But the League of Nations existed to prevent future wars by guaranteeing the security of its various members. This meant guaranteeing their existing frontiers and, since Germany was bitterly dissatisfied with the terms granted her, this meant that the League was in its early years something very like an alliance against Germany. Americans were determined that their intervention in the war of 1914 was not to be taken as a precedent. They had no intention of fighting another European war about problems with which they were not vitally concerned and so, when they saw what the League was becoming, they refused to have anything to do with it. Their withdrawal made the League more than ever like an alliance against Germany.

The League, however, was intended by its founders to be something much more important than this. They thought that it, or bodies connected with it, could undertake a general supervision of important questions which affected all or many nations and for which there had been no machinery before the war. Thus arrangements were made to protect the rights of 'national minorities', people belonging to one nation who were compelled to live inside another nation-state. Similar arrangements were made to safeguard various 'backward races' which were to be governed by one or other of the great powers. When Syria was given to France and Palestine to England, for example, they were given not as colonies but as 'mandated areas' to be governed for the benefit of their inhabitants under the guarantee of the League of Nations. An International Labour Office was also set up with the hope that it would be able to secure uniform, good conditions of labour in all countries. These promising experiments had some success, but not nearly as much as they would have done if the League had not seemed to be primarily an alliance against Germany.

III. The Post-War Years

Four black years. The first four years of peace were black years for four reasons. First, there was an acute economic crisis in central Europe caused partly by the effects of the war-time blockade of Germany and partly by the demand for 'reparations' or compensation for damage done. A year of hunger and starvation was followed by several years of inflation, when German money lost all its value and the savings of the German middle class disappeared. Secondly, there were two years of civil war in Russia in which France and England took a hand against the Communist government. The success of the Communists caused a momentary danger of a successful invasion of central Europe where many starving workmen would have welcomed the Russian armies which were finally turned back by the Poles outside Warsaw. Thirdly, the relations of France and Germany were worse than ever. The French first vainly tried to persuade the Germans of the Rhineland, which was occupied by French and English troops, to set up an independent state of their own under French protection. In 1923 they went further and invaded the Ruhr valley, Germany's richest industrial district, which was occupied for some months. Apparently France could only feel secure if Germany was absolutely powerless.

The coming of Fascism. Fourthly, representative government was abolished in Italy in favour of a nationalist dictatorship under Benito Mussolini. In 1922 he organised a march on Rome as a demonstration against the government. There was little organised force behind Mussolini and a strong government would have had little difficulty in dealing with the situation. But the Italian government was not strong and simply capitulated before the threat. For a time Mussolini was content to allow parliament to continue, provided that it gave up its power. Gradually, however, he established a new Fascist constitution side by side with the old parliamentary one, and finally he abolished the old forms of government altogether, retaining only the king as a figurehead.

Some people think that the feeble surrender of the parliamentary leaders in 1923 was a sufficient proof that Italy stood in need of a revolution. Certainly the Italian parliament had frequently been a corrupt, and never an inspiring body. It was in 1922 deeply discredited by the inglorious Italian record during the war, by its relative failure at the peace conference and by its total failure to find a solution for the serious post-war industrial unrest.

Mussolini offered Italy a fresh start. He promised Italy strong government which would be above the corruption of party politics; he promised her people discipline and her youth training which would fit them to be the heirs of Rome. Finally he offered her a new cure for industrial unrest. Mussolini himself had been a socialist before the war and he promised to undertake a strict state control of industry in the interests of the worker as well as of the general community. Private ownership was to remain, but it was to be taught its duty to the nation and to its own employees. Not all these promises were kept, but at least Mussolini gave Italy a new, if rather hysterical, self-respect. The Fascist revolution was really the product of post-war unrest acting on a tradition of fifty years of shabby government.

The Locarno period. Towards the end of 1923 a welcome change came over Europe. The next five years were a time of increasing prosperity. The economic crisis in central Europe was ended by a reasonable temporary settlement of the reparations problem, devised by an American banker. The Russian danger to Europe had already been removed in 1920 by the victory of Poland in the Russo-Polish war. In 1921 a New Economic Policy had been introduced which left some room for private enterprise, and European countries began again to trade with Russia. Before long it was apparent also that Fascist Italy was ready for the time being to be a good neighbour.

The relations of France and Germany steadily improved until in 1925 the Locarno treaties were signed by which both countries freely recognised the existing frontiers as permanent while Britain undertook to guarantee either state against an attack from the other. This arrangement was soon followed by the admission of Germany into the League of Nations and by a new and more favourable settlement of the old reparations problem. Men began to talk of an early evacuation of the Rhineland by its foreign garrisons—this was actually carried out in 1930—and a disarmament conference was called to discuss the possibility of a general agreement about the size of armies. During this period both France and Germany enjoyed considerable prosperity. In France the damage caused by the war was made good, clean new villages taking the place of the dirty, destroyed ones; in Germany American loans were used to finance great housing developments.

The Great Depression. Europe had been full of hope in 1929; in 1939 she was almost without hope. The change for the worse began when a great economic crisis started in America in 1929. Germany's recent prosperity disappeared overnight for it had been almost entirely financed by American loans. The reparations problem again appeared to worry statesmen since the temporary solutions, which had worked when trade was good, proved far too heavy for a time when trade was bad. Unemployment became general and affected the well-educated middle classes as well as the manual workers. England, which was caught by the depression in 1931, had reserves of strength on which to fall back; Germany had none. The inflation of 1920–22 had robbed her people both of their savings and of their courage. Despair became universal and out of that despair came in 1933 the Nazi revolution.

The Nazi Movement. In the early days after the collapse of Germany in 1918 there were many wild political movements led by men who were more like romantic conspirators than serious statesmen. They quarrelled violently among themselves, but were alike in believing that they alone knew the way out of the chaos and despair which had overwhelmed the strongest European state. Some of these movements were led by men who had been orthodox socialists, some by men who had been orthodox conservatives. They shared an outright rejection of the old social order and of all existing parties, and they all boasted an open belief in revolutionary violence as the way to reach their different ideals. It may seem curious that many young men joined first one, then another of these groups, swinging quickly from extreme Right to extreme Left and back again. This is explained by the fact that men who had served in the war found it much easier to agree on their dislikes, and on the way to express them, than on what they wanted to see done. Among these movements there was one called the National Socialist German Labour Party (Nazis) founded by an ex-corporal, Adolf Hitler. Among the other early leaders, two—Hermann Goering and Josef Goebbels—were later to achieve world notoriety. As times mended, most of these romantic movements faded into insignificance and, ten years after the end of the war, this fate seemed to be overtaking the Nazi party. It got a second start in the great depression. Hitler blamed his four bugbears—the peace of Versailles, the 'swindle' of democratic government, the Jews and the Communists—for the ever increasing distress. He offered deeds as well as words. The street battles in which night after night his Brown Shirts fought armed Communist gangs served a triple purpose. They attracted the publicity without which the Nazis could not thrive; they discredited the moderate government which was unable to suppress them; and they built up a body of fanatical, disciplined political gangsters upon whose loyalty Hitler could base the new state, the Third German Empire, which he was determined to found. It is significant that the Nazi song, which became almost a second national anthem, was about this street fighting and was called the Horst Wessel song after its author, a particularly unsavoury gangster who was killed in one of these early affrays.

In 1932 the Nazi party won more votes than any other party in a general election, although it polled little more than a third of all the votes recorded. Early in 1933 Hitler became Chancellor in a coalition government, and set to work rapidly to turn Germany into a single-party dictatorship with the aid of his private army of Brown Shirts and their more highly trained and disciplined élite, the black-shirted S.S. The most adroit and pervasive propaganda, supervised by Goebbels, was used to persuade all Germans who would listen that they stood at the beginning of a new, heroic future. It gave them what they most needed—fresh courage and self-confidence. Meanwhile the minorities, who could not be converted, or who were not eligible for conversion, could be suppressed. Throughout Germany concentration camps were set up. In them, without any form of trial, there were confined and often tortured the various enemies of Hitlerism. They fell into three main groups. There were Jews, whose crime was the unavoidable one of being Jews; there were Christians and democrats, who refused to accept quietly a dictatorship which tried to direct men's consciences and opinions; and there were Communists, who had no objection in principle to Hitler's methods and who would indeed have cheerfully used very similar ones to attain their own ends. To many foreigners it seemed strange that so many decent Germans should apparently remain unaware that many thousands of their fellow countrymen were being persecuted with a calculated and scientific cruelty which had perhaps never been equalled in earlier centuries.

The Russian Dictatorship. Meanwhile the hope roused by the New Economic Policy that Communist Russia would gradually mellow into a friendly, if eccentric, democratic power gradually disappeared. Lenin's death in 1924 was followed by a tense struggle for power between Trotsky and Stalin. Trotsky, the man who had raised the Communist armies which won the civil war, believed that socialism in one country alone was impossible, and that the first duty of Communist Russia was to foster revolution in other lands. Stalin, a man little known outside Russia, was the secretary of the Communist party, and the leader of those who put the Russian revolution first and its duty to the international revolution second. The quarrel ended with the exile of Trotsky, who took refuge in Mexico where many years later he was murdered by political opponents. Stalin proceeded to carry through a rapid industrial revolution in Russia, putting most of her resources into long-term capital investments, such as the hydro-electric scheme on the river Dnieper, and as little as possible into the production of things which people could eat, wear and enjoy in the present. The original weakness of the Communists had been greatest in the country, where the peasants had supported the revolution without any intention of becoming socialists, because it gave them the squire's land. The Communists therefore started a policy of industrialised farming by setting up enormous, mechanised 'collective farms' in which the individual peasant holdings were merged. This was strongly resisted by many of the peasants and only accomplished at the expense of very great suffering—a major famine which killed many hundreds of thousands, if not millions, of men and women, and the driving of countless others into distant labour camps which were as barbarous, if not quite as scientific, in their horrors as Hitler's concentration camps. Opposition inside the Communist party was dealt with by exhaustive purges of the party membership. The most spectacular feature of these purges was the series of treason trials in which the prisoners made no attempt to defend themselves but, on the contrary, made abject confessions of guilt and expressions of repentance. The whole process, which saved neither the life nor the liberty of the accused, was completely bewildering to western people.

IV. The Return of War

Pacific and Mediterranean. The establishment of the Nazi dictatorship in Germany began a new period in European history. The initiative passed from the western democracies to the dictatorships of central and eastern Europe. Each year the existing state of affairs was changed to meet the wishes of the dictatorships against the futile opposition of the western powers. These changes were brought about either by force or by the threat of force. Taken together they amounted by 1939 to a completely new territorial settlement, quite as drastic as the one imposed by the western powers at Versailles twenty years before.

The changes fall naturally into three groups: those in the Far East brought about by Japan; the Mediterranean changes for which Italy was responsible, and the north European changes carried out by Hitler. It was Japan which started the period of violent change when she invaded Manchuria and took it from China. After a short breathing space this completely successful adventure was followed by an invasion of China proper. This time the Chinese were able to maintain a more effective resistance although they were soon driven inland from the more prosperous coastal regions. The conduct of the war caused relations between Japan and Britain to become very difficult, and there were times when the only alternatives seemed to be a British withdrawal from China or an Anglo-Japanese war.

Meanwhile Italy had been engaged in three major adventures in the Mediterranean area. Her first move was the conquest of Abyssinia, an undertaking which she was encouraged to attempt by the failure of the League of Nations to take any steps to prevent Japan's conquest of Manchuria. Abyssinia, however, was not left quite as much to its fate as Manchuria had been. Ineffective protests were supplemented this time by equally ineffective attempts to prevent Italy from obtaining the raw materials without which modern war cannot be carried on. These 'sanctions', as they were called, failed because the countries which adopted them merely gave up selling to Italy but did not stop other countries doing so.

The most important result of the 'sanctions' period was to persuade Italy that the western powers, France and England, were her enemies, and that she had better establish a firm friendship with Germany.

After her conquest of Abyssinia Italy turned her attention to Spain. There is no doubt that an increasing number of Spaniards were bitterly dissatisfied with the influence of the advanced socialists in the government of the republic, which had been established in 1931 after the abdication of Alfonso XIII. Italy used this dissatisfaction for her own ends. When a military rebellion began it was given all the help that Italy could afford. A large Italian army was sent to Spain; Italian submarines helped to blockade the Spanish republican ports and Italian naval and air bases were established in the Balearic Islands. All this help was given openly but unofficially—Italy herself remained a member of the international 'Non-intervention Committee' whose duty was to prevent the great European powers taking part in the war. A fair amount of German help was also given to the Spanish rebels under General Franco, while Russian war material was sent to the republican government. After a long struggle the Spanish war ended with a complete victory for General Franco.

Italy at once turned her attention eastwards and on Good Friday, 1939, suddenly invaded Albania. Ever since the European War of 1914 Albania had been a more or less obedient Italian sphere of influence; what western weapons she had came from Italy, and it is not surprising that her king was unable to resist the Italian invasion. The net result of the Spanish and Albanian adventures was greatly to increase Italy's importance as a Mediterranean power, to frighten the smaller states such as Greece and Turkey and to bring them considerably closer to France and England.

Rake's progress. While Mussolini was engaged in building what he hoped would be a new Roman Empire in the Mediterranean, Hitler was busily occupied in tearing up the Versailles Treaty. There were five stages in this process before the final stage which caused the second European war to break out in September, 1939.

First, Germany re-fortified the Rhineland in defiance of the Locarno treaties. Then she brought about by force the annexation of Austria, against the opposition of England and France, but probably with the good-will of the majority of Austrians. Next she took from Czecho-Slovakia its German-speaking provinces. England and France tried to prevent this by every means short of war, but in the end they had reluctantly to agree at Munich in September, 1938. In this way Hitler managed to bring back into Germany almost all those Austrian Germans whom Bismarck had expelled from the German Confederation in 1866 as the price of building the German Empire.

The Munich turning point. These first three stages were carried out peaceably not only because the western powers did not want war, and were not ready for it, but also because large numbers of people thought that there was some justice in the German case. It was possible to argue that Hitler was merely correcting the mistakes of the Versailles treaties, though no doubt in a very undesirable way. When Mr Chamberlain flew back from Munich in September, 1938, and claimed that he had brought back 'peace with honour', he was in fact mistaken. The actual arrangements made at Munich were honourable enough; Mr Chamberlain's mistake lay in thinking that Hitler intended to keep them.

A few months showed that Hitler was not and probably never would be satisfied. He seized Memel from Lithuania, and absorbed the purely Slav parts of Czecho-Slovakia into Germany. Neither annexation was resisted by the western powers. There was, however, a difference between their inaction after Munich and their inaction when Hitler was fortifying the Rhineland, incorporating Austria or dissecting Czecho-Slovakia. Before Munich the western powers had been quiet because they were not certain that war would have been justified; after Munich they were inactive merely because for a time the opportunities for resistance were unfavourable. This was a difference which Hitler never grasped. It accounts for his invasion of Poland. No doubt he expected another bloodless victory, particularly after he had secured in the last days of August, 1939, an agreement with Soviet Russia. This miscalculation of how the western powers would behave in the end undid all the changes which Hitler made in the map of Europe in his first years of power. Many of them were distasteful to Englishmen and Frenchmen, but they had been accepted, however reluctantly. Nobody proposed to go to war to restore the old order. But once war came all these issues were reopened and vast additional changes became possible. The result of the second German war was never likely to be a return to the Versailles settlement.

The Second German War. This second German war lasted six years, compared with the four years of the first, and ultimately involved even more of the world. Once again sweeping German victories in the early days almost destroyed the western powers; once again American intervention in the middle period more than redressed the balance of force and in the end brought about the overthrow not only of the German armies but also of the German state. The stakes, however, were considerably greater than in 1914; the struggle marked by even more dramatic changes of fortune, and its end by still harsher melodrama. German armies reached the Pyrenees, the Channel, the North Cape, the Caucasus and the borders of Egypt; but the tide ebbed as rapidly as it had flowed. In 1945 Hitler and Goebbels destroyed themselves in an air-raid shelter as Russian troops stormed devastated Berlin. A few months later listeners in their homes in England heard on the wireless sentence of death actually being passed in Nürnberg on Goering and other German leaders as war criminals. Not long after, the same listeners, waiting to hear that the sentences had been carried out, were surprised by the announcement that Goering himself, in spite of the most elaborate precautions, had managed to commit suicide and thus escape the hangman.

Mussolini threw in his lot with Germany after her early victories only to find his armies completely out-classed and to hear the British Prime Minister, Mr Winston Churchill, truthfully describe his country as 'the soft under-belly of the Axis'. The Italian collapse before British and American invasion was the first stage in the overthrow of the central powers. Here, too, melodrama marked the end. Mussolini's overthrow by a palace revolution, his arrest, his rescue by a German aerial cutting-out expedition and, in the end, his lynching by his fellow countrymen make the last days of Fascism read like a boy's adventure story.

The lurid epic of the rise and fall in German fortunes is paralleled by the sudden expansion and almost equally sudden contraction of the Japanese empire which occurred between the treacherous sinking of the American Pacific fleet in Pearl Harbour in time of peace and the dropping of the first atomic bombs by American bombers on Hiroshima. The Japanese conquests, brief though they were, carried her fleets and armies through Malaya and Burma almost into India, and south through the East Indies to within striking distance of Australia.

Twelve years after the outbreak of war it seems that the greatest immediate beneficiary of the collapse of the German and Japanese empires has been Soviet Russia. Her story too, has been one of fantastic changes of fortune. The sudden German-Russian agreement of 1939 gave Russia the eastern half of Poland. The Baltic states of Esthonia, Lithuania and Latvia were then rapidly reabsorbed into Russia by the Hitlerian technique of peaceful penetration, and Finland was defeated in war. At this point Hitler suddenly turned on Russia, invaded her and seemed likely to destroy for ever the Communist dictatorship. Once overthrown, it is most unlikely that it could ever have been restored. But the fighting at Stalingrad marked the furthest limit of Hitler's success and to-day Russia has not only regained the advances she made in 1939 and 1940, but has erected a ring of dependant Communist states farther west. Poland, compensated at the expense of Germany for her losses to Russia on the east, Czecho-Slovakia, Rumania, Hungary and Bulgaria are all to-day dutiful Russian dependencies. Jugo-Slavia, too, is a Communist state, but one which is now on bad terms with the rest of the Communist world. In China in 1949 Communist armies overran the whole country. The next year they invaded South Korea, but instead of the easy victory they expected they were met by opposition from a United Nations force in which the United States played the largest part. Farther south in South East Asia Communist guerillas waged effective war against the established governments. Against the rising tide of Communism, however, is to be set the decision of the United States not to fall back as in 1919 on a policy of isolation, but to play the major part in the leadership of the forces which still believe in free and responsible government. The effects of this are to be seen in the Korean War and in the creation in the West of the North Atlantic Treaty Organisation.

A final judgment. The complete change in the political outlook since 1919 and again since 1929, 1939 and 1945 shows how dangerous it is to attempt any sort of general judgment on the work of the nineteenth century. Parliamentary government, for which its heroes struggled, seemed stronger than ever before in 1919; in 1951 it seems weaker than at any time since the seventeenth century. There have been, however, two great changes in politics since the French Revolution about which we can be fairly certain. First, there has been an enormous extension of the authority of the state. Whether we live under a dictatorship or a democracy, the state regulates far more of our lives than it did of our ancestors'. Secondly, the old autocratic kings have disappeared entirely. The last traces of divine right have gone. Louis XVI was King of France by the grace of God, because he was his father's son, because—simply—France belonged to him. The Tsar Nicholas II of Russia and Kaiser William II of Germany ruled by the same authority. No ruler in Europe to-day makes this his claim to power. Mussolini and Hitler controlled, and Stalin still controls, the lives and liberties of their people more completely than any of their predecessors did, but by a different title. Modern dictators rule in the name of those whom they govern, or at least they claim and pretend that this is so. We may, perhaps, say that the war of 1914 completed the work of the French Revolution by sweeping away the last absolute, hereditary monarchy. It left undecided, however, the struggle between parliamentary government, as it is practised in England and France to-day, and plebiscitary dictatorship, as practised in France by Napoleon and, in a much harsher form, in Hitler's Germany and in the Communist world. Nor was this struggle settled by the war of 1939.

APPENDIX

APPENDIX. The Revolutionary

Assembly or Government	What it was
1. The States-General 1789 which became the National Assembly 1789–91.	Lack of money caused Louis XVI to call the States-General, an old-fashioned parliament in which nobles, clergy and commons sat in separate houses. After it had been decided that they should sit together it was called the National Assembly.
2. The Legislative Assembly 1791–2.	When the new constitution was finished the National Assembly was replaced by the Legislative Assembly, a much weaker body, because none of the members of the old Assembly were allowed to be elected to it.
3. The Convention 1792–5.	A much stronger body than the Legislative because it contained experienced M.P.'s from both the former assemblies. It was elected by universal suffrage.
4. The Directory 1795–9.	The Convention drew up a new Constitution providing that two thirds of the members of the new assembly must have been members of the Convention. This prevented a possible Royalist restoration. The new assembly elected a government of five called the Directory. The assembly was divided into two Houses of Parliament, one third of the members to retire each year.

Governments of France

What it did	Parties
1. Abolished feudal privilege, e.g. tax exemption, game laws, etc. and made all Frenchmen equal before the law. 2. Made Church a department of State and thus offended Pope and clergy. 3. Set up Constitutional Monarchy; but the flight of the king made men think of a republic. 4. Stopped M.P.'s being ministers and thus prevented strong government being formed which might have saved France from war and the Terror. 5. Restricted votes to middle class. 6. Divided France into modern departments instead of old provinces. 7. Started issue of 'assignats': paper money not backed by gold.	Absolute Monarchists (the queen, etc.) *v.* Constitutional Monarchists (Lafayette, Mirabeau).
1. Declared war on Austria. N.B. Girondins wanted war to spread revolution to other countries. Jacobins opposed war because they feared revival of monarchy or military dictatorship. 2. Allowed power to pass to Commune which organised revolution of August 10 and persuaded assembly to suspend king. 3. Tolerated September Massacres.	Constitutional Monarchists (Lafayette) *v.* Moderate Republicans (Brissot).
1. Drew up Republican Constitution. 2. Executed king and thus helped to bring England into the war. 3. Destroyed Girondins and started Reign of Terror. 4. Established Committee of Public Safety, thus giving France a strong government for first time since 1789. 5. Conquered Belgium and Holland and suppressed civil war in France. 6. Passed Law of Maximum fixing bread prices and foreshadowed social(ist) as well as political revolution. 7. Ended Reign of Terror and set up moderate Republican Government (the Directory).	(*a*) Moderate Republicans (Girondins) *v.* Extreme Republicans (Jacobins). (*b*) Hébert *v.* Danton and Robespierre. (*c*) Danton *v* Robespierre. (*d*) Robespierre *v.* Barras.
1. Employed Napoleon to win victories in Italy. 2. Set up new republics in conquered territories. 3. With help of army destroyed revived royalists in 1797. 4. Got rid of extreme Republicans by force, 1798. 5. Tolerated corruption and profiteering on one hand, while extreme poverty led to Baboeuf's first communist movement. 6. Lost Napoleon's conquests while he was in Egypt.	Royalists and real Republicans *v.* Barras (nominal Republican) supported by Army.

THE REVOLUTIONARY AND NAPOLEONIC WARS

1. This diagram shows the relative parts played by England, Austria, Prussia and Russia in the struggle against the French Revolution and Napoleon. Each year in which one of the allies was at war is shown by a soldier or a sailor in the appropriate line. They hold a flag aloft in years when their country was successful, downwards when their country was defeated. The most important peace treaties are indicated by an asterisk.

2. Notice how much longer England was at war than any of the other allies. It should, however, be remembered that from 1794 to 1808 the war was almost entirely naval.

3. Austria was, after England, France's most consistent enemy. Notice, however, the paralysing effect on her of Marengo, Austerlitz and Wagram.

4. The almost complete absence of Russia and of Prussia (after1795) from the early stages of the wars was due to their greater interest in the partition of Poland, and, in the case of Russia, to a Turkish war.

5. Spain and Piedmont were also enemies of France. Spain fought against her until 1795, then for a time on her side, and finally during the Peninsular War played a large part in defeating Napoleon. Piedmont, though a small state, was important at the beginning from her geographical position.

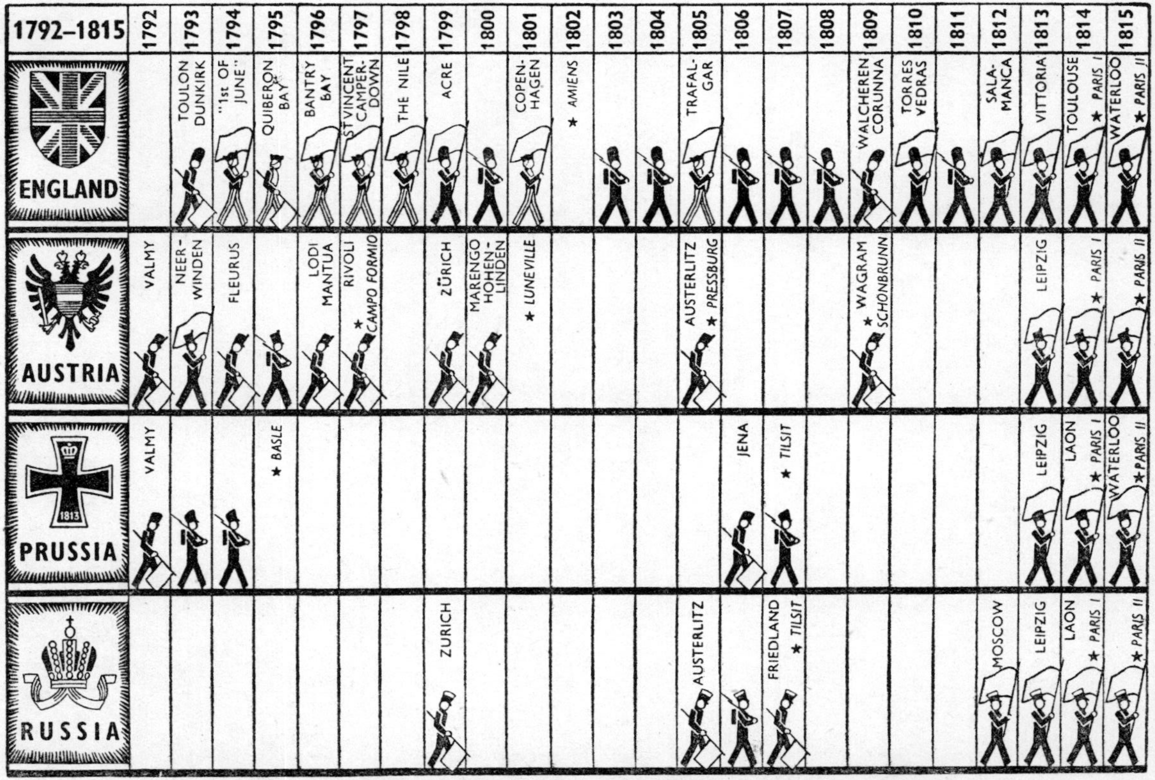

THE REVOLUTIONARY AND NAPOLEONIC WARS

DICTIONARY OF POLITICAL TERMS

ABSOLUTE GOVERNMENT, ABSOLUTE MONARCHY, ABSOLUTISM
The government of a country where there is no legal check on the ruler (usually the king) giving what orders and making what laws he likes.

ACT (of Parliament)
A law. A bill (q.v.) becomes an act when it has been approved (technically 'read') three times by both Houses of Parliament and received the royal assent.

ALLIANCE
An arrangement between two or more countries, usually in the written form of a treaty (q.v.), to support one another. It may be a general agreement to support one another in all circumstances or limited to particular occasions, e.g. when one of the allies is attacked by another state. In the latter case it is called a defensive alliance.

AMBASSADOR
The representative of one country at the capital of another important country. In the case of a less important country the foreign representative is called a minister (q.v.). Thus Great Britain is represented in France by an ambassador and in Holland by a minister.

ANTI-SEMITISM
Opposition to the influence and power of the Jews.

ARBITRATION
A method of settling disputes, either between nations or between different sections of a nation, e.g. miners and mineowners, by which both parties agree to refer the dispute to a third party and to accept his decision.

ARISTOCRACY, ARISTOCRAT
A social class, to which only birth or a king's favour gives admission, and which enjoys or claims special respect, privileges and/or power. Thus (1) before 1789 the French aristocracy were excused many taxes; (2) in the eighteenth century the government of England was aristocratic, i.e. it was controlled by aristocrats. Aristocracy comes from two Greek words meaning 'rule of the best people'.

AUTOCRACY, AUTOCRAT
A system of government in which all power is concentrated in the hands of one man, responsible to himself alone.

BELLIGERENT RIGHTS
When one government is at war with another it is entitled, if it can, to prevent neutrals from giving certain kinds of help to the other side, e.g. it may declare a blockade. The neutral powers recognise that under these con-

ditions the belligerents may do things to the neutral's property which otherwise would be piracy or an act of war. It is, therefore, important for a government making war to secure a recognition from the neutrals of its 'belligerent rights'. Such recognition is not usually given to rebels against a recognised government.

Bill

In England a suggested law or act of parliament (q.v.). The proposal is put before Parliament which can alter or 'amend' it.

Bolshevik, Bolshevism

An extreme form of socialist or socialism (q.v.). The word comes from the Russian for 'more'. In 1903 the Russian Socialist party divided into two groups: the minority (Mensheviks) wanted to co-operate with the middle class, the majority (Bolsheviks) wanted revolutionary action. Bolshevism and Communism (q.v.) now have nearly the same meaning.

Bourgeois, Bourgeoisie

Generally, the middle class; but used in three different senses: (1) all people who are not either landed gentlemen or manual workers, (2) the class who enjoy power because they own factories or businesses as opposed to the old aristocracy whose power was based on landowning, (3) as an adjective to describe dull, comfortable middle-class life.

Cabinet

The body of ministers (q.v.) governing Great Britain. It may also be used of any country which has responsible government (q.v.). All the ministers are bound to support all the decisions of the cabinet. This is described as the 'collective responsibility' of the cabinet.

Capital, Capitalist

A capitalist is one who owns capital, i.e. the means of production such as factories, machines or money with which he employs people to work for him.

Caucus

The people forming and controlling the organisation of a political party (q.v.).

Centre, Left, Right

Words describing political parties. The representatives in the French revolutionary assemblies sat in a semicircle facing the chairman, and not on opposite sides (supporters and opponents of the government) as in England. Those who wanted the biggest changes sat on the chairman's left, those who wanted to keep things as they were or to restore them to their old position sat on the chairman's right, those who wanted moderate changes sat in the centre. Thus, roughly, Right = Conservative; Centre = moderate or (to-day) Liberal; Left = radical or (to-day) socialist. Most European parliaments are still arranged on this semicircular plan. In Germany the Centre party was the Roman Catholic party.

CHAMBER OF DEPUTIES
The French name for the equivalent of our House of Commons. It is also used in many other countries.

CHANCELLOR
The name for the chief minister, or head of the government, in Germany and many other central and east European countries. In England the Lord Chancellor is the head of the law, not the government.

CLERICALISM
The influence of the church, especially the Roman Catholic church, in politics. Anti-clericalism is, therefore, the opposition to this influence.

COALITION
(1) A government formed by representatives of different political parties, e.g. Lloyd George's Coalition Government 1916–1922.
(2) The name given to the successive alliances against France during the Revolutionary and Napoleonic Wars.

COMMISSAR
The present Russian equivalent of a cabinet minister.

COMMUNE
Three different meanings: (1) the smallest unit in French local government = roughly, town or village council; (2) the local government of Paris at the time of the great French Revolution; (3) 'The Commune' = the name given to the French socialist government which seized power for a time in Paris in 1871.

COMMUNISM, COMMUNIST
Originally the same as socialism (q.v.). Thus Marx's 'Communist Manifesto' of 1848 might equally well have been called the 'Socialist Manifesto'. Since the war it has been applied to those extreme socialist parties which are in agreement with Russia.

CONSERVATISM, CONSERVATIVE
A person, political party or belief aiming at keeping things as they are or restoring them to their former condition. The objects of conservatism, therefore, differ from country to country and from generation to generation, e.g. European conservatives in 1820 were opposed to nationalism (q.v.); to-day they encourage it.

CONSTITUTION, CONSTITUTIONAL GOVERNMENT, CONSTITUTIONAL MONARCHY
In politics 'constitution' has two meanings: (1) in general, the form of government of any country, e.g. 'the French constitution before 1789 was an absolute monarchy' (q.v.); (2) in particular, a definite scheme for the government of a country giving its people (or some of them) a right to be consulted through a parliament. Only countries which have a constitution in the latter sense are described as having a constitutional government or monarchy. Constitutions may be written, like the American, or unwritten (i.e. based on tradition) like the British.

Customs Union

An arrangement by which two or more countries agree (1) not to impose any taxes on goods taken from one member's territory to another's and (2) to impose the same taxes on goods sent to any member's territory from an outside country.

'De Facto' and 'De Jure' Governments

One country may recognise the government of another country 'de facto', i.e. as being the actual government without recognising it 'de jure', i.e. as being the rightful government. It may continue to recognise 'de jure' a government which has lost all actual power. This distinction is important in cases of civil wars and revolutions.

Democracy, Democrat

That form of society in which everybody has an equal share through their votes in the choice of a government and in which all men are treated equally by the government irrespective of their race, religion or wealth.

Deputy

The usual continental name for what we should call a member of parliament.

Dictator, Dictatorship

A form of absolute government (q.v.) in which the ruler is not a king.

Diplomacy, Diplomat

Diplomacy is the business of managing the relations between different countries by peaceful means. Thus diplomatic relations are broken when a war starts. A diplomat is a person, e.g. an ambassador, professionally engaged in this business.

Election, Direct and Indirect

Direct election is the system by which the ordinary citizen votes directly for the representative or official to be chosen. Thus in England M.P.'s are directly elected. Indirect election is the system by which the ordinary citizen votes for people whose duty it is to choose the representative or official concerned. Thus in France senators (q.v.) are indirectly elected; the ordinary citizen chooses the equivalent of county councillors and the county councillors choose the senator.

Émigré

A political refugee. Originally used of the French nobles who went abroad at the outbreak of the Revolution.

Entente

An agreement between two or more countries, similar to but less formal than an alliance (q.v.).

Entrepreneur

A business man who collects capital from the public (the shareholders) and uses it to develop some new industry or business.

Executive, Judiciary, Legislative

The authority of the state (q.v.) is divided into three branches. The Executive carries on the ordinary work of ruling, usually proposes new laws and always sees that they are carried out by bringing offenders before the Judiciary whose business it is to decide whether the laws have been broken and to decide disputes between citizens. The Legislative makes laws. Thus, roughly, the Executive = the government; the Judiciary = the Law Courts; the Legislative = Parliament.

Fascism

Fascism has two meanings: (1) the present Italian system of government; (2) any roughly similar system of government, e.g. the National Socialist system in Germany. Fascism differs from (*a*) Conservatism (q.v.) because it endeavours by government interference to improve the position of the wage earner at the expense of the private capitalist; (*b*) Socialism (q.v.) because it is strongly nationalist (q.v.) and allows private property; (*c*) Democracy (q.v.) because it replaces parliamentary or responsible government (q.v.) by unquestioning obedience to a dictator.

Federal State; Unitary State

A federal state (q.v.) is one which is made up of a number of smaller, separate governments deciding on their own responsibility all their local affairs and leaving to the central or federal government only those matters which concern the whole country. A unitary state is one in which the central government has complete control of everything, although it may allow some things, e.g. public health, housing, to be managed locally under its supervision. England and France are unitary states; Australia and Switzerland federal.

Franchise

The right to vote.

Free Trade

The system under which a country imposes no taxes on foreign imported goods without imposing an equal tax on similar goods produced at home. The opposite to Protection (q.v.).

Government, The

See under Cabinet and Executive. It is most important not to confuse the government with Parliament, although in England when we elect a parliament we are in practice indirectly electing also a government.

Group System, The

In certain countries, e.g. France to-day, Germany before Hitler, there are so many political parties that none can secure an absolute majority of members of parliament. Governments, therefore, have to rely for support on coalitions of parties or groups. This state of affairs is known as the group system as opposed to the 'Two Party System'.

Imperialism

Means generally the desire to found or extend an empire. It is used to-day normally to describe the effort to secure an empire in economically unexploited regions where a market can be found for the manufactured goods of the home country and from which she can fetch raw materials.

Indemnity

(1) In internal politics a pardon for a wrong act, usually a political one, e.g the Act of Indemnity in Charles II's reign pardoning those who had supported Cromwell.

(2) In foreign affairs a sum of money paid by the defeated country in a war to the victorious country as part of the terms of peace.

Intelligentsia

The highly educated part of a nation. Often used as a term of abuse.

Judiciary (*see under* Executive)

Junker

The great landowners of Prussia east of the Elbe. They were exceedingly conservative and so the word is sometimes used for any class of conservative landowners.

Left (*see under* Centre)

Legislative (*see under* Executive)

Liberal, Liberalism

When spelled with a small 'l' liberal means one who believes in the greatest possible freedom for the individual—particularly freedom (*a*) to carry on his business as he likes; (*b*) to express in speech or writing what views he likes; (*c*) to have a share by his vote in the government of his country. When spelled with a capital 'L' Liberal is the name of a political party, usually a middle-class party, which tries to secure these privileges.

Limited Monarchy

A form of government in which the power of the king is limited by a constitution (q.v.).

Middle Class (*see under* Bourgeoisie)

Minister

(1) A member of the government, a cabinet minister—usually the head of a department, e.g. the Minister of Labour. (2) A diplomatic representative at the capital of one of the less important countries, an inferior kind of ambassador.

Monarchy (*see under* Absolute Monarchy, Constitutional Monarchy, Limited Monarchy)

Nationalism, Nationalist, Nationality

A political belief which may take three forms: (1) a belief that people of one nationality ought not to be governed by people of another; (2) a belief that all people of one nationality should form one state (q.v.); (3) a belief that one's own nation is so much better than some or all other nations that it ought to rule them. It is difficult to define nationality but it may be said that people of one nationality always speak the same language, usually have a common historical tradition and frequently belong to only one race. But people may speak the same language (English and Americans), have a common historical tradition (Norwegians and Danes) and be of the same race (Belgians and Dutch) and yet belong to different nationalities. The only test is to find out what nationality people think they are and accept that.

PARLIAMENT

Parliament (technically 'The King in Parliament') is sovereign (q.v.) in England. The House of Commons, which can always override the House of Lords, is directly elected by all men and women over twenty-one. It makes the laws, decides on the taxes and can turn out a government of which it disapproves by passing a vote of censure. The name parliament is often loosely used—in this book and elsewhere—for the legislative (q.v.) of any country.

PARTY

An organisation of people holding the same views, who try to carry them out by securing the election of a sufficient number of M.P.'s to secure a majority in parliament so that the party may form a government or cabinet (q.v.). In England, where parties are strong, all M.P.'s belong to one of two or three parties and have to do what the party leaders tell them. Thus when we vote for a member of parliament we are really choosing which party we should prefer to win a majority and form a government.

PLEBISCITE

A direct vote of the people of a country or district held in exceptional circumstances to decide an important question, e.g. whether Napoleon should be Emperor or to what country the Saar should belong.

PLUTOCRACY, PLUTOCRAT

A state of society in which power belongs to men simply because they are rich.

POGROM

A Russian word meaning 'destruction'. Used in politics for an organised massacre, usually of Jews.

PRESIDENT

The head of a republic, elected either directly by the people, e.g. in the U.S.A., or indirectly by the parliament, e.g. in France. The president may, as in France, have no more power than the King of England or, as in the U.S.A., be the real head of the government like the British prime minister.

PROGRESSIVE

A name often used by political parties which wish to make changes, e.g. Liberal or Socialist parties.

Proletariat
The working class who own no capital and have nothing to sell but their skill or labour.

Protection
The system by which goods imported from abroad are taxed so that the home producers may have 'protection' or an advantage. The opposite of Free Trade (q.v.).

Radical, Radicalism
When spelled with a small 'r', one who wants to make great changes (from the Latin radix=root). In this sense both Socialist and Fascist parties are radical. When spelled with a capital 'R' it is the name of a political party whose object is or was to make great changes. It sometimes happens, however, that the name survives after the changes have been made and the people who wanted them are satisfied and have become conservatives. Thus the French Radicals to-day are really conservatives.

Reactionary
One who wants to put the clock back politically, an extreme conservative.

Referendum
A direct vote of the citizens to decide some particular question such as prohibition. It differs from a plebiscite in that, in the countries where it is in use such as Switzerland, it is part of the normal constitution.

Reichstag
The German word for parliament.

Reparations
The name given to the indemnity (q.v.) which Germany was to pay to the allies under the Treaty of Versailles (1919). Its object was to repair the damage done by the war.

Representative Government, Responsible Government
Representative Government exists where the citizens are consulted through a parliament which makes the laws. Responsible government exists only where the citizens through parliament can get rid of a government or minister of whom they disapprove. It is possible to have representative government without responsible government, e.g. in Germany before the war, but not vice-versa.

Republic
A state (q.v.) of which the head is the representative of the people, elected constitutionally.

Responsible Government (*see under* Representative Government)

Right (*see under* Centre)

SENATE
The equivalent in foreign parliaments of the House of Lords. Its members may either be nominated by the president, king or prime minister or else elected, directly or indirectly. In federal states they are usually chosen in equal numbers to represent each of the smaller governments.

SEPARATION OF POWERS
The system by which the executive, legislative and judiciary are each independent of the other and of equal status, as in the U.S.A. In England and most other countries either the executive or the legislative is usually supreme.

SOCIALISM, SOCIALIST
A society in which the means of production (e.g. mines, farms, factories), distribution (e.g. railways) and exchange (e.g. shops, banks) are owned by the whole community and not by private individuals. It is usually said that in a socialist state everybody would do the work for which they were best fitted and everybody would receive whatever they needed, as far as the resources of the community went.

SOVEREIGN
(1) The person or body in a state (q.v.) which has supreme power. (2) A title of honour given to a king whether really sovereign in the first sense or not.

SOVIET
The Russian word for 'council'. In politics it was used particularly for the councils of representatives of soldiers, workmen and peasants, which were first formed in the 1905 revolution and which carried through the 1917 one. The official title of Russia is the 'Union of Socialist Soviet Republics'.

STATE
A political organisation which has sovereign powers, i.e. which is not dependent on any other body. Thus Great Britain is a state, but London or Lancashire is not. The word 'country' is often used to mean 'state', but it is possible for a country not to be one, e.g. Scotland is a country but not a state, 'Italy' in 1848 was a country but not a state.

SUFFRAGE
The right to vote.

'TAMMANY'
A loose word for any corrupt force in politics which works by bribing the voters directly or indirectly. It is the name of the headquarters of the Democratic Party in New York.

TAXATION
The method by which states raise money. A tax is a sum of money, calculated in different ways, which has to be paid either directly by the citizen to the government (direct taxation) or on the sale of certain goods (indirect taxation).

Treaty

A written agreement between two or more governments either settling past difficulties, e.g. a treaty of peace, or deciding future action, e.g. a treaty of alliance, a disarmament treaty.

Unitary State (*see under* Federal State)

Veto

The right to prevent a proposed law coming into force either for a time or altogether. In some countries the head of the state, the king or the president, has this right.

Working Class (*see under* Proletariat)

BIOGRAPHICAL DICTIONARY

Abd-el-Kader (*c.* 1807–1883) p. 130
Algerian leader in struggle against French. Influenced by Mehemet Ali's Egyptian reforms. Kept up struggle for fifteen years until he surrendered to Marshal Bugeaud in 1847. At first interned but released by Napoleon III, pensioned and decorated with Legion of Honour.

Abdul Hamid (1842–1918) pp. 263, 327, 328
Sultan of Turkey 1872–1909. Nicknamed in England 'Abdul the Damned' for his cruelty. After Congress of Berlin (1878) he changed traditional pro-English policy for pro-German one. Responsible for Armenian massacres in 1890. Visited by Kaiser in 1898 and fell wholly under German influence. In 1908 Young Turks forced him to grant constitution. He tried to revoke it and was deposed in 1909.

Aberdeen, Lord (1784–1860) pp. 146, 147
British diplomatic representative with allied armies 1813–1814; foreign secretary 1828–1830 and 1841–1846. Prime Minister 1852–1855.

Albert, Prince Consort (1819–1861) p. 264
Married Queen Victoria, 1840.

Alexander I, Tsar of Russia (1777–1825) pp. 45, 61, 64, 78–80, 83, 85, 92, 93, 101, 103, 109, 116, 120, 133, 151, 272, 274, 279, 280
Tsar 1801–1825, succeeding his murdered father. Opposed Napoleon until at Tilsit 1807 they agreed to share Europe between them. Continental System broke their friendship and led to Moscow Campaign 1812 after which Alexander took lead against Napoleon. Founded Holy Alliance 1815. This was originally liberal but after 1819 Alexander took lead in suppressing liberalism everywhere.

Alexander II, Tsar of Russia (1818–1881) pp. 259, 261–265, 272–275, 280, 281, 334
Tsar 1855–1881 when murdered by Nihilists. Reorganised country after Crimean War, freeing serfs (1861) and developing local government. Then turned reactionary, brutally suppressing Polish rebellion 1867. Great expansion in central Asia, but attempts to secure Constantinople checkmated by Congress of Berlin 1878. Basis of policy: friendship with Prussia.

Alexander III, Tsar of Russia (1845–1894) pp. 332–334
Tsar of Russia 1881–1894. Thorough and efficient reactionary. Devoted to Pan-Slav ideals. Not so friendly with Prussia as Alexander II, and gradually drew closer to France. Considerably extended Russian possessions in Asia.

Alexander of Battenberg, Prince of Bulgaria (1857–1893) pp. 325, 326
First Prince of Bulgaria from which he was removed by Russia.

Alfonso XII, King of Spain (1857–1885) p. 283

ALFONSO XIII, King of Spain (b. 1886) pp. 283, 376
Posthumous son of Alfonso XII. Came of age 1902 and ruled until 1931 when expelled by republican revolution.

AMADEUS, King of Spain (1845–1890) p. 283
Son of Victor Emmanuel of Italy. Elected King of Spain 1870 and tried to rule constitutionally, but resigned and returned to Italy 1873.

ANTONELLI, CARDINAL (1806–1876), p. 216
Secretary of State to Pope Pius IX and chief leader in struggle for temporal power of Pope against Italian nationalism. 1847, drafted constitution for Papal States; 1848, arranged Pope's flight from Rome; 1850, on his return persecuted liberals; 1860, tried to restore Bourbons to Naples; arranged for French garrison in Rome; after Italian occupation of Rome (1870) confined to foreign affairs.

ARABI PASHA (1839–1911) pp. 292, 297, 298
Egyptian soldier and politician. Dismissed for embezzlement by Ismail Pasha 1875; anti-foreign military rising made him dictator 1881, but overthrown by British 1882 through bombardment of Alexandria and battle of Tel-el-Kebir. Exiled to Ceylon but returned 1901.

ARNDT, ERNST MORITZ (1769–1860) p. 82
German patriot. In 1793 at Jena University fell under Fichte's influence; became Professor of History and wrote books against Napoleon. Fled to Sweden and then Russia where he composed patriotic songs and helped Stein to organise struggle against Napoleon. In 1819 forbidden to lecture because suspected of liberalism.

ARTOIS, Comte d' (*see* Charles X)

AUGEREAU, Marshal (1757–1816) pp. 59, 61
Humble birth, served in French, Russian and Prussian armies. Commanded division in Napoleon's first Italian campaign, helped in *coup d'état* of 1797; made marshal 1804; served at Jena and Eylau; in Catalonia (Spain) gained reputation for cruelty; joined Louis XVIII in 1814 but rejoined Napoleon in Hundred Days; after Waterloo lost title and pension.

BAKUNIN, MICHAEL (1842–1876) p. 196
Russian anarchist. Resigned commission in Imperial Guard because of treatment of Poles. In exile in France, Switzerland, Germany. Handed over to Russia and sent to Siberia 1855 but escaped to England 1861. Defeated by Marxist party in International 1872.

BALZAC, HONORÉ DE (1799–1850) p. 143
French novelist; admirer of Walter Scott and friend of Victor Hugo who said his greatest qualities were 'observation and imagination'. Lived irregular life and wrote his books largely after midnight.

BARRAS, PAUL FRANÇOIS (1755–1829) pp. 31, 33, 34, 38
French politician. Spent early life in French India. Became revolutionary leader and helped to organise fall of Robespierre. A member of the Directory 1795–1799. Patronised Napoleon. Left politics after fall of Directory. A thoroughly corrupt man.

BAZAINE, Marshal (1811–1888) p.239
French soldier. Entered army as private, made reputation in Algeria, Spanish Carlist wars, Crimea and 1859 campaign in Italy. In command in Mexican expedition 1862–1867. Surrendered Metz too easily in Franco-Prussian War. Sentenced to twenty years' imprisonment for this by court-martial but escaped to Italy in 1872.

EUGÈNE DE BEAUHARNAIS, Viceroy of Italy (1781–1824) pp. 62, 68, 85
Josephine's son and Napoleon's step-son. Accompanied Napoleon on first Italian campaign and to Egypt. Made Viceroy of Italy 1805: a good and tolerant administrator. Served at Wagram and in Moscow campaign.

BEETHOVEN, L. VAN (1770–1827) p. 81
German musician. His nine symphonies revolutionised writing of music. After 1798 greatly handicapped by deafness which became almost complete. He still went on composing, however. A keen liberal, greatly stirred by French Revolution.

BERESFORD, Marshal (1768–1854) p. 100
Took part in defence of Toulon 1793, served in Egypt *v.* French 1799–1803; with Sir John Moore at Corunna, fought through Peninsular War, reorganising Portuguese troops, and stayed there until 1819.

BERNADOTTE, Marshal (1763–1844) pp. 43, 59, 83, 85
Served with distinction on northern front during early French revolutionary wars. One of Napoleon's original marshals, but relations always strained especially after Jena (1806), where he took no part in battle, and Wagram (1809), for which he claimed all the credit. Offered and accepted position of heir to childless King of Sweden 1810. Devoted all his energies to taking Norway from Denmark, which he did (1815) with help of allies whom he had helped against Napoleon. Became Charles XIV King of Sweden and Norway 1818.

BERRI, Duc de (1778–1820) pp. 98, 136
Younger son of Charles X of France. Served with émigré army 1792–1797, then with Russian. Came to England 1801 and married Englishwoman—marriage dissolved for political reasons 1814. Assassinated in Paris 1820.

BERTHIER, Marshal (1753–1815) p. 60
Chief of Staff of Napoleon's Grand Army. Joined Bourbons 1814. Committed suicide 1815.

BESSIÈRES, Marshal (1768–1813) p. 60
Served as captain in Napoleon's first Italian campaign and took part in most of his subsequent wars. A great cavalry leader.

BISMARCK, OTTO VON (1815–1898) pp. 192, 220–230, 239, 247–253, 262, 264–266, 281, 282, 289, 301, 348–350, 358

BLANC, LOUIS (1811–1882) pp. 143, 173, 174, 196, 377
French socialist leader. In *L'Organisation du Travail* 1839 expounded his socialist system. Organised national workshops 1848. In exile in London 1849–1870, where he wrote history of French Revolution. Elected member of French National Assembly 1871 and in 1879 secured pardon for imprisoned supporters of the Paris Commune.

BLÜCHER, Field-Marshal (1742–1819) p. 82
In Swedish army at fourteen, captured by Prussians and joined their army. Leader of war party in Prussia 1805–1806, fought bravely at Jena. Commanded Prussian armies in War of Liberation. His victory at Laon 1814 led to Napoleon's first abdication. Defeated at Ligny 1815 but arrived at Waterloo after forced march in time to help Wellington.

BONAPARTE, JEROME (1784–1860) pp. 62, 69
Napoleon's fourth brother. Married American wife, but marriage dissolved 1806. Made King of Westphalia. After Waterloo lived in Italy and Switzerland until Second Empire when returned to France and made president of Senate.

BONAPARTE, JOSEPH (1768–1844) pp. 62, 68, 74, 75
Napoleon's eldest brother. Made career as diplomat negotiating Treaty of Lunéville with Austria and of Amiens with England. King of Naples 1806, King of Spain 1808, where he was complete failure.

BONAPARTE, LOUIS (1778–1846) pp. 62, 67, 68
Napoleon's third brother, and father of Napoleon III king of Holland, 1806–1810 but removed for slackness in enforcing Continental Systems.

BONAPARTE, LUCIEN (1775–1840) p. 43
Napoleon's second brother. A keen Jacobin, helped to organise and carry out *coup d'état* of 1799. Relations with Napoleon subsequently bad. Captured by English ship on way to America 1810.

BOUILLÉ, Marquis de (1739–1800) p. 15
French general and governor of France's eastern provinces at outbreak of Revolution. Sternly suppressed military risings at Metz and Nancy. Arranged for flight of king and queen to his army 1792. When it failed went to Russia and then England.

BOULANGER, General (1837–1891) pp. 341, 342
French soldier, served in Italy 1859; the Colonies and Franco-Prussian War. War Minister 1886. Attempted to become Dictator, but fled to Brussels and committed suicide 1891.

BRIAND, ARISTIDE (1862–1932) p. 341
French politician responsible for resuming friendly relations with Germany after European War, 1914. Eleven times prime minister of France.

BROGLIE, DUC DE (1821–1901) p. 245
French writer and royalist politician. Elected to Academy, 1862; ambassador in London, 1871–1872; responsible for Thiers' resignation and MacMahon's election as president, 1873. Prime Minister in 1873 and 1877—forced to resign both times by Gambetta, republican leader.

BRUNSWICK, Duke of (1735–1806) pp. 18–20
Succeeded to Dukedom 1780 and ruled as model benevolent despot. Commanded Austrian and Prussian armies *v.* France 1792; and Prussian army against Napoleon 1806. Defeated and mortally wounded at Jena.

BUGEAUD, Marshal (1784–1849) p. 130
Served in Napoleon's armies 1804–1815. In retirement until 1830, after which he made reputation by breaking strikes and suppressing riots. Responsible for French conquest and colonisation of Algeria.

BYRON, Lord (1788–1824) pp. 103, 108, 118–120, 279
English poet. Famous at twenty-four with publication of *Childe Harold's Pilgrimage*. Owing to separation from wife left England for good 1816. Lived in Italy until 1823, then helped Greeks in War for Independence. Died at Missolonghi 1824. A liberal in politics both domestic and foreign.

CADOUDAL, GEORGES (1771–1804) pp. 57, 58
French royalist leader during Revolution. Took part in La Vendée rising. Conspired against Napoleon, caught and executed 1804.

CANNING, GEORGE (1770–1827) pp. 106, 108–114, 117–122, 254
English statesman. Under-Secretary for Foreign Office 1796–1799; Foreign Secretary 1807–1809, when quarrelled with Castlereagh and fought duel with him. Foreign Secretary again 1822 (on Castlereagh's death) to 1827—Prime Minister 1827.

CAPODISTRIAS, Count (1776–1831) p. 118
Greek patriot; 1800 ran semi-independent republic in Ionian Isles; after Treaty of Tilsit 1807 joined Alexander of Russia's service. After 1813 Tsar's most trusted adviser, resigned when Alexander became reactionary. Elected President of Greek Republic 1827 but failed to control bandit leaders. Assassinated 1831.

CARLOS, Don of Spain (1788–1855) pp. 146, 147
Spanish conservative leader and pretender to the throne. In prison in France 1808–1814. Heir to throne until King Ferdinand, his brother, altered succession in favour of his own daughter Isabella. Carried on civil war against her 1834–1839 when, defeated, he fled to France.

CARLYLE, THOMAS (1795–1881) pp. 13, 32
Scottish historian and writer.

CARNOT, LAZARE (1753–1823) pp. 27, 31, 35, 38
Trained as engineer officer. Member of Committee of Public Safety, responsible for organisation of republican armies. A member of the Directory until 1797 when he fled abroad. Returned in 1799 and, as minister of war 1800, reorganised army. Retired 1801.

CARRIER, JEAN BAPTISTE (1756–1794) p. 29
Jacobin leader, formerly a country lawyer. 1793, punished La Vendée rebels at Nantes by mass executions by drowning—'noyades'. 1794, took part in attack on Robespierre, but soon after was executed as terrorist.

CASTLEREAGH, Viscount (1769–1822) pp. 84, 88, 92, 93, 99, 103, 108, 109
English statesman. Largely responsible for Act of Union with Ireland 1800, but resigned owing to failure to grant Catholic Emancipation. Minister of War 1807–1809, when he quarrelled with Canning and fought duel with him. Foreign Secretary 1812–1822, when he committed suicide.

CATHERINE II (The Great), Empress of Russia (1729–1796) p. 271
Ruled alone after her husband's death in 1762. Great patron of French ways. During her reign Russia expanded at expense of Turkey and Poland, whose partitions she organised.

CAVAIGNAC, General (1802–1857) pp. 174, 175
Engineer officer and French republican leader. Served in Algeria during Louis Philippe's reign, in 1848 returned to Paris and crushed socialist rising. Arrested by Louis Napoleon 1851 but soon released to live in retirement.

CAVOUR, Count (1810–1861) pp. 210–217, 220–222, 258, 259

CÉZANNE, PAUL (1849–1906) p. 338
One of the most influential French artists in the early days of 'Post-Impressionism'. Used colour largely to accentuate form.

CHAMBERLAIN, HOUSTON STEWART (1855–1927) p. 290
Son of British admiral, who became naturalised German. Taught theory of Aryan superiority in *Foundations of 19th Century*, published 1890.

CHAMBERLAIN, JOSEPH (1836–1914) pp. 287, 288, 290
English statesman. Extreme radical leader in parliament until he quarrelled with Gladstone over Irish Home Rule 1886. Colonial Secretary 1895–1903, when resigned to preach Protection. Retired from politics 1906.

CHAMBERLAIN, NEVILLE (1869–1940) pp. 287, 377.
British Prime Minister 1937–1940 until succeeded by Churchill.

CHAMBORD, Comte de (1820–1883) pp. 243, 244
Grandson of Charles X and son of Duc de Berri, who was murdered in 1820. Extreme royalist pretender to French throne.

CHARLES, Archduke (1771–1847) pp. 42, 77
Best of the Austrian generals against French Revolution and Napoleon. Drove French out of Germany 1796 and 1799 and defeated them in Italy 1805. Reorganised Austrian army after Austerlitz and drew battle of Aspern with Napoleon 1809, though defeated by him at Wagram same year. Rest of his life in retirement.

CHARLES X, King of France (1757–1836) pp. 9, 57, 130, 132, 136–138, 172, 243, 280
King of France 1824–1830.

CHARLES ALBERT, King of Piedmont (1798–1849) pp. 102, 154, 155, 159, 160, 162
During revolution of 1821 granted constitution as regent for his cousin Charles Felix, who disowned him, and defeated him with Austrian help at Novara. Became King in 1831. In 1848 granted constitution and declared war on Austria, but defeated and abdicated.

CHARLES FELIX, King of Piedmont (1765–1831) pp. 102, 154
King 1821–1831.

CHRISTIAN IX, King of Denmark (1818–1906) p. 224
Became King in 1863 and lost Schleswig-Holstein next year to Prussia and Austria.

CHURCHILL, WINSTON (b. 1874) p. 378.
British Prime Minister 1940–1945 and 1951–

CLEMENCEAU, GEORGES (1841–1929) pp. 343, 344
French radical, anti-clerical and nationalist leader. One of chief supporters of Dreyfus. As Prime Minister 1906–1909 devoted himself to Franco-British Entente. Bitter enemy of Germany. As Piime Minister 1917–1920 restored broken French morale and largely responsible for Treaty of Versailles.

COBDEN, RICHARD (1804–1865) p. 233
English politician. Founded Anti-Corn Law League 1838; arranged commercial treaty with France 1859–60.

COCKERILL, JOHN (1790–1840) p. 186
Great industrialist, owning large iron works at Liége with branches in France, Germany and Poland. Particularly concerned in continental railway development. Son of William Cockerill, founder of the firm.

CODRINGTON, Admiral Sir EDWARD (1770–1851) p. 121
Served in channel during Revolutionary Wars, commanded *Orion* at Trafalgar. In command of French, Russian and English ships which destroyed Turkish and Egyptian Fleets at Navarino 1827.

COMBES, ÉMILE (1835–1921) p. 340
Principal French anti-clerical leader.

CROMER, EARL OF (1841–1917) p. 297
As British financial representative in Egypt 1876–1879 exposed Ismail's methods and forced his abdication. From 1883 to 1907 he virtually ruled Egypt, reforming finances and making the country prosperous.

DANTON, GEORGES JACQUES (1759–1794) pp. 18–22, 26, 27, 30

DARWIN, CHARLES (1809–1882) pp. 180, 181
Sailed on *Beagle* as naturalist 1831–1836. Published *Origin of Species* 1859 and *Descent of Man* 1871. Principal teacher of theory of 'evolution by natural selection'.

DAUDET, ALPHONSE (1840–1897) p. 338
French novelist and journalist, sometimes compared with Dickens. Author of *Lettres de Mon Moulin* 1869 and *Tartarin de Tarascon* 1872.

DAVOUT, Marshal (1770–1823) pp. 59, 61
Educated with Napoleon at Military School at Brienne, accompanied him to Egypt. Fought brilliantly at Austerlitz and Jena. Organised army for Moscow campaign. Napoleon's Minister of War in 1815.

DEÁK, FRANCIS (1803–1876) p. 319
Hungarian statesman and lawyer. Elected to Hungarian Parliament 1833. Moderate leader trying to reconcile Hungarians to Habsburg Emperors. In 1848 he urged Hungary to be moderate, but after suppression of Kossuth's rebellion he was court-martialled. He was, however, acquitted. Largely responsible for negotiating 'Ausgleich' 1867.

DECAZES, ÉLIE (1780–1860) p. 135
A lawyer under Napoleon, became Louis XVIII's principal and most moderate adviser. He was most important minister from 1816 to 1820 when forced to resign after murder of Duc de Berri. Joined Louis Philippe in 1830.

DELCASSÉ, THÉOPHILE (1852–1923) p. 300
French Foreign Minister at time of Fashoda incident 1898; arranged Entente with England 1904 but resigned owing to difficulties with Germany over Morocco 1905. One of the founders of French colonial empire.

DESAIX, General (1768–1800) p. 44
French noble who supported revolution. Made reputation 1796 as commander of Moreau's rear-guard in retreat from Germany. With Napoleon in Egypt. Killed at Marengo, for which victory he was largely responsible.

DISRAELI, BENJAMIN, Earl of Beaconsfield (1804–1881) pp. 193, 197, 224, 263–266, 287
Wrote *Coningsby* 1844 and *Sybil* 1846. Prime Minister 1868 and 1874–1880. Purchased Suez Canal shares 1875; attended Congress of Berlin 1878.

DREYFUS, ALFRED (1859–1935) pp. 342–344
French Jewish officer falsely accused of espionage.

DUMOURIEZ, General (1739–1823) pp. 22, 23, 57
Served in Seven Years' War; imprisoned in Bastille and elsewhere 1771–1774; supported Revolution and joined Girondins 1792. Largely responsible for declaration of war, checked enemy at Valmy, conquered Netherlands, but defeated at Neerwinden 1793. Tried to persuade troops to march on Paris; but, failing, fled to Austrians. Finally settled in England 1804.

DUPLEIX, JOSEPH-FRANÇOIS (1697–1763) p. 39
The maker of the French empire in India.

EDWARD VII, King of England (1841–1910) p. 347
Succeeded 1901.

ENGELS, FRIEDRICH (1820–1896) pp. 194, 195
Son of German cotton spinner, sent to Manchester 1842 where father had factory. Met Marx in Paris 1844 and started lifelong friendship. Engels lived mainly in Manchester and London and helped Marx in his socialist work.

D'ENGHIEN, Duc (1772–1804) pp. 57, 58
Son of last Prince of Condé. 'Emigrated' after Fall of Bastille 1789 and fought with Royalists until 1799. Kidnapped from Germany and executed by Napoleon 1804.

ENVER PASHA (1882–1922) p. 327
Leader of Young Turk movement, 1908. Commander of Turkish armies, 1914–1918. Killed in fighting against Soviet Russia after having been in its pay.

EUGÉNIE, Empress (1826–1920) pp. 233, 236, 239
Daughter of Spanish nobleman, lived in France after 1834; married Napoleon III 1853. Used her influence against liberalism. Lived in England after 1870.

FALK, ADALBERT (1827–1900) p. 251
Prussian statesman.

FERDINAND I, King of the Two Sicilies (1751–1825) pp. 101, 102
Opposed French Revolution and was turned out of Naples, but able to keep Sicily thanks to support of Britain which, however, insisted on his granting a constitution. After 1815 restored to Naples. Completely under Austrian influence. Abolished constitution but forced to restore it owing to military rising 1820 which he later suppressed 1821 with Austrian help. Constitution abolished. A corrupt and cruel king.

FERDINAND II, King of the Two Sicilies (1810–1859) pp. 159, 161, 162, 214
King 1830–1859. Cruel and despotic, his reign was marked by frequent conspiracies. Nicknamed 'Bomba' for the way he suppressed the 1848 revolution by bombardment. His treatment of political prisoners was denounced by Mr Gladstone in 1851 in his *Letters to Lord Aberdeen.*

FERDINAND I, Emperor of Austria (1793–1875) p. 169
Emperor 1835–1848 when forced to abdicate in favour of nephew Francis Joseph. Utterly incompetent.

FERDINAND VII, King of Spain (1784–1833) pp. 75, 99, 100, 110, 146

FERDINAND (of Saxe-Coburg), King of Bulgaria (1861–1938) pp. 326, 328
Elected Prince of Bulgaria 1887; took part unsuccessfully in Balkan Wars 1912–1913. Joined Germany in European War 1915 in hope of getting grievances redressed. Abdicated 1918.

FERRY, JULES (1832–1893) p. 341
French republican leader. Opposed Franco-Prussian War. Prominent in colonial expansion—arranged occupation of Tunis 1881—and in anti-clerical movement.

FICHTE, JOHANN GOTTLIEB (1762–1814) pp. 81, 150
German philosopher. His early work was praised by Kant. 1807–1808, delivered his patriotic *Addresses to the German People* against Napoleon. 1810, first Rector of new Berlin University. 1814, died of fever caught from wife who was nursing in war hospitals.

FOUCHÉ, JOSEPH (1763–1820) pp. 29, 56–58, 136
Trained as a priest, but joined revolutionary party and started atheist campaign 1792–4. Helped to overthrow Robespierre. Helped Napoleon in *coup d'état* 1799 and served with short breaks as Minister of Police until 1815. Tried to make friends with Bourbons but failed and had to live in exile after 1816.

FRANCE, ANATOLE (1844–1924) p. 338
Most important French novelist from 1881 until his death. A radical in politics, took great part in Dreyfus case.

FRANCIS II, King of Naples and Sicily (1836–1894) pp. 214, 215
Became King in 1859, driven out of Sicily and Naples by Garibaldi, 1860–1. Nicknamed 'Bombino'.

FRANCIS FERDINAND, Archduke (1863–1914) pp. 321, 322, 361

FRANCIS JOSEPH, Emperor of Austria (1830–1916) pp. 169, 170, 213, 264–266, 316–321, 323
Emperor of Austria 1848–1916.

FRANCO, General (b. 1892) p. 376
Head of the Spanish state since 1936.

FREDERICK OF AUGUSTENBURG (1829–1880) pp. 224–226
German claimant to Duchies of Schleswig and Holstein, 1863.

FREDERICK THE GREAT, King of Prussia (1712–1786) pp. 2, 61

FREDERICK WILLIAM III, King of Prussia (1770–1840) pp. 90, 151
King of Prussia during its greatest degradation (at time of Jena) and during the subsequent recovery. He opposed, but was too feeble to stop, the reforms of Stein which made that recovery possible.

FREDERICK WILLIAM IV, King of Prussia (1795–1861) pp. 162–165, 221, 222
King of Prussia 1840–1861 (insane after 1857).

GAMBETTA, LÉON (1838–1882) pp. 239, 241, 243–245, 340

GARIBALDI, GIUSEPPE (1807–1882) pp. 161, 162, 210, 214–218, 222, 238
In exile in South America 1836–1847. Fought for Charles Albert 1848 and defended Roman Republic 1849. Fought for Piedmont in Alps 1859; conquered Sicily and Naples 1860. Tried to overthrow Pope and take Rome 1862 and 1867. Also served in Seven Weeks' War *v.* Austrians 1866 and in Franco-Prussian War 1870 *v.* Prussia. Given Freedom of City of London 1864.

GAUGUIN, PAUL (1848–1903) p. 338
French artist famous for pictures of Tahiti where he settled 1891.

GEORGE IV, King of England (1762–1830) pp. 109, 121
Prince Regent 1811; King 1820.

GIOBERTI, VINCENZO (1801–1852) pp. 154, 155
Italian priest and political thinker. Piedmontese royal chaplain, in exile 1833–1847. President of Piedmont Chamber of Deputies 1848; in exile in Paris 1849–1852. Advocated federation of free Italian states under Pope's presidency.

GLADSTONE, WILLIAM EWART (1809–1898) pp. 214, 262–264, 324
Prime Minister 1868–74, 1880–1885, 1886, 1892. Denounced Bomba's government in Naples 1851; conducted campaign against Bulgarian Atrocities 1876–1880; advocated Home Rule for Ireland 1885; denounced Armenian massacres 1895.

GOEBBELS, JOSEF (1897–1945) pp. 372, 378.
Early German Nazi leader. Hitler's propaganda chief from 1929. Committed suicide with wife and seven children in Berlin 1945.

GOERING, HERMANN (1893–1946) pp. 372, 378.
German Nazi leader. Airman in 1914 war winning German V.C. Responsible for building up German Air Force. Condemned to death at Nürnberg as war criminal, but committed suicide on eve of execution.

GOETHE, JOHANN WOLFGANG VON (1749–1832) p. 81
German poet, dramatist and philosopher. Invited to Weimar, small central German state, 1774 and became an important government official there. Intensely interested in all reforms. 1786 visited Italy. 1792 took part in invasion of France. Formed lifelong friendship with Schiller 1794. His greatest works were his short lyric poems and the epic *Faust*.

VAN GOGH, VINCENT (1853–1890) p. 338
Dutch painter who lived and worked in France, principally at Arles. Suffered from periodical madness. Gift of vivid colour and swirling line.

GORDON, General CHARLES (1833–1885) p. 298
British soldier. Served in Crimean War 1855; commanded Chinese army 1860–1863; as governor-general tried to suppress slave trade in Sudan 1874–1879. Sent to evacuate Khartum threatened by Mahdi. Besieged there and killed 1884–1885.

GORTSCHAKOFF, Prince (1798–1883) pp. 265, 266
Russian Minister of Foreign Affairs from 1856 to 1882. Managed by Prussian alliance to make good Russia's losses in Crimean War.

GRIMM, The Brothers (1785–1863 and 1786–1859) p. 150
Edited German fairy stories 1812–1815, laid foundation of science of folk lore; later edited old German laws and published great historical German grammar. Both brothers made librarians at Göttingen University 1829 and Professors at Berlin 1840.

GUIZOT, FRANCOIS (1787–1874) pp. 141–143, 146, 147, 172
French statesman and historian.

HARDENBERG, Prince KARL AUGUST VON (1750–1822) pp. 82, 90, 133
Prussian statesman. Negotiated Treaty of Basel 1795. Centre of opposition to Napoleon. Made Chancellor 1810 and reorganised Prussian administration for War of Liberation. Represented Prussia at Paris 1814, Vienna 1815 and subsequent international congresses.

HAUSSMANN, Baron GEORGES (1809–1891) pp. 233, 234
Born in Paris of German extraction. Made Prefect of the Seine (Paris) 1853 and started rebuilding of city. Sewers built, boulevards constructed, parks such as Bois de Boulogne laid out. Served as Bonapartist deputy in French parliament under Third Republic.

HAYDN, FRANZ JOSEF (1732–1809) p. 2
Austrian composer of Slav race. Conductor of Prince Esterhazy's private orchestra in Vienna 1760–1790. During this time he composed 5 masses, 12 operas, 40 quartets, 100 symphonies and overtures, etc. 1792 gave Beethoven lessons. Composer of Austrian National anthem (tune: 'Glorious things of Thee are spoken, Zion, city of our God').

HAYNAU, General (1786–1853) p. 170
Entered Austrian Army 1801. Responsible for suppressing Hungarian revolt 1849.

HÉBERT, JACQUES RENÉ (1757–1794) pp. 29, 30
Corrupt French revolutionary leader. Edited *Le Père Duchesne* newspaper. Influential leader of Commune. Started atheistic Worship of Reason. Denounced by Robespierre and guillotined 1794.

HEGEL, GEORG WILHELM FRIEDRICH (1770–1831) p. 150
German philosopher. Professor at Berlin 1818 and Rector of the University 1830. Died of cholera 1831.

HENRY, Colonel (1848–1898) pp. 342, 343
French Officer concerned in Dreyfus case.

HITLER, ADOLF (1889–1945) pp. 372, 373, 375–380.
German Chancellor 1933. Dictator until his death in 1945. Founded Nazi party 1920; attempted revolution 1923 and imprisoned. Wrote *Mein Kampf* (My Struggle) in prison.

HUGO, VICTOR (1802–1885) pp. 143, 232
French Romantic poet, dramatist and novelist. Exiled in 1840 for Bonapartist propaganda and spent nearly twenty years in Guernsey. Author of *Les Misérables* and *Notre Dame de Paris*.

HUMBOLDT, KARL WILHELM VON (1767–1835) p. 82
Prussian statesman and thinker. Reorganised educational system after Jena and founded University of Berlin, 1809.

IBRAHIM PASHA (1789–1848) pp. 119–121
Son or adopted son of Mehemet Ali. Conquered Mecca and Medina 1816–1818; reorganised army with French help 1820; commanded expedition against Greeks 1824–1828; conquered Syria 1831; forced to evacuate it 1841.

ISABELLA II, Queen of Spain (1830–1904) pp. 146, 147, 283
Succeeded to throne 1833; after Carlist wars declared of age when thirteen years old. Perpetual palace intrigues and bad government led to republican revolution 1868. Fled to France and renounced throne in favour of son 1870.

ISMAIL PASHA (1830–1895) pp. 296, 297
Son of Ibrahim Pasha. Educated in Paris. Ruler of Egypt 1863–1879 when forced to abdicate, after which he lived as a prisoner in Turkey.

JAURÈS, JEAN (1859–1914) p. 341
French socialist leader and historian. Assassinated on outbreak of war of 1914.

JELLACIC, Count JOSEF (1801–1859) pp. 168–170, 319

JOHN VI, King of Portugal (1769–1826) pp. 100, 110, 111

JOSEPHINE, Empress of the French (1763–1814) pp. 34, 77, 134
Her first husband, Vicomte de Beauharnais, executed during French Revolution. Married Napoleon 1796, divorced 1809.

JOURDAN, Marshal (1762–1833) pp. 35–37
Served in American War of Independence. Won battles of Wattignies (1793) and Fleurus (1794). Military adviser to Joseph Bonaparte in Naples 1806; joined Bourbons 1814; supported revolution of 1830.

KANT, IMMANUEL (1724–1804) p. 81
Greatest modern philosopher. Lived and died at Königsberg in Prussia, where he was professor. His greatest work, *The Critique of Pure Reason*, appeared in 1781.

KARA GEORG (*c.* 1766–1817) p. 325
Serbian leader and founder of Karageorgevich dynasty. Son of a peasant, brutal and ill-tempered, but a soldier of genius. Trained as sergeant in Austrian army, deserted and led Serbian rebellion 1804–1813. In exile 1813–1817, when returned to Serbia and was murdered in his sleep. Said to have killed in anger 125 men.

KIPLING, RUDYARD (1865–1936) p. 290
English writer.

KITCHENER, Lord (1850–1916) pp. 298, 299
British soldier. Reconquered Sudan 1892–1896. Made War Minister 1914, drowned on voyage to Russia 1916.

KORAIS, ADAMANTAIOS (1748–1833) p. 118
Greek patriot and scholar. Improved modern Greek dialects and made them nearer to ancient language. Lived in Paris after 1788 and interested Europe in Greek independence.

KÖRNER, THEODOR (1791–1813) p. 83
Wrote light comedies for Vienna theatres; served as volunteer in War of Liberation; 1813 killed in battle. Wrote many patriotic songs.

KOSSUTH, LOUIS (1802–1894) pp. 168–170
Hungarian patriot. Imprisoned by Metternich 1837. Leader of Hungarian revolutionary party 1847–1849. Lived in exile after failure of war of independence.

KOTZEBUE, AUGUST VON (1761–1819) p. 151
German dramatist and Russian government official. 1817, acted as Russian spy against German liberals; murdered by student 1819.

LAFAYETTE, Marquis de (1757–1834) pp. 4, 10, 11, 18, 139
Became American Major-General in American War of Independence. Dismissed from French army for liberalism 1788. Deputy in States-General 1789 and commander of National Guard. Planned to restore king's power 1792, fled to enemy but imprisoned as revolutionary. Returned to France 1799 but lived in retirement.

LAFITTE, JACQUES (1767–1844) p. 138
Paris banker and organiser of the July Revolution, 1830. President of the Chamber of Deputies, 1843.

LAMARTINE, ALPHONSE (1790–1869) pp. 173, 175
French romantic poet. Entered Chamber of Deputies during Louis Philippe's reign and played important part as republican leader during Second Republic, returning to literature 1851.

LAMENNAIS, FÉLICITÉ (1782–1854) p. 183
French liberal Roman Catholic leader, condemned by Pope. Sentenced to one year's imprisonment for article 'Le Pape et le Gouvernement', 1840. Took part as republican deputy in Second Republic. Subsequently translated Dante.

LANNES, Marshal (1769–1809) pp. 59, 60
Of humble birth, served in French army of Pyrenees 1792–1794. Promoted general by Napoleon in First Italian campaign; helped to carry out *coup d'état* 1799; saved Napoleon at Marengo 1800, promoted marshal 1804; mortally wounded at Aspern 1809.

LENIN, VLADIMIR ILYICH (1870–1924) pp. 197, 367, 374
Son of school inspector, elder brother hanged for plotting to murder Alexander III. 1892, became lawyer; 1894, began socialist propaganda; arrested 1895, and banished to Siberia; 1905, became leader of Bolshevik (revolutionary) wing of socialist party; 1906–1917, in exile; smuggled by Germany to Russia after first 1917 revolution; organised second Bolshevik revolution, head of government until his death.

LEO XIII, Pope (1810–1903) pp. 251, 340
Pope 1878–1903. Made reputation as administrator in old Papal States; created Cardinal by Pius IX 1853; violently opposed loss of temporal power 1870; persuaded French Roman Catholics to recognise republic 1894.

LEOPOLD I, King of the Belgians (1790–1865) p. 145
King of the Belgians 1831–1865. Son of the Duke of Saxe-Coburg and uncle of Queen Victoria who often consulted him. Served as a general in the Russian Army against Napoleon. 1816 settled in England. Refused the crown of Greece 1830; elected King of the Belgians 1831 after their rebellion against Holland. 1832 married Louis Philippe's daughter thus conciliating France.

LEOPOLD II, King of the Belgians (1835–1909) p. 295
King of the Belgians 1865–1909, son of Leopold I. Founded International African Association 1876; 1878 employed H. M. Stanley to organise settlement of Congo basin. 1884 secured recognition of the Congo Free State by Europe. Personally responsible for terrible abuses of Congo government; international campaign made him appoint enquiry in 1904 which confirmed their existence.

LEOPOLD OF HOHENZOLLERN (1835–1905) p. 227
German candidate for Spanish throne 1870.

LEOPOLD II, Holy Roman Emperor (1747–1792) p. 15
Holy Roman Emperor 1790–1792. An 'enlightened despot'. As Grand Duke of Tuscany introduced many liberal reforms but quarrelled with Roman Catholic church 1770–1790. In 1791 jointly with Prussia issued Declaration of Pillnitz promising to intervene in France if called on to do so.

DE LESSEPS, FERDINAND (1805–1894) pp. 233, 342
While vice-consul at Alexandria 1832 inspired by memorandum on Suez Canal written for Napoleon I. 1854 obtained concession for canal, built by him 1859–1869. Undertook Panama canal 1879, but work stopped in 1888. 1893, sentenced to imprisonment for corruption concerned with Panama Canal Company.

LINCOLN, ABRAHAM (1809–1865) pp. 204, 205
President of the United States 1860–1865. Son of Kentucky farmer; self-educated. 1856, became leader of Republicans (i.e. anti-slavery party) in Illinois; 1860, elected president of U.S.A. 1861, southern (slave) states seceded and American Civil War began. 1863, Lincoln proclaimed emancipation of slaves throughout United States. 1865, last southern army surrendered; same year Lincoln assassinated in Washington theatre.

LIST, FRIEDRICH (1789–1846) p. 152
German economist. Lived in U.S.A. 1825–1832. Coal found on his land there made him financially independent. Later American consul at Leipzig. Advocated railway construction and the Zollverein, preaching national self-sufficiency. Lost fortune and committed suicide, 1846.

LIVINGSTONE, Dr (1813–1873) p. 294
1840 went to Africa as missionary doctor. After 1852 mainly concerned with exploring Zambesi. From 1863 devoted himself to exposing slave trade. Met on Lake Tanganyika 1871 by Stanley who had been sent to find him. Buried in Westminster Abbey 1874.

LLOYD GEORGE, DAVID (b. 1863) pp. 253, 300
British cabinet minister 1905–1922 and Prime Minister 1916–1922.

LOUIS XIV (1638–1715) pp. 2, 3, 5, 89
King of France 1643–1715.

LOUIS XV, King of France (1710–1774) p. 5
King of France 1715–1774.

LOUIS XVI, King of France (1754–1793) pp. 3, 7, 10, 15, 17, 22, 51, 56, 140
King of France 1774–1792. 1774–1781, vain attempts at financial reform. 1789, States-General met, Louis hesitated between support of, and opposition to, revolution. Carried off to Paris Oct. 1789. In 1791 King and Queen fled to Varennes but were recaptured; Louis took oath as constitutional king. April 1792, France declared war on Austria; August, Louis imprisoned; September, deposed; December, tried. Executed January 1793.

LOUIS (XVII), Dauphin of France (1785–1795) pp. 27, 33

LOUIS XVIII, King of France (1755–1824) pp. 85, 86, 132–136, 138, 141
King of France (1795), 1814–1824. Brother of Louis XVI, known as Comte de Provence, fled to Belgium 1791. Lived in England 1807–1814. Restored 1814 and after Waterloo. Ruled constitutionally until his death. Too sensible for extreme royalists.

LOUIS PHILIPPE, King of the French (1773–1850) pp. 130, 138–142, 144–147, 153, 171–173, 175, 232, 239, 243, 280
Son of Philippe 'Egalité', Duc d'Orleans. In republican army at battle of Valmy 1792. In exile 1797–1817. King of the French 1830–1848. Died in London 1850.

LUGARD, Lord (1858–1945) p. 289
British colonial administrator and reformer. High Commissioner of N. Nigeria 1900–1906; Governor of Nigeria 1912–1918.

LUTZOW, FREIHERR VON (1782–1834) pp. 83, 150
Resigned from Prussian army 1808 in disgust at Treaty of Tilsit; rejoined 1811. Organised 'free corps' of volunteers to harry French rear during War of Liberation. 1815, taken prisoner by French at Ligny, but escaped on day of Waterloo. 1834, challenged Blücher, fruitlessly, to duel for insults about his cavalry's behaviour at Ligny.

LYAUTEY, Marshal (1854–1934) p. 289
Entered French army 1875 and spent nearly all his service in North Africa until 1894. Resident-General and Commissioner in Morocco 1912–1916 and 1917–1925. Created Marshal 1921. The greatest figure in French imperial history.

MACMAHON, Marshal (1808–1893) pp. 238, 244, 245, 340
French soldier and politician of Irish descent. Served in Algeria and the Crimea. Made Marshal and Duke of Magenta 1859 after Italian campaign. Defeated and surrendered at Sedan 1870 in Franco-Prussian War; commanded siege of Paris against the Commune 1871. President 1873–1879. Worked for Royalist restoration but checkmated on 'seize mai' 1877.

MAHDI, The (1848–1885) p. 298
Fanatical Sudanese leader; claimed to be descendant of Mahomet. Rebelled against Egyptians 1881. Defeated several expeditionary forces under European leaders; besieged Gordon in Khartum 1884–5.

MAHMUD II, Sultan of Turkey (1785–1839) pp. 119, 128, 129
Sultan 1808–1839. During his reign Greece, Algeria and Egypt were lost to Turkey.

MANET, ÉDOUARD (1832–1883) p. 338
Most distinguished French painter of Impressionist school. Carried on struggle against Salon, French Royal Academy, 1861–1867. Served as gunner in Franco-Prussian war. Reputation finally established 1873.

MANIN, DANIELE (1804–1857) pp. 159, 162
Italian patriot of Jewish birth. Led opposition to Austrians in Venice. As President of Venetian Republic defended city against Austrians 1849. Exiled to Paris. Converted from republicanism to support of Piedmontese royal family.

MARCHAND, Captain (1863–1934) pp. 298, 299
French explorer and soldier. In charge at Fashoda, 1898. Commanded brigade in European War, 1914.

MARIE ANTOINETTE, Queen of France (1755–1793) pp. 8, 9, 15, 16, 30
Daughter of Maria Theresa of Austria. Married Louis XVI 1770 and became Queen 1774. Devoted to pleasure and often interfered in politics to support unworthy favourites. During revolution led extreme royalists and showed great courage.

MARIE LOUISE, Empress of the French (1791–1847) pp. 77, 88, 134
Habsburg princess and Napoleon's second wife. Refused to go to Elba with him. Made ruler of Parma 1815 and governed well until 1831, then harshly.

MARMONT, Marshal (1774–1852) pp. 59, 67, 76
Met Napoleon at siege of Toulon 1793 and served with him in Italy and Egypt. Governor-General of Dalmatia 1805–1810 when transferred to Spain. Went over to Bourbons 1814 and followed Louis XVIII into exile during Hundred Days.

MARX, KARL (1818–1883) pp. 194–198, 242

MASSÉNA, Marshal (1756–1817) pp. 42–44, 60
Greatest soldier among Napoleon's marshals. Served in first Italian campaign. Saved France at battle of Zürich 1799 and siege of Genoa 1799. Retired from active service 1811. Refused to serve on Ney's court-martial.

MAUPASSANT, GUY DE (1850–1893) p. 338
French novelist and poet. Under Flaubert's influence 1873–1880. Began slowly to go mad after 1886.

MAXIMILIAN, Archduke, Emperor of Mexico (1832–1867) p. 237
Brother of Emperor Francis Joseph. Married daughter of King of the Belgians 1857. Persuaded by Napoleon III to become Emperor of Mexico 1863; captured and shot by Mexicans 1867. News drove his wife insane. A liberal-minded man with scientific tastes.

MAZZINI, GIUSEPPE (1805–1872) pp. 148, 153–155, 160, 161, 194, 210, 214–216, 222

MEHEMET ALI (1769–1849) pp. 119, 124, 127–130, 146, 224, 296, 298

MENELIK II, Emperor of Abyssinia (1844–1913) p. 301
On father's death imprisoned 1855 by usurper, but escaped. Failed to secure throne until 1889 on death of usurper's son. Defeated and massacred Italians at battle of Adowa 1896. Built railway to coast and began to open up country.

METTERNICH, Prince (1773–1859) pp. 83, 84, 88–93, 99, 101, 103, 110–112, 120, 133, 135, 145, 147, 148, 151–155, 158, 159, 163, 166, 168, 171, 194, 221, 270, 279.

MIGUEL, Dom (of Portugal) (1802–1866) p. 110

MILLERAND, ALEXANDRE (1859–1943) p. 341
French statesman. Originally a socialist he gradually became more conservative, helping to break great railway strike in 1910. Succeeded Clemenceau as Prime Minister 1920 and elected President same year. Continued to intervene as conservative in politics and consequently forced to resign 1924 after radical victory in elections.

MILOSH OBRENOVITCH I (1780–1860) p. 325
Leader in revolt against Turks 1807–1815. Made Prince of Serbia 1817. Extended his territory 1830. Forced by Russia to abdicate 1839 and lived in exile until recalled as prince in 1858.

MIRABEAU, Comte de (1749–1791) pp. 13–15, 20, 49

MOLTKE, HELMUTH VON (1800–1891) pp. 222, 227, 238
Born a Danish subject. Translated Gibbon's *Decline and Fall of Roman Empire* 1832; served in Turkish army 1835–1838. Chief of Prussian General Staff 1858–1888. Greatest strategist of later nineteenth century.

MONROE, JAMES, President of United States (1758–1831) pp. 106, 112, 237
Virginian gentleman. Served in War of Independence, and later as diplomatist in France and England. Elected President 1816. Proclaimed 'Monroe Doctrine' 1823.

MOREAU, General (1763–1813) pp. 36, 37, 40, 43, 44, 57, 58
French republican lawyer discovered as soldier by Carnot. Served mainly in northern campaigns and won battle of Hohenlinden 1800. Possible rival of Napoleon. Exiled to America. Served with allies in War of Liberation, mortally wounded at battle of Dresden 1813.

MOZART, WOLFGANG AMADEUS (1756–1791) p. 2
Greatest German operatic composer. Court Composer to Emperor at Vienna 1787. Started 'Requiem' 1791 but died before finished. Buried in pauper's grave.

MURAT, Marshal (1767–1815) pp. 34, 43, 59–62, 68
Napoleon's brother-in-law and greatest cavalry leader. Made King of Naples 1808, failed to take Sicily from British. 1814 thought of joining Austria but left it too late. Captured and shot 1815 on attempt to reconquer Naples.

MUSSOLINI, BENITO (1883–1945) pp. 370, 371, 376, 378, 380
Son of a blacksmith. Edited socialist paper before 1914. Founded Fascists after war and became dictator 1922. Declared war 1940. Overthrown and arrested 1943 but rescued by German parachutists. Captured by Italians while trying to escape 1945 and shot. Body taken to Milan and publicly exposed to insults.

NAPOLEON I, Emperor of the French (1769–1821) pp. 12, 16, 29. Chaps. 6 to 12. pp. 88, 89, 91–93, 102, 106, 108, 116, 119, 128, 134, 136, 140, 150, 232, 254, 257, 282, 298, 340, 362

NAPOLEON II, King of Rome (1811–1832) pp. 67, 85
Son of Napoleon I and Marie Louise. Remained with mother on parents' separation.

NAPOLEON III (Louis), Emperor of the French (1808–1873) pp. 62, 103, 161, 172, 175–177, 191, 212–217, 220, 225, 228, 232–239, 243, 245, 257, 259
Son of Louis Bonaparte, and Napoleon I's nephew. Took part in Carbonaro rising in Italy 1831. Bonapartist pretender after Napoleon II's death. Planned unsuccessful revolt at Strassburg 1836 and at Boulogne 1840 for which sentenced to life imprisonment but escaped 1846. Exile mostly spent in London. President of France 1848. Prince President 1851. Emperor 1852–1870.

NAPOLEON, LOUIS (Prince Imperial) 1856–1879 p. 243
Son of Napoleon III. Entered Woolwich after Franco-Prussian War. Killed in British army in Zulu war 1879.

NELSON, HORATIO (1758–1805) pp. 39, 45, 59, 65, 116, 121
English admiral at the Nile (1798), Copenhagen (1801), and Trafalgar (1805).

NEY, Marshal (1769–1815) pp. 59, 79, 85, 134, 135
Nicknamed by Napoleon 'the bravest of the brave' after the battle of Friedland. Held army together by personal courage on retreat from Moscow. Accepted Bourbons 1814 but rejoined Napoleon 1815. Shot by Bourbons after Waterloo.

NICHOLAS I, Tsar of Russia (1796–1855) pp. 120, 170, 256–260, 272, 275, 280
Emperor 1825–1855. 'Decembrist' rising in 1825 made him hate reform—the 'Iron Tsar'. Fought Turkey 1828 and secured Treaty of Adrianople 1829. Negotiated Treaty of Unkiar Skelessi 1833 and crushed Polish rebellion. Firm friendship with Austria 1835, crushed Hungarian rebellion *v.* Austria 1849. Annoyed at Austria's ungrateful neutrality in Crimean War 1854.

NICHOLAS II, Tsar of Russia (1868–1918) pp. 332, 334–337, 380
Tsar 1894–1917. Murdered 1918.

NIGHTINGALE, FLORENCE (1820–1910) p. 259
Studied nursing in Germany and France. Reorganised hospital at Scutari during Crimean War. With government help founded homes for London hospital nurses.

ORSINI, FELICE (1819–1858) p. 212
Italian revolutionary conspirator. Imprisoned for life 1844 but set free by Pius IX. Took part in Roman Republic 1849. Again imprisoned but escaped 1854. Tried to murder Napoleon III 1858 and executed.

OTTO (of Bavaria), King of Greece (1815–1867) p. 122
King 1832–1862, when deposed by military rebellion.

OUDINOT, General (1791–1863) p. 161
Son of one of Napoleon's marshals. Served in later Napoleonic wars. Commanded French expedition to restore Pope Pius IX 1849. Opposed Louis Napoleon and retired.

PALMERSTON, Viscount (1784–1865) pp. 129, 130, 146, 147, 224
British Foreign Secretary 1830–1840 and 1846–1851; Prime Minister 1855–1858 and 1859–1865.

PARIS, Comte de (1838–1894) pp. 243, 244
Grandson of Louis Philippe and heir apparent after 1842. In exile 1848–1871 and after 1886. Present at early battles of American Civil War and then organised Lancashire Cotton Famine Fund. Compromise with Bourbon pretender 1873, made royalism unpopular by supporting Boulanger.

PEDRO I, Emperor of Brazil (1798–1834) p. 111
Regent in Brazil 1821, Emperor 1822–1831. Abdicated and returned to Europe. Invaded Portugal 1832 to support liberal party. Acted as Regent 1834.

PERIER, CASIMIR (1777–1832) p. 146
Parisian banker and statesman. His moderate policy avoided European war over Belgian revolution 1830–1831.

PHILIP II, King of Spain (1527–1598) p. 144
King of Spain 1556–1598.

PICASSO (b. 1881) p. 338
A Spanish painter who settled in Paris. Greatly influenced by Cézanne. One of the original creators of Cubism who, however, later abandoned that style.

PICHEGRU, General (1761–1804) p. 57
Brilliant French revolutionary general on northern front 1794–1795. Fled to London 1797 after unsuccessful conspiracy. Shared in Cadoudal conspiracy 1803, arrested and found dead 1804.

PICQUART, Colonel (1854–1914) pp. 342, 343
French officer concerned in Dreyfus case, subsequently minister of war.

PILSUDSKI, Marshal (1867–1935) p. 282
Organised Polish legion during European War, 1914. Successfully defended Poland against Russia, 1919–1920; virtual dictator of Poland from 1926 to his death.

PISSARRO, CAMILLE (1831–1903) p. 338
French Impressionist painter, particularly famous for his pictures of Paris. Gauguin's first teacher.

PITT, WILLIAM (1759–1806) pp. 22, 23, 43
English Prime Minister 1784–1801 and 1804–1806.

PIUS IX, Pope (Pio Nono) (1792–1878) pp. 155, 159, 160, 162, 183, 184, 216, 250, 251
Pope 1846–1878. At first mildly liberal but events of 1848 and Roman Republic 1849 turned him rigidly conservative. Lost two-thirds Papal States 1859; retained remainder only through French garrison. Issued encyclical 'Quanta Cura' 1864 denouncing liberalism and modern civilisation. Called Vatican Council which proclaimed Pope's infallibility 1870. Lost all Papal States except Vatican Palace same year.

POLIGNAC, Prince de (1780–1847) pp. 111, 112, 130, 138
Imprisoned 1804–1813 for share in Cadoudal plot. Leader of extreme reactionary party after Restoration. His policy responsible for fall of Charles X.

PRIMO DE RIVERA (1870–1930) p. 283
Spanish soldier and politician; dictator under Alfonso XIII from 1923 to 1930.

RADETZKY, Marshal (1766–1858) pp. 160, 162, 168, 169
Served in Austrian armies throughout revolutionary and Napoleonic wars, after 1809 as chief of staff. On retired list 1829–1834 when given command in Italy. Saved Austrian Italy 1848–1849. Died in harness 1858.

RICHELIEU, Duc de (1766–1822) p. 136
French statesman. During revolutionary and Napoleonic wars served in Russian army. Created modern city of Odessa while governor there. Louis XVIII's chief minister after Hundred Days. Secured early withdrawal of army of occupation and France's admission to Quadruple Alliance.

ROBESPIERRE, MAXIMILIEN (1758–1794) pp. 21, 26, 30–33

ROLAND, Madame (1754–1793) p. 30
A leader of the Girondins; executed 1793.

ROON, ALBRECHT VON, Field–Marshal (1803–1879) pp. 222–224, 226–228
Prussian War Minister 1859–1871.

ROSSI, Count (1787–1848) p. 160
Italian statesman. Professor of Law at Bologna 1812. Lived in Geneva 1815–1832, was naturalised and member of Geneva parliament. Became naturalised Frenchman 1834 and French ambassador at Rome 1845. Became Pius IX's chief minister 1848 but assassinated that year.

ROTHSCHILD, JACOB (1792–1868) pp. 233, 235
The youngest of the original brothers Rothschild. Settled in Paris 1815 and provided capital for many early French railways.

ROUSSEAU, JEAN JACQUES (1712–1778) pp. 4, 23
French writer whose book, *The Social Contract*, had great influence in preparing French Revolution.

RUDOLF, Crown Prince of Austria (1857–1889) p. 321
Son of Francis Joseph: intelligent and liberal minded. Married daughter of King of the Belgians 1881. Committed suicide at Mayerling 1889.

RUSSELL, Sir WILLIAM (1820–1907) p. 259
War correspondent in Crimea, Indian Mutiny, American Civil War, Seven Weeks' War, Franco-German War, Zulu War and in the Sudan.

SALISBURY, Lord (1830–1903) pp. 265, 266, 324
British Foreign Secretary 1878–1880; Prime Minister and Foreign Secretary 1885–1886, 1886–1892, 1895–1902 (resigned foreign secretaryship 1900).

SAMSON, CHARLES HENRI (b. 1739) (better spelling Sanson) pp. 31, 32
Succeeded his father as public executioner of Paris 1778, accused of royalism 1789 but acquitted. Handed over his office to his son 1795. Date of death unknown.

SCHARNHORST, GERHARD VON (1755–1813) p. 82
Reorganised Prussian army after Tilsit, arranging short-service system to pass as many men through the ranks as possible.

SCHILLER, FRIEDRICH VON (1759–1805) p. 81
While a medical student aged eighteen wrote first verse play, *The Robbers*. Wrote histories of Dutch War of Independence and of Thirty Years' War 1788–93. Became friend of Goethe and returned to poetry 1794. Published great historical plays in verse 1799–1804—*Wallenstein*, *Mary Stuart*, *William Tell* and *Joan of Arc*.

SCHUBERT, FRANZ (1797–1828) p. 150
German composer and perhaps the greatest song writer. Set many of Goethe's poems to music. Died in Vienna where he had worked.

SCHURZ, Major-General CARL (1829–1906) pp. 204, 205
Took part in 1848 revolution in Bonn, fled to Switzerland but returned 1850 to help a friend escape from prison. Lived in Paris and London until he emigrated to U.S.A. 1852. Friend of Abraham Lincoln. Became Major-General during American Civil War, later Senator, cabinet minister (reforming American civil service) and newspaper editor.

SCHWARZENBERG, Prince (1800–1852) pp. 169, 170, 319
Austrian statesman. Restored Habsburg power after disasters of 1848, calling in Russia to crush Hungary. By the Convention of Olmütz (1850) restored the old constitution of the German Confederation. A stronger character than, and quite as unscrupulous as, his master Metternich.

SHAFTESBURY, Lord (1801–1885) pp. 193, 253
British social reformer, largely responsible for factory acts.

SHELLEY, PERCY BYSSHE (1792–1822) p. 108
English poet. Lived latterly in Italy, drowned in Gulf of Spezzia.

SIEYÈS, Abbé (1748–1836) p. 43
Well-known priest before Revolution. Elected to States-General 1789 on strength of pamphlet 'What is the Third Estate?' Responsible for most of the republican constitutions. Kept quiet during Terror—'j'ai vécu'. Helped Napoleon organise 1799 *coup d'état*, then retired. In exile 1814–1830.

SMITH, Sir SIDNEY (1764–1840) p. 116
English admiral whose defence of Acre checked Napoleon 1799.

SOULT, Marshal (1769–1851) pp. 59, 60
Served throughout early revolutionary wars on northern front. Distinguished himself at Austerlitz 1805, later served in Spain and south France against Wellington. In politics a 'Vicar of Bray'. The last of Napoleon's marshals to die.

STALIN, JOSEF (b. 1879) pp. 374, 380
Russian revolutionary leader and dictator. Repeatedly imprisoned before 1917. Secretary of the Communist party from 1922. Reached supreme power in 1926.

STAMBULOFF, STEFAN (1854–1895) pp. 325, 326
Bulgarian statesman.

STANLEY, ARTHUR PENRHYN (1815–1881) p. 255
Dean of Westminster, 1863.

STANLEY, H. M. (1840–1904) pp. 294, 295
British explorer. Sent by newspaper to find David Livingstone 1869. Explored course of Congo 1874–1877. Sent out by Leopold, King of the Belgians, to establish Congo Free State 1879. Attended Berlin Conference 1884. Elected liberal M.P. 1892.

STEIN, FREIHERR VON (1757–1831) pp. 82, 90
Prussian civil servant in charge of commerce in Westphalia 1796. Minister of Commerce 1804. 1807–1808 reorganised Prussia, extended local government, freed serfs. Dismissed by Napoleon's orders. 1812 visited Russia and helped to arrange coalition against Napoleon. Regretted reaction after 1815.

STOLYPIN, PETER (1863–1911) p. 336
Russian prime minister 1906–11 when murdered. A rigid conservative.

STRATFORD DE REDCLYFFE, Lord (1786–1880) (Stratford Canning) pp. 254–257, 260, 266
British diplomatist. 1st secretary at embassy in Constantinople 1808, acting ambassador 1810–1812, ambassador 1825–1829, 1831 and 1844–1858.

SUVOROFF, Field-Marshal (1729–1800) pp. 40, 42
Made reputation in wars against Turkey 1773–1790. Commanded Russian army in Italy with great success 1799 and led heroic march across the St Gotthard and the Alps. Returned to St Petersburg and died in disgrace. The greatest Russian general.

TALLEYRAND, CHARLES MAURICE DE (1754–1838) pp. 91, 132, 133, 136, 141
French revolutionary politician. Bishop of Autun 1789–1791 when he resigned and was banned by Pope. French foreign minister 1797–1807. Resigned after Treaty of Tilsit and, though keeping seat on council, began secretly to oppose Napoleon. 1814 persuaded allies to restore Bourbons. Louis XVIII's prime minister for a short time, later ambassador in London.

TEWFIK PASHA (1852–1892) p. 297
Succeeded his father, Ismail Pasha, as ruler of Egypt when latter deposed 1879. Maintained friendly relations with England, accepting control.

THIERS, ADOLPHE (1797–1877) pp. 129, 130, 138, 141, 142, 146, 172, 232, 239, 243, 338
Strong supporter of 1830 revolution. Served in various offices under Louis Philippe 1832–1836 and 1840. Organised opposition during last years of reign. A conservative republican 1848–1852. Leader of opposition to Napoleon III 1863–1870. Organised Third Republic 1871, crushed Commune, forced to resign presidency 1873. Wrote histories of French Revolution and Napoleon.

TIRPITZ, Grand Admiral VON (1849–1930) p. 349
German admiral. Chief of Naval Staff 1892, minister of marine 1897. Responsible for Germany's big navy. Resigned 1916. Entered parliament 1921 as Nationalist deputy.

TOGO, Admiral (1847–1934) p. 312
Japanese admiral. Studied in England 1871–1878. Completely destroyed Russian fleet in Russo-Japanese War 1904.

TROTSKY, LEO (1877–1940) p. 374
Russian revolutionary. Exile in Siberia. Communist Foreign Minister 1917–1918; War Minister 1918–1924; exiled 1928; murdered in Mexico 1940.

VENIZELOS, ELEUTHERIOS (1864–1936) p. 328
Greek statesman. Took part in risings in Crete 1889 and 1897. Head of Cretan government most of the time from 1898 to union with Greece 1909. Prime Minister of Greece 1910–1913 when he quarrelled with king. Deposed king and joined allies 1917. After the war republican leader sometimes in power and sometimes in exile.

King of Piedmont 1849. King of United Italy 1861.

Succeeded 1837.

French admiral. A noble who sympathised with revolution. Failed (1) at battle of Nile, (2) to arrange for Napoleon's invasion of England and (3) at Trafalgar. Committed suicide.

French writer.

French statesman. English father and himself Cambridge rowing blue but naturalised Frenchman. French foreign minister 1877, attended Congress of Berlin 1878, prime minister 1879. Ambassador in London 1883–1893.

Prussian soldier. Succeeded Moltke as chief of staff; commanded international force sent to suppress Boxer movement in China, 1900.

British natural scientist.

Accompanied father on 1814 campaign. Helped to crush liberal movements 1849. Regent of Prussia 1858, King 1861, appointed Bismarck 1862, proclaimed German Emperor 1871.

Succeeded 1888 and dismissed Bismarck. Lived in Holland after abdication 1918.

President of U.S.A. 1913–1921. Brought America into war 1917; laid down '14 points' as war aims 1918, attended Peace Conference and helped to form League of Nations 1919. Failed to persuade U.S.A. to join League.

Austrian field-marshal. Served through Napoleonic wars. Suppressed revolts in Prague 1848 and besieged Vienna rebels. Occupied Budapest 1849. Relieved of command and retired.

Son of George III. Unsuccessful commander in Low Countries during early stages of revolutionary wars.

YPSILANTI, Prince ALEXANDER (1783–1828) p. 117
Leader in movement for freeing Balkan Christians from Turks. Served in Russian army, 1812. Chosen as leader by Greek liberals; led Rumanian revolt, 1821 and fled to Austria on its failure.

ZOLA, EMILE (1840–1902) pp. 338, 343
French novelist. Originally a clerk but made fortune from series of novels 1871–1876. Leader of the naturalistic school, his novels are crowded with closely observed characters. Prominent supporter of Dreyfus and for a time forced to take refuge in England because of his pro-Dreyfus writings.

LIST OF TREATIES

In this appendix the principal European treaties between 1789 and 1919 involving changes to territory are arranged alphabetically. It is prefaced by chronological lists of the principal treaties dealing with various important problems or areas.

Revolutionary and Napoleonic Wars:

Basel (1795); Campo Formio (1797); Lunéville (1801); Amiens (1802); Pressburg (1805); Schönbrunn (1) (1805); Tilsit (1807); Schönbrunn (2) (1809); Abo (1812); Vienna Settlement (1815).

Eastern Problem:

Tilsit (1807); Bucharest (1) (1812); London (1) (1827); Adrianople (1829); Unkiar-Skelessi (1833); London (3) (1841); Paris (3) (1856); San Stefano (1878); Congress of Berlin (1878); London (4) (1913); Bucharest (2) (1913); Versailles (1919).

German and Austrian Problems (after 1815*):*

Vienna Settlement (1815); Convention of Olmütz (1850); Convention of Gastein (1865); Prague (1866); The Ausgleich (1867); Frankfurt (1871); Versailles (1919).

Colonial Problems (after 1880*):*

Berlin Conference (1884); Anglo-French Entente (1904); Portsmouth (1905); Algeciras Conference (1906); Anglo-Russian Entente (1907); Versailles (1919).

ABO 1812, p. 83

Occasion Preparation for war between Russia and Napoleon.

Parties Russia, Sweden.

Terms (1) Sweden recognised Russia's conquest of Finland from her.
(2) Russia undertook to help Sweden annex Norway from Denmark.
(3) Sweden undertook to help Russia against Napoleon.

Duration (1) The union of Russia and Finland lasted until 1917.
(2) The union of Sweden and Norway was accomplished in 1815 and lasted until 1905 when Norway became independent.

ADRIANOPLE 1829, p. 122

Occasion Greek War of Independence and Russo-Turkish War 1828–1829.

Parties Turkey, Russia.

Terms (1) Virtual independence of Greece as proposed by Treaty of London 1827 recognised by Turkey.

(2) Russia's hold on mouths of the Danube increased.

(3) Moldavia and Wallachia (Rumania) given virtual independence under Russian protection.

(4) Special privileges for Russians in Turkey—their own law courts.

Duration (1) Greece became a fully independent kingdom in 1832.

(2) The territorial provisions were unaltered until the Crimean War.

ALGECIRAS CONFERENCE 1906, p. 300

Occasion The Kaiser's visit to Tangier 1905.

Parties The Great Powers.

Terms (1) The sovereignty of the Sultan of Morocco was recognised.

(2) France to control the Bank of Morocco.

(3) France and Spain to organise police forces in the ports.

Duration Until Germany recognised full French protectorate over Morocco 1911 (after Agadir incident) in return for territory in the Congo.

AMIENS 1802, pp. 45–47

Occasion Nelson's victory at Copenhagen; end of Second Coalition.

Parties France, England.

Terms (1) France surrendered to their old rulers the Papal States, Naples and Egypt; and recognised independence of Ionian Islands.

(2) England restored all her conquests in the colonies except Ceylon (taken from Holland) and Trinidad (taken from Spain).

(3) England agreed to give Malta back to the Knights, but did not.

Duration Until war between England and France was renewed 1803.

ANGLO-FRENCH ENTENTE 1904, pp. 268, 299, 300, 311, 349, 360

Occasion Fears of German Expansion.

Parties England, France.

Terms (1) France recognised England's special interests in Egypt, i.e. her virtual protectorate over Egypt.

(2) England recognised France's special interests in Morocco.

Duration Still in force.

ANGLO-RUSSIAN ENTENTE 1907, pp. 268, 349, 360

Occasion Fears of German Expansion, particularly Berlin-Baghdad Railway.

Parties England, Russia.

Terms (1) Russia recognised English influence to be supreme in Afghanistan.

(2) Persia was divided into three sections: a Russian sphere of influence in the north and a British in the south, divided by a neutral zone.

Duration Until the fall of the Tsarist government.

AUSGLEICH, The 1867, pp. 319, 320

Occasion Hungarian Nationalist demands after Austria's defeat in 1866.

Parties Austria, Hungary.

Terms [N.B. The Ausgleich was not a treaty between two states but a compromise between two parts of the Austro-Hungarian Empire.]

(1) Francis Joseph to rule as Emperor in Austria and King in Hungary.

(2) The two countries to be completely separate except that:

(3) common ministries for Foreign Affairs, War and Finance;

(4) separate parliaments in Austria and Hungary, but delegations from each to discuss the business of the common ministries.

Duration Until the end of the empire 1918.

BASEL (or Bâle) 1795, p. 35

Occasion End of First Coalition; Third Partition of Poland.

Parties France, Prussia, Spain.

Terms (1) Prussia gave France her provinces west of the Rhine.

(2) Spain gave France part of San Domingo (in West Indies).

Duration As far as Prussia was concerned until the first Treaty of Schönbrunn 1805.

BERLIN CONFERENCE 1884, pp. 294, 295

Occasion Stanley's exploration of the Congo; Germany's new interest in colonies.

Parties England, France, Germany, Belgium, Portugal.

Terms Africa south of the Sahara was partitioned into 'spheres of influence' between 1884–1886.

(1) England took Nigeria, the Gold Coast, Kenya, Uganda and Rhodesia.

(2) France took most of the rest of West Africa as far south as the Congo, and Madagascar.

(3) Germany took South West Africa, Togoland, the Cameroons, and Tanganyika.

(4) Belgium took the Congo Free State.

(5) Portugal took Angola and Mozambique.

Duration Until 1914.

BERLIN, CONGRESS OF 1878, pp. 265–267, 322–324, 327

Occasion The Treaty of San Stefano 1878.

Parties Turkey, England, France, Austria, Prussia, Russia, the Balkan States.

Terms (1) Russia gained Batum and Kars from Turkey.

(2) Russia regained southern Bessarabia (touching the Danube) which she had lost in 1856.
(3) The independence of Serbia, Montenegro and Rumania was recognised and their boundaries slightly enlarged.
(4) Bulgaria was reduced to half its size under the Treaty of San Stefano and divided into two principalities.
(5) Austria occupied Bosnia and Herzegovina which remained in Turkish ownership.

Duration (1) The two Bulgarian principalities were united in 1885.
(2) Greece acquired Thessaly from Turkey 1881.
(3) Austria annexed Bosnia and Herzegovina in 1908 and held them until 1918.

BUCHAREST (1) 1812, p. 116
Occasion Russo-Turkish War of 1806–1812.
Parties Turkey, Russia.
Terms (1) Russia gained Bessarabia and the control of the Danube Delta.
(2) The Serbians were given self-government.
Duration Until the Russo-Turkish War 1828.

BUCHAREST (2) 1913, p. 329
Occasion The second Balkan War 1913.
Parties Turkey, the Balkan States.
Terms (1) Turkey regained Adrianople lost in first Balkan War.
(2) North and Central Macedonia to Serbia.
(3) South Macedonia to Greece.
(4) Bulgaria had to be content with East Macedonia.
Duration Until European War 1914–1918.

CAMPO FORMIO 1797 (sometimes written as Leoben), pp. 37, 40
Occasion Napoleon's first Italian Campaign.
Parties France and Austria.
Terms (1) The Austrian Netherlands were given to France.
(2) Austria recognised Cisalpine Republic (French protectorate) formed out of Austrian Lombardy and small North Italian duchies.
(3) Venice was divided: France took the western strip and the Ionian Islands, Austria the rest.
(4) Austria took Salzburg and part of Bavaria.
Duration Until the formation of the Second Coalition 1798.

FRANKFURT 1871, pp. 239–243
Occasion Franco-Prussian War.
Parties France, Germany.
Terms (1) France gave Germany Alsace (except Belfort) and Eastern Lorraine.
(2) France undertook to pay Germany a heavy indemnity.
(3) German troops to stay in France until indemnity was paid.
Duration Until the European War 1914.

GASTEIN, CONVENTION OF, 1865, p. 225
Occasion The Schleswig-Holstein Question.
Parties Prussia, Austria.
Terms Prussia was to occupy Schleswig, Austria Holstein.
Duration Until 1866.

LONDON (1) 1827, p. 120
Occasion Greek War of Independence.
Parties England, France, Russia.
Terms The three Powers agreed to force Turkey:
(1) to grant an armistice;
(2) to establish Home Rule for Greece.
Duration Until England withdrew from the Eastern Question after the battle of Navarino 1827.

LONDON (2) 1839, pp. 146, 360
Occasion Declaration of Belgian Independence 1831.
Parties Belgium, Holland, England, France, Prussia, Austria, Russia.
Terms (1) Belgium was recognised to be an independent kindom.
(2) The neutrality of Belguim was recognised and guaranteed by each of the powers.
[N.B. This treaty marks the recognition by Holland of the arrangements agreed to by the Great Powers in London in 1831.]
Duration Until the European War 1914.

LONDON (3) 1841, p. 130
Occasion Last stage of Mehemet Ali's wars.
Parties England, France, Russia, Prussia, Austria, Turkey, Egypt.
Terms (1) Mehemet Ali gave back Syria and Crete to Turkey, but kept Egypt.
(2) The Dardanelles were closed to all warships under guarantee of the five great powers: thus destroying Treaty of Unkiar-Skelessi.
Duration Until the Crimean War 1854.

LONDON (4) 1913, p. 329
Occasion First Balkan War 1912.
Parties Turkey, the Balkan States, the Great Powers.
Terms (1) Turkey lost Albania, Macedonia, most of Thrace.
(2) New state of Albania set up by Great Powers.
(3) Failure to agree on partition of Macedonia.
Duration Until Second Balkan War 1913, but Albania then continued in existence.

LUNÉVILLE 1801, p. 44
Occasion The victories of Marengo and Hohenlinden (1800).
Parties France, Austria.

Terms (1) Confirmed the terms of the Treaty of Campo Formio 1797.
(2) Recognised the Rhine as the French eastern frontier.
(3) Recognised the French protectorates over Holland (Batavian Republic) and Switzerland (Helvetic Republic).
Duration Until formation of Third Coalition 1805.

OLMÜTZ, CONVENTION OF, 1850, p. 165
Occasion The German and Austrian Revolutions of 1848–1849.
Parties Prussia, Austria, the other German States.
Terms (1) The Prussian Union (a suggested Confederation of German States under Prussian leadership) abandoned.
(2) The old German Confederation of 1815 under Austrian leadership re-established.
(3) But Austria refused admission to the Prussian Zollverein (Customs Union).
Duration Until the Seven Weeks' War 1866.

PARIS (1) and (2) (*see under* Vienna Settlement)

PARIS (3) 1856, pp. 259, 261
Occasion The Crimean War.
Parties England, France, Turkey, Russia, Piedmont.
Terms (1) Turkey promised good treatment of Christian subjects.
(2) Russia abandoned claim to interfere in Turkey to protect Christians.
(3) Russia gave up protectorate over Moldavia and Wallachia (now Rumania).
(4) Black Sea neutralised, i.e. no warship to enter it or arsenals on its shores.
Duration Until Russia again established Black Sea fleet 1870.

PORTSMOUTH (U.S.A.) 1905, pp. 312, 336
Occasion Russo-Japanese War 1904.
Parties Russia, Japan.
Terms (1) Russia surrendered Port Arthur to Japan.
(2) Russia gave back Manchuria to China.
(3) Japan gained control of Korea which she annexed 1910.
Duration Until the end of the rule of the Tsars.

PRAGUE 1866, p. 226
Occasion The Seven Weeks' War.
Parties Prussia, Austria, the smaller German States, Italy.
(1) Prussia annexed Hanover, Schleswig-Holstein and several smaller German States.
(2) All Germany north of the Main was formed into a North German Confederation under Prussia.

(3) The old German Confederation was abolished and Austria ceased to be part of 'Germany'.
(4) Austria surrendered Venetia to Italy.

Duration (1) The North German Confederation lasted until the Franco-Prussian War 1870
(2) The exclusion of Austria from Germany lasted until 1938.
(3) Venetia still belongs to Italy.

PRESSBURG 1805, p. 61

Occasion The Austerlitz Campaign.
Parties France, Austria.
Terms (1) Austria surrendered Venice and Dalmatia (part of Jugo-Slavia to-day) to France.
(2) Austria surrendered Tyrol etc. (her Alpine provinces) to Bavaria.
(3) Austria surrendered her lands in south-west Germany to Baden and Würtemberg.
Duration Until Austria again declared war on Napoleon in 1809.

SAN STEFANO 1878, pp. 264–266

Occasion Russo-Turkish War of 1877.
Parties Turkey, Russia.
Terms Points 1, 2 and 3 were roughly the same as those of the Treaty of Berlin 1878.
(4) A great Bulgarian state was established including all Macedonia, stretching from the Danube to the Aegean Sea.
Duration Until the Congress of Berlin 1878.

SCHÖNBRUNN (1) 1805, p. 61

Occasion After Austerlitz as a result of Prussia's belated threat of joining Austria.
Parties France, Prussia.
Terms (1) Prussia forced to accept Hanover from Napoleon, who took it from England. In this way Napoleon hoped to make Prussia and England enemies.
(2) Prussia forced to exclude English ships from her harbours.
Duration Until the outbreak of the Jena campaign 1806.

SCHÖNBRUNN (2) 1809 (also called Vienna), p. 77

Occasion The Wagram Campaign.
Parties France, Austria.
Terms (1) Austria surrendered the Dalmatian Provinces to Napoleon (parts of Jugo-Slavia to-day).
(2) Austria surrendered Western Galicia to Grand Duchy of Warsaw and Eastern Galicia to Russia.
(3) Austria gave back Tyrol, Salzburg etc. (her Alpine provinces) to Bavaria.
(4) Austria joined Continental System.
Duration Until Austria joined in the War of Liberation 1813.

TILSIT 1807 pp. 54, 61, 64, 72, 74, 77, 78, 83, 116.

Occasion The Jena (1806) and Friedland (1807) Campaigns.

Parties France, Russia, Prussia.

Terms (1) Prussia surrendered all provinces west of Elbe and her share of second and third partitions of Poland.

(2) Prussia was to pay a large indemnity and join Continental System.

(3) Russia promised to mediate between France and England. If this failed she would join Continental System and force Sweden, Denmark and Portugal to go to war with England.

(4) Russia was allowed to take Finland from Sweden (which she did) and Moldavia and Wallachia (Rumania) from Turkey (which she failed to do).

(5) Russia gave France the Ionian Islands.

(6) Russia and Prussia recognised the new Napoleonic kingdoms, e.g. Westphalia, Grand Duchy of Warsaw.

Duration (*a*) Russia: until the Moscow Campaign 1812.

(*b*) Prussia: until outbreak of War of Liberation 1813.

UNKIAR-SKELESSI 1833 pp. 124, 129.

Occasion Mehemet Ali's conquest of Syria.

Parties Turkey, Russia.

Terms The Dardanelles to be closed to the warships of every country except Russia.

Duration Until the Treaty of London 1841.

VERSAILLES 1919 pp. 191, 240, 261, 365–369, 375–378

Occasion End of European War (1914–1918).

Parties The principal countries of the world.

[N.B. A large number of separate treaties are grouped together in this summary.]

Terms

A. Germany. Lost (1) Alsace-Lorraine to France.

(2) Schleswig to Denmark.

(3) Her Polish Provinces to Poland.

(4) Memel to Lithuania.

(5) Her colonies divided under 'mandate' between France, Belgium, Japan, England and the British Dominions.

(6) Forced to pay enormous reparations.

(7) Forced to allow foreign occupation of Rhineland.

(8) Forced to disarm.

B. Austria-Hungary. Disappeared as a state.

C. Turkey. Lost (1) Her remaining European provinces except Constantinople.
(2) Part of Asia Minor (to Greece).
(3) Her Arab provinces (independent states).

D. Bulgaria. Lost territory to Greece and Rumania.

E. Russia. Lost (1) Finland.
(2) Her other Baltic Provinces.
(3) Poland.

F. The following new states were established or revived:
Finland, Latvia, Esthonia, Lithuania } cut off from Russia.
Poland: cut out of Russia, Germany and Austria-Hungary.
Czecho-Slovakia, Austria, Hungary } cut out of Austro-Hungarian Empire.
Irak, Arabia, Syria (French Mandate), Palestine (British Mandate), Trans-Jordania (British Mandate) } cut out of Turkey.

G. The following states had their territory greatly increased:
France (from Germany and Turkey).
England (from German colonies and Turkey).
Italy, Rumania, Serbia (now Jugo-Slavia) } from Austria-Hungary.
Greece from Turkey.

H. The League of Nations membership was originally open to all countries except the defeated powers and Russia. The United States refused to join. All the excluded powers joined later but some have resigned as have other countries. Objects:
(1) to keep peace by providing machinery for peaceful settlement of disputes where agreement can be reached. No machinery for compulsory alteration of unjust treaties.
(2) to secure international cooperation to improve health and labour conditions etc.

Organisation (1) An Assembly of all member-states meeting once a year; (2) a council of the Great Powers plus representatives of some of the smaller powers meeting more frequently; (3) the International Labour Office and the Permanent Court of International Justice which work independently of, but in connection with the League.

Duration (1) Virtually all the non-territorial clauses of the treaties, e.g. reparations, disarmament, had ceased to apply by 1938.
(2) Most of the territorial changes remained in force until 1939 but
(*a*) Germany regained the Saar (from the League of Nations) 1935 and annexed Austria and part of Czecho-Slovakia 1938.
(*b*) Turkey reconquered the Greek part of Asia Minor 1922–1923 (Treaty of Lausanne).

VIENNA SETTLEMENT 1815, pp. 88–98, 104, 133, 137, 144, 170, 265, 276, 279.
Occasion The end of the Napoleonic Wars.
Parties All the European Powers.
[N.B. A large number of separate treaties are grouped together in this summary.]
Terms
A. France and England (1) First Treaty of Paris 1814.
(*a*) France kept her frontiers of 1792 plus part of Savoy.
(*b*) England restored all colonies except Tobago, St Lucia and Mauritius (from France), Malta (from the Knights of St John), Cape Colony and Ceylon (from Holland).
(*c*) France promised to abolish slave trade.
(*d*) France allowed to keep most art treasures taken by Napoleon.
(*e*) No indemnity imposed on France.
(*f*) Napoleon sent to Elba.
(2) Second Treaty of Paris 1815 as first treaty except
(*a*) Savoy surrendered by France.
(*b*) France to surrender art treasures.
(*c*) France to pay war indemnity.
(*d*) Foreign army of occupation until indemnity paid.
(*e*) Napoleon sent to St Helena.
B. Austria. (*a*) Gave up Austrian Netherlands (Belgium) and lands in south-west Germany.
(*b*) Regained Lombardy and added Venetia. Austrian princes in central Italian states of Parma, Modena, Tuscany.
C. Prussia. (*a*) Gave up most of her Polish lands except those gained in First Partition.
(*b*) Took instead lands in west Germany (on the Rhine) and Central Germany.

D. Russia. (*a*) Took most of Poland (virtually all the Grand Duchy of Warsaw).
(*b*) Kept Finland (taken from the Swedes).

E. Sweden. (*a*) Took Norway from Denmark.
(*b*) Finally lost Finland to Russia.

F. Holland. Took the Austrian Netherlands (Belgium).

G. 'Germany.' A German Confederation under Austria's Presidency was established to take the place of the old Holy Roman Empire and the Confederation of the Rhine. It had little real power.

H. 'Italy.' Remained a 'geographical expression' with no political existence.

J. Other Countries. No important changes. The old rulers were restored.

K. Alliances. (1) Holy Alliance—Principal countries, except England and France, joined together through their kings to uphold 'Christian principles'. Ineffective idea of Alexander of Russia. Pp. 92, 106, 109, 112.
(2) Quadruple Alliance—England, Austria, Prussia and Russia agreed to hold periodical meetings to discuss matters of common interest—the Congress System. Originally intended to prevent Napoleonic restoration. Later became general reactionary organisation. Pp. 92–94, 98, 120.

Duration (1) French frontiers remained unchanged until 1859 (Italy) and 1870 (Germany).
(2) Belgium separated from Holland 1831.
(3) Austrian frontiers remained unchanged until 1859.
(4) Prussian frontiers remained unchanged until 1864.
(5) Russia's western frontier remained unchanged until 1917.
(6) Norway separated from Sweden 1905.
(7) 'Italian' frontiers remained unchanged until 1859.
(8) France admitted to Quadruple Alliance 1818.
(9) England withdrew from Congress System 1822.

VILLAFRANCA, Armistice of, 1859, pp. 213, 214
Occasion The Magenta and Solferino Campaign.
Parties France, Austria and, against her will, accepted by Piedmont.
Terms (1) Lombardy (capital Milan) given to Piedmont.
(2) Venetia and Mantua kept by Austria.
Duration Until the Seven Weeks' War 1866.

INDEX

[*For references to people and treaties see the Biographical Dictionary and List of Treaties—treaties not involving changes to European frontiers are, however, indexed here, e.g. Treaty of Nanking.*]